Praise for th ▍▍▌▍▌▌▍ edition of *Corrupted Science*

"You [may] remember how much fun Grant's *Discarded Science* was and scoop this up in search of chewy anecdotal goodness. The tone this time is considerably darker, though, as Grant ascribes to malice what cannot be adequately explained by stupidity. His comparison of the current administration's assault on scientific truth and inquiry goes even further than Christopher Mooney's *The Republican War on Science*."

—*USA Today*

"I just finished the book last night, and my highly unscientific verdict—amazing! . . . The writing is clear and witty, and the content seems (to the best of my knowledge) spot on. The few places that I found myself in minor disagreement with Grant's comments, he followed them up and proved me depressingly wrong. . . .

"*Corrupted Science* is an excellent and highly readable book about fraud and ideological fallacy in science, and serves both as an introduction and a reference for those interested in learning more about the tenuous thread by which hangs rationality."

—*Skulls in the Stars*

Praise for John Grant's other works on science

CORRUPTED SCIENCE

FRAUD, IDEOLOGY, AND POLITICS IN SCIENCE

JOHN GRANT

SEE SHARP PRESS ◆ TUCSON, ARIZONA

For information contact:

    See Sharp Press
    P.O. Box 1731
    Tucson, AZ 85702

    www.seesharppress.com

Grant, John.
    Corrupted Science : Fraud, ideology, and politics in science (revised & expanded) /
John Grant ; Tucson, Ariz. : See Sharp Press, 2018.
    Includes bibliographical references and index.
    448 p. ; 23 cm.
    ISBN 978-1-947071-00-1

CONTENTS

Part 2: The Political Corruption of Science

Dedication to the First Edition

To Keith Barnett (1938–2006), the best big bro any little bro could have hoped for

Dedication to the Second Edition

To Carl Fink, for countless words of encouragement over the years

INTRODUCTION

"Science is a set of rules that keep the scientists from lying to each other."

—Kenneth S. Norris, cited in *False Prophets* (1988) by Alexander Kohn

People in general believe what they're told. That seems hardwired into the human brain. One can understand why this should be so: verbal communication presumably came about as a means of exchanging vital information among tribal members ("the food's over there") and with members of other tribes ("if you try to take my food I'll kill you"). Speech would have had little survival value for tribal peoples—and indeed little purpose—unless the information it contained were true. We can guess that the invention of the lie came some while after speech was in widespread use—and a devastating invention it must have been. Even today we naturally tend to believe what we're told—skepticism is an educated response, not an instinctive one, as demonstrated by the ease with which parents can fool small children with tall tales. Similarly, most of us tell the truth almost all of the time.*

In general it's essential for the smooth functioning of society that this be so. Consider the simple social interaction in which you ask a passing stranger for directions. Society would soon weaken if strangers habitually gave false directions, or if tourists habitually disbelieved the directions given to them by strangers. Of course, sometimes strangers do quite deliberately give false directions, either because they don't know the answer and have an infantile dislike of displaying ignorance, or in the misguided belief that it's funny. Similarly, travelers sometimes disbelieve genuine directions because "they don't make sense."

---

* Hence the effectiveness of the old joke about politicians:
Q: How can you tell if he's lying?
A: His lips move.

The deliberate giving of false directions can be regarded as a small-scale demonstration of the inherent flaw in our natural assumption of truth-telling. If one party, almost always the teller, disobeys the tacit rules of the game, the other is exceedingly vulnerable. Hence the effectiveness of false propaganda, as exemplified today by broadcasters such as Rush Limbaugh and the pundits of Sinclair Broadcast and Fox News: they can tell whatever fibs they like, secure in the knowledge that a high percentage of the audience will believe what they say; further, since those members of the audience who perceive the lie will soon go elsewhere for their information, the "credulosity quotient" of the remaining audience tends to rise.

This particular dishonest gambit is, of course, not a new development: it can be traced back through most of recorded history. False tales of the disgusting licentiousness of Cleopatra were circulated in Rome to shape the citizens' attitudes toward the Egyptian queen. Lies concerning the sexual appetites and extravagance of Marie Antoinette contributed to the onset of the French Revolution. And we all know the devastating human costs of the early-twentieth-century antisemitic propagandist forgery, *Protocols of the Learned Elders of Zion.**

The same vulnerability in our social structure is of course exploited alike by the hoaxer, the forger, the fraud artist, the trickster, the prankster, the dissembling politician, the televangelist, the propagandist and the straightforward liar. Sometimes, as per the *Protocols,* the efforts of these assorted crooks have enormously damaging consequences; political and media denials of imminent catastrophic climate change, for example, may spell the collapse of human civilization.

Most hoaxes and frauds are outside the purview of this book, in which we're concerned only with the sciences, but that still leaves us plenty to play with. We'll start with scientists themselves—scientists who for one reason or another have felt driven to fake things—before turning to those who've corrupted science not from within but from without: ideologues (religious and otherwise), powerful corporations and of course political regimes, three of which I'll discuss in the final three chapters of the book.

The political corruption of science is not new. Since the dawn of the twentieth century there seems to have been an upsurge in ideology-based

---

* This was actually a reworking of *Dialogue in Hell Between Machiavelli and Montesquieu* (1865) by Maurice Joly. Joly's original was a satirical attack on Napoleon III. All the anonymous forgers of the Protocols did was change a few names and details and pretend it wasn't a satire.

distortion of public science by those in political power. The three classic cases over the past century are Hitler's Germany, Stalin's USSR, and twenty-first-century America. The third of these is treated in slightly more detail here than the other two not just because it is current but because its ideological spread has arguably been on the widest scale, affecting not one or two sciences but many. It is also, because of its assault on climate science, potentially the most globally dangerous in human history.

Let me stress something. The selection of these three case studies should not be read to suggest that recent US administrations are comparable to the Hitler and Stalin regimes other than in their corruption and suppression of science. The emphasis is (alas!) essential because more than one amateur online reviewer of *Corrupted Science*'s first edition made the outrageous claim—I assume with the intention of deflecting criticism they couldn't otherwise counter—that I was attempting to equate the Bush II administration with Nazi Germany.

It's not just those in power who're responsible for the political corruption of science. An allied factor is the quaint perversion of democratic thought that seems to believe scientific truth can be determined by popular vote—or at least by opinion polls. The playwright Nathaniel Lee supposedly jibed, "They called me mad, and I called them mad, and, damn them, they outvoted me." In terms of sanity, "their" vote may have been of note, in that sanity is largely defined as conformity to society's norm; but in other areas of science a voting exercise—most especially if most of the voters are ignorant of or poorly informed about the issue at hand—is meaningless. Reality is not affected by human preferences, and most certainly does not obey human wishes. Reality simply *is*.

Regarding This New Edition

In 2006 the UK publisher AAPPL released a book of mine called *Discarded Science*, which was a fairly lighthearted stroll through those scientific ideas, from the sensible-but-wrong to the outright silly, that have fallen by the wayside, sometimes almost immediately (if not quicker than that) and sometimes only after centuries. The book sold surprisingly well, and so naturally AAPPL's Cameron Brown welcomed the idea of a follow-up.

That follow-up, published in 2007, was the first edition of this book. *Corrupted Science* was in part a genuine companion to *Discarded Science*, in that it concerned itself with false science, although this time science that had been either deliberately falsified or at least molded by the ideological or other preconceptions of scientists or the scientific community. The other main component of Corrupted Science was my response to the increasing tendency of some parts of the media and some of our elected public servants to deliberately corrupt the public understanding of science. Although the most obvious manifestation of the political falsification of science was the then US government's pretense that the science of climate change was as yet unsettled (a falsehood of suicidal stupidity), there was distortion aplenty in other fields of public science from bogus sex education to the classroom expurgation of evolution.

Prominent politicians were telling flat-out lies about science, and the media was too frightened, too afraid of being portrayed as taking sides, to call out the lies. Powerful corporations were quite openly paying to promote those lies, buying not just some of our politicians but also certain elements of our supposed news media.

I was far from the only person concerned about this usually profit-based war on science—you'll find plenty of others mentioned in these pages. Luckily Cameron Brown at AAPPL felt similarly, and he provided the book stalwart support; he even, at the coaxing of writer Jeff VanderMeer, gave it a relaunch a year or so after its first publication, with Jeff kindly devoting much time and effort to gain it additional publicity.

Time passed. A new administration arrived in the White House and, while its attitude toward science was far from perfect, it was a huge improvement on what had gone before. Now it was countries like Canada, under the Stephen Harper administration, and Australia, under Tony Abbott, that were the developed world's science pariahs.

Years passed. Although *Corrupted Science* fared well, and received more notice than its predecessor, it became dated. Cameron decided it was time for him to fold AAPPL and retire to a life of horse-riding country squiredom.

But the political assault on science never really went away, although it was held in check for a number of years. The political anti-science warriors were deprived of direct power, so they simply bided their their time. Meanwhile, the corporate corruption of science hadn't stopped at all.

The ascendancy of Donald Trump to the presidency, and the ushering-in of an era of "alternative facts," marked the onset of a new and far more serious attack on the sciences in the US than anything that had gone before. With scientifically illiterate and often aggressively science-denying individuals being appointed to major science posts within the administration, it seemed obvious to me that a new and very extensively revised edition of *Corrupted Science* was necessary. Luckily publisher Chaz Bufe of See Sharp Press agreed with me, which is why you're able to read the book you're holding.

Some parts of this book have received only a minor overhaul since the first edition: updating, correction, the introduction of relevant recent cases. Other parts have been very extensively revised, most notably the final chapter, on the political corruption of science in the US in the twenty-first century. I decided that the rather brief discussion in the original edition of the corporate corruption of science needed expansion into a full-length chapter of its own; it became the longest chapter in the book, and by a fair margin.

Revisiting *Corrupted Science* has been at once an invigorating and a depressing experience—invigorating, because it's a book that holds a special place for me, and depressing, because in my optimistic way I'd hoped its content would by now have come to be of solely historical interest: *Those were the days when politicians and pundits lied about science and large numbers of people believed the lies, but we've grown up a bit since then.*

Unfortunately, we haven't.

1

FRAUDULENT SCIENTISTS

The prominent UK surgeon Paul Vickers remarked in a 1978 lecture:

What the public and we [doctors] are inclined to forget is that doctors are different. We establish standards of professional conduct. This is where we differ from the rag, tag and bobtail crew who like to think of themselves as professionals in the health field.

High words, and arrogant ones, but colored in hindsight by the fact that just a few years later, in 1983, Vickers, in association with his mistress, was charged with—and later convicted of—murdering his wife Margaret, using "professional conduct" to do so: he poisoned her with an anti-cancer drug so that her death was initially attributed to natural causes.

Arrogance and hypocrisy are recurrent themes, although far from the only ones, in considering fraudulent scientists: it is as if some get so engaged in their own world that the external one becomes a sort of secondary reality, one in which events have an almost fictional status and where consequences need not be considered.

But there are plenty of other motivations for fraudulence in science.

In 1993 the physicist John Hagelin—three times US presidential candidate for the Natural Law Party—gathered together in Washington, DC, some 4,000 transcendental meditators with the aim of reducing the violent-crime rate in that city by 25 percent. Unfortunately, the police records showed that Washington's murder rate for that year actually rose. Or did it? In 1994 Hagelin announced he'd done a *proper* analysis of the figures and, sure enough, they showed a 25 percent decline.

On the face of it, this might seem like an instance of the fraudulent abuse of statistics, but was that really the case? It seems unlikely. Far more probable is that Hagelin, unable to believe the results of his experiment, quite unconsciously read into the statistics what he wanted to see there.

The borderlines between fraud, self-deception, gullible acceptance of the fake, and the ideological corruption of science can be very blurred. In theory none of them should happen; in practice, all too often, they all do, sometimes in combination. Where does one draw the line between, say, self-deception and ideological corruption? The latter may be deliberate and self-serving, but it can equally well be a product of the same desire to have reality obey one's wishes that seems to have driven Hagelin to derive an "alternative" message from those crime statistics.

All of the categories of scientific falsification are dangerous: people can die of them—either directly, as with fraudulent cancer cures, or indirectly, as when distortion of science about supposed racial differences reinforces irrational prejudices and leads to violence between peoples. The Nazis had their scientific "proof" that the Jews were subhuman, and so murdered them by the million. In eighteenth- and nineteenth-century North America the white man had "proof" of the subhuman nature of black and red races, so enslaved one and waged a campaign of near-genocide on the other. The pseudoscience of eugenics not only helped fuel the Nazis' murderous spree but also, before that spree shocked sense into people, looked well set to initiate a culling in North America of the mentally ill, the socially inadequate, and, of course, those of "lesser" races.

We live in an age when the falsification of science, in particular the ideological corruption of science, has reached a new level of importance—the very survival of our species may be threatened by it.

Not all falsifications seem of such significance—although one could make the case that false belief in itself does considerable damage by way of a sort of intellectual pollution, a brain rot that hampers all our other efforts at progress, or by generating enough "noise" that genuine knowledge becomes obscured. What may start as a relatively harmless hoax or fraud can be compounded through human gullibility or self-deception to the point where it assumes an importance far beyond anything the original perpetrator could have conceived.

On a small scale, this is what happened in the late sixteenth century in one of the oddest spats in the history of science, the infamous case of the Silesian Boy. This child was born on December 22, 1585, and was discovered, when his teeth grew in, to be possessed of a golden molar. How could science explain this?

The best-regarded hypothesis of the time was produced by Professor

Jacob Horstius of Helmstädt University, author of the definitive *De Au-reo Dente Maxillari Pueri Silessii* (1595). Horstius's hypothesis managed to conflate astrology with a medical belief widely current at the time: the notion that, if a pregnant woman became covetous of something she saw, the next time she touched herself on her body she would generate an appropriate birthmark on the corresponding bodily part of her as-yet-unborn infant. Horstius claimed the astrological conditions at the time of the Silesian Boy's birth (under the sign of Aries with the planet Saturn in conjunction) were so favorable that the boy's body began to produce not bone (as teeth were thought to be) but gold. The gold had manifested as a tooth because the boy's mother, while carrying him, must have coveted something golden she'd seen and not long afterward stuck her finger in her mouth. Horstius went on to theorize, obscurely, that the tooth was a sign from Heaven that the Turks would be kicked out of Europe.

Another book, this time by Regensburg physician and alchemist Martin Ruland, emphatically seconded Horstius's hypothesis. Others felt driven to write volumes in rebuttal, among them Duncan Liddel, who pointed out an elementary flaw in the astrology of Horstius's proposal: the sign of Aries falls in spring, not in December. The Coburg chemist Andreas Libavius produced a book about not the phenomenon or the theory but the controversy itself.

It finally occurred to someone that it might be a good idea to examine the boy and his tooth. It was at this point that it was discovered the whole affair was a hoax: the parents had jammed gold leaf over the child's molar.

Not many years later, precisely in order to avoid such embarrassments, Francis Bacon set forth his version of the scientific method, advocating the collection of empirical evidence before you advance hypotheses. For centuries people accepted the principle . . . but carried on much as before, believing what they wanted to believe and finding the "proof" where they could. It's something we still do.

The idea that at least some blind people can distinguish different colors by touch dates back at least to the eighteenth century; there's mention of it in Boswell's *Life of Johnson* (1791). At that time it was believed minuscule differences in surface texture were responsible for the appearance of different colors; although these variations were too fine to be detected by most people, the nonvisual senses of the blind were known to be often more astute than those of the unimpaired. In different form, the idea re-emerged in the nineteenth century, this time the underpinning being the notion that the different colors generated slightly different temperatures.

One practitioner of the art in England was teenager Margaret M'Avoy, who was all the rage in 1816. A peculiarity of her uncanny ability which puzzled people was that it would function only in the light; in darkness her fingers were just as blind as anyone else's. Her devotees helpfully explained that this was because everything looks equally black in darkness.

We wouldn't be so easily fooled today, would we?

In the 1960s, when the fad for parapsychology was at a peak in the west, reports emerged from the USSR of various women who could "see" with their fingertips or other extremities. Several of the practitioners were soon exposed as frauds, and the credulous reports in western media tapered off rather abruptly when something very obvious was pointed out: stage magicians had been performing precisely similar feats for generations. Yet, where those reports were retracted at all, it was with relatively little fanfare, and so it's still widely believed the women's abilities were genuine.

The notion of "the emotional life of plants" can be traced to the polygraph expert Cleve Backster, the founder in New York, after a career with the CIA, of the Cleve Backster School of Lie Detection. In 1968 Backster published a paper in the *International Journal of Parapsychology*, "Evidence of a Primary Perception in Plant Life," which claimed that, by in essence hooking up a polygraph to them, he'd been able to show plants possessed a primitive form of ESP: they displayed a reaction to the destruction nearby of living cells. Money poured in to help him establish the Backster Research Foundation, whose express purpose was to investigate further the ESP abilities of plants. The initial results from these researches were very positive, and it seemed Backster and his team had made a great breakthrough. However, other researchers couldn't replicate the experimental results, and soon the claims were debunked. This did not stop the publication of several bestselling books on the subject of plant psychology—notably *The Secret Life of Plants* (1989) by Peter Tompkins and Christopher Bird. It seems Backster's claims may have been the product of the fairly common phenomenon whereby perfectly rational researchers can unwittingly, and despite all self-imposed safeguards against bias, skew their results to favor their preconceptions. Yet some of the relevant books are still in print.

The two latter examples lead to another important tile in the falsification-of-science mosaic: the role of the print and broadcast media, which are almost always eager to trumpet sensational claims of the extraordinary and then, when the claims are shown to be bunkum, are near-criminally negligent in acknowledging the fact. This is frequently compounded in the modern era by the perversion that has grown up of the old idea of journalistic balance: the new *faux*-balance seeks to find a midway point not between two reality-based viewpoints but between a reality-based viewpoint (right or wrong) and one that is demonstrably false. The result in the minds of the audience—and, who knows, perhaps in the minds of the journalists too—is a fallacious perception that facts are somehow subject to debate. The attitude that everyone's opinion is equally valid, no matter their level of ignorance or expertise—and certainly no matter what the reality actually *is*—is lethally dangerous in some areas of science.

BABBAGE'S REFLECTIONS

In 1830 Charles Babbage, best remembered today for his early work on the computer, published *Reflections on the Decline of Science in England, and On Some of Its Causes*. In this little book's Chapter V there's a subsection on the genesis of fraudulent science that's as valid now as it was then—it's almost a textbook-in-miniature of how fraud, deliberate or unconscious, can arise within the sciences.

> There are several species of impositions that have been practiced in science, which are but little known, except to the initiated, and which it may perhaps be possible to render quite intelligible to ordinary understandings. These may be classed under the heads of hoaxing, forging, trimming, and cooking.
>
> OF HOAXING. This, perhaps, will be better explained by an example. In the year 1788, M. Gioeni, a knight of Malta, published at Naples an account of a new family of Testacea, of which he described, with great minuteness, one species, the specific name of which has been taken from its habitat, and the generic he took from his own family, calling it Gioenia Sicula. . . . He gave figures of the animal, and of its parts; described its structure, its mode of advancing along the sand . . .
>
> The editors of the *Encyclopedie Methodique*, have copied this description, and have given figures of the Gioenia Sicula. The fact, however, is, that no such animal exists, but that the knight of Malta, finding on the Sicilian shores the three internal bones of one of the species of Bulla [*Bulla lignia*] . . . described and figured these bones most accurately, and drew the whole of the rest of the description from the stores of his own imagination.
>
> Such frauds are far from justifiable; the only excuse which has been made for them is, when they have been practiced on scientific academies which had reached the period of dotage. . . .
>
> FORGING differs from hoaxing, inasmuch as in the latter the deceit is intended to last for a time, and then be discovered, to the ridicule of those who have credited it; whereas the forger is one who, wishing to acquire a

reputation for science, records observations which he has never made. This is sometimes accomplished in astronomical observations by calculating the time and circumstances of the phenomenon from tables. The observations of the second comet of 1784, which was only seen by the Chevalier D'Angos, were long suspected to be a forgery, and were at length proved to be so by the calculations and reasonings of Encke. The pretended observations did not accord amongst each other in giving any possible orbit. . . .

Fortunately instances of the occurrence of forging are rare.

TRIMMING consists in clipping off little bits here and there from those observations which differ most in excess from the mean, and in sticking them on to those which are too small; a species of "equitable adjustment," as a radical would term it, which cannot be admitted in science.

This fraud is not perhaps so injurious (except to the character of the trimmer) as cooking, which the next paragraph will teach. The reason of this is, that the average given by the observations of the trimmer is the same, whether they are trimmed or untrimmed. His object is to gain a reputation for extreme accuracy in making observations; but from respect for truth, or from a prudent foresight, he does not distort the position of the fact he gets from nature, and it is usually difficult to detect him. He has more sense or less adventure than the Cook.

OF COOKING. This is an art of various forms, the object of which is to give to ordinary observations the appearance and character of those of the highest degree of accuracy.

One of its numerous processes is to make multitudes of observations, and out of these to select those only which agree, or very nearly agree. If a hundred observations are made, the cook must be very unlucky if he cannot pick out fifteen or twenty which will do for serving up.

Another approved [recipe], when the observations to be used will not come within the limit of accuracy, which it has been resolved they shall possess, is to calculate them by two different formulae. The difference in the constants employed in those formulae has sometimes a most happy effect in promoting unanimity amongst discordant measures. If still greater accuracy is required, three or more formulae can be used. . . .

In all these, and in numerous other cases, it would most probably happen that the cook would procure a temporary reputation for unrivalled accuracy at the expense of his permanent fame. It might also have the effect of rendering even all his crude observations of no value; for that part of the scientific world whose opinion is of most weight, is generally so unreasonable, as to neglect altogether the observations of those in whom they have, on any occasion, discovered traces of the artist. In fact, the character of an observer, as of a woman, if doubted is destroyed. . . .

That last observation is pretty ripe. Babbage's long-term colleague in his unsuccessful attempts to create a computer was Ada Lovelace. She wrote what can be regarded as the first computer program, and is thus generally accepted as a more significant figure in the history of computing than Babbage himself. There were plenty who doubted her character.

CHEATING

Even very distinguished scientists are not immune to the temptations of fraud. For some 1,500 years the western world regarded the geocentric cosmology of Ptolemy as the last word on the subject, and he was greatly admired for the way in which he had confirmed his theories by experiment. Even long after the Copernican Revolution made his cosmology outmoded, Ptolemy was still held in high regard *as a scientist*. It was only in the twentieth century that astronomers began to become more skeptical about some of his results: they seemed almost too good to be true . . . and in fact, on closer examination, they were. Further, it seemed improbable that he could have made some of the claimed observations at all. Once his stated results were fully analyzed, it was evident that a lot of his observations would make better sense had they been performed from about the latitude of the island of Rhodes. Ptolemy worked in Alexandria, but the great observational astronomer Hipparchus had worked in Rhodes a few centuries before him. Rather than make observations of his own, it seems Ptolemy spent his time in Alexandria's great Library cribbing many of Hipparchus's results and claiming them as his own.

The clincher came when modern researchers calculated the exact time of the autumnal equinox in the year 132 CE. Ptolemy recorded that he had observed it very carefully at 2:00 pm on September 25; in fact the equinox occurred that year at 9:54 am on September 24. Ptolemy was attempting to prove the accuracy of the determination Hipparchus had made of the length of the year; using as his base a record Hipparchus had made of observing the moment of equinox in 146 BCE, 278 years earlier, Ptolemy simply multiplied Hipparchus's figure for the year's length by 278. Unfortunately for Ptolemy's credibility, Hipparchus's figure was slightly off—hence the 28-hour disparity in 132 CE. And it clearly wasn't just that Ptolemy adjusted the time of the observation to suit his theory; it was that he never bothered to make the observation. Had he found the equinox arrived over a day ahead of schedule he'd have been in a position to make an *even better* calculation of the year's length than Hipparchus had—and, Ptolemy being Ptolemy, this would have been an achievement he'd have crowed about.

Ptolemy wasn't the only cosmologist to cheat in this fashion. Galileo Galilei certainly never conducted some of the most important experiments he reported having done when investigating gravity; those experiments would not have worked using the materials available at the time, as was evidenced by the fact that many of his contemporaries were puzzled when they couldn't replicate his results.*

John Dalton, famed as the propounder of the atomic theory of matter, is known to have fudged his experimental data in order to support that theory,** and even Sir Isaac Newton fiddled some of the mathematics in his *Principia* to present a more convincing case for his theory of universal gravitation. As Richard S. Westfall put it in *Never at Rest: A Biography of Isaac Newton* (1980), "Having proposed exact correlation as the criterion of truth, he took care to see that exact correlation was presented, whether or not it was properly achieved. . . . [N]o one can manipulate the fudge factor so effectively as the master mathematician himself."

Another significant figure who may have faked what he's best known for is Marco Polo, whose ghostwritten account of his travels in the orient, *Description of the World* (c1298), contains some curious lacunae: there's no mention of chopsticks, foot-binding or even the Great Wall of China. At the same time, Polo did mention certain other Chinese innovations unknown to Europeans of his day, such as paper money and the burning of coal. What seems very possible is that he heard about these and other items on the grapevine via the traders with whom his family did business, and constructed the rest of his account around them. We may never know the truth of the matter. Another possibility is that his ghostwriter, Rustichello da Pisa—to whom Polo supposedly dictated his memoir while they were in prison together in Genoa—invented most of the *Description* out of whole cloth, tossing in, for the sake of verisimilitude, bits and pieces of genuine information he'd heard from Polo. Or maybe Rustichello did his best based on a somewhat incoherent narrative from Polo?***

---

* The story of his dropping weights from the Leaning Tower of Pisa was a later invention, so that's at least one fictitious experiment about which he didn't lie.

** An aside: Dalton was profoundly red–green colorblind, and came to believe this was because the inside of his eyeballs must contain a blue stain—a blue filter would, he reasoned, block off the red and green light. He decreed that after his death his eyes should be dissected to prove or disprove his hypothesis. The dissection was duly done, and his hypothesis disproved. Even so, red–green colorblindness is still often called Daltonism.

*** We do know, though, that another account of exploration was fraudulent, even though it was given great credence in its day. This was *The Travels of John Mandeville* (c1371), and described the eponymous knight's journeys through much of the Near and Far East, where he encountered, so the book related, dog-headed humans, cyclopean giants, the famous

The German physiologist Ernst Heinrich Haeckel is best known today for his long-defunct Biogenetic Law—the notion that "ontogeny recapitulates phylogeny" (i.e., that the physiology of the developing vertebrate embryo mimics the evolutionary development of its species). The importance of the Biogenetic Law in scientific history is frequently overstated; it has often been claimed, for example, that it played a major part in the evolutionary thinking of Charles Darwin, although a closer look reveals that Darwin carefully made no mention of the Biogenetic Law even while acknowledging Haeckel's other, valid work in vertebrate physiology. Long after the hypothesis had lost all credibility, in the mid-1990s the UK embryologist Michael Richardson happened to notice that a particular diagram by Haeckel seemed curiously, well, wrong. This led Richardson and colleagues to conduct a more thorough examination, at the end of which they concluded Haeckel had extensively doctored his drawings of developing embryos, the main evidence supporting his Biogenetic Law, so as to fit the theory.

Louis Pasteur deceived in at least two of his famous demonstrations, those concerning the vaccination of sheep against anthrax and the inoculation of humans against rabies, in 1881 and 1885 respectively. In the first instance he pretended the vaccines had been prepared by a method of his own devising when in fact he'd had to fall back on a rival method devised by a colleague, Charles Chamberland. The fabrication in the second instance involved the experimental work Pasteur did, or claimed he did, with dogs before his first, successful inoculation of a human against rabies. That the rabies sufferer, Joseph Meister, was cured was strictly thanks to luck, largely unsupported by any canine experiments of Pasteur's. On the other hand, without the treatment Meister would almost certainly have died anyway, so presumably Pasteur convinced himself he was taking a legitimate risk. In both instances the motive for Pasteur's dishonesty appears to have had its roots in his desire always to present himself as the master-scientist.

TAKING CREDIT

Sir Charles Wheatstone was famed for his experiments in electricity; his name remains well known because of the Wheatstone bridge, that staple

---

vegetable lamb, and many more marvels besides. The book was not an entire fiction—the geography in it is as accurate as one could expect at the time—but there's no clear evidence that Sir John Mandeville himself ever existed outside the anonymous author's imagination.

of the school science laboratory. One has to wonder if Wheatstone himself was actually responsible for it, going by an incident that occurred in the 1840s. The Scottish inventor Alexander Bain is little remembered today, but he seems to have been possessed of the most astonishingly fertile brain, most of whose formidable powers he turned toward the then exciting new field of electricity. Among his countless inventions were various electric clocks, telegraphy systems and railway safety systems, a way of synchronizing remote clocks, the insulating of electric cables, the earth battery, and even the fax machine; he constructed a functional version of the last of these, the "electrochemical telegraph," to transmit images between the stations of Glasgow and Edinburgh.*

Developing such devices cost money, and Bain didn't have much of it. The editor of *Mechanics Magazine*, hugely impressed by Bain's work and sympathetic to his financial lack, arranged an introduction to Wheatstone. Bain came to London in 1840 and visited the distinguished scientist, demonstrating various items, including his electric clock. The older man was dismissive: these were wonderful inventions indeed, but no more than toys; the future of electricity lay elsewhere. Very fortunately, Bain ignored the advice and applied for patents anyway, because just a few months later Wheatstone demonstrated to the Royal Society . . . the electric clock, which he claimed to have invented himself!

Wheatstone then tried to have Bain's various patents struck down; he was unsuccessful, but set up the Electric Telegraph Company, making pirated use of several of Bain's inventions. His downfall came when he sought government funding for the company. The House of Lords held an inquiry at which Bain appeared as a witness. Wheatstone's theft was revealed, and the company was forced to make financial restitution to the Scot and acknowledge him, not Wheatstone, as the true inventor of several of the devices upon which its business relied.

Of course, in the US there was no House of Lords to see justice done, and Bain's inventions were stolen wholesale. He spent the latter years of his life embroiled in interminable legal actions trying to get some measure of compensation.

Edward Jenner, the "Father of Vaccination," similarly claimed credit that was not his due—but got away with it. According to all the standard histories, Jenner realized that immunity to the killer disease smallpox might

---

* Bain's "fax machine" certainly worked, but with the technology of the day could have been little more than a curio.

be conferred by inoculating people with the milder strain of the disease, cowpox, which commonly infected, but did not seriously discommode, farmers and milkmaids, who were well known to be less susceptible to smallpox than the general population. In 1796 Jenner inoculated an un-witting eight-year-old, James Phipps, with cowpox and then a few weeks later with smallpox, revealing that the cowpox had indeed made Phipps immune to the more serious disease. This was and is generally accepted as a magnificent medical achievement. In truth it was also an astonishing lapse of medical ethics: for all Jenner knew, the experiment could have killed the lad.

Or is that so? Jenner seems certainly to have been aware that over twen-ty years earlier, in 1774, the Dorset farmer Benjamin Jesty had successfully performed an exactly similar action on his family—in his case not as a scientific experiment but as an act of desperation, because the county was in the midst of a smallpox epidemic. Yet Jenner never acknowledged Jesty's precedence, even while accepting a large reward from a grateful Parlia-ment (in 1802); neither did his medical contemporaries, and almost with-out exception neither has history.

More recently, Alexander Fleming became famed for his discovery of penicillin and thus opener of the door for the host of antibiotics that have saved so many lives around the world. In fact, although Fleming discov-ered penicillin—in 1928, through accidental contamination of one of his culture dishes—and although he did name it and performed some desul-tory experiments with it as a possible bacteria-killer, he decided it was of little therapeutic value and moved on to other concerns.

A sample of penicillin was sent to the pathology department at Oxford University for use in an experiment; the stuff proved useless for that par-ticular experiment, but was kept in case it might be handy later. A new professor took over the department, Howard Florey; he knew about pen-icillin but, like Fleming, assumed it had little medical potential. Not so Ernst Chain, a young Jewish refugee from Hitler's Germany. A biochemist, Chain was fascinated when he came across a reference to Fleming's discov-ery and investigations.

Fleming had discarded penicillin because of its instability: it was effec-tive in killing bacteria, but only briefly. Assisted by Florey, his professor, Chain extracted a stable version, and the age of antibiotics was born. It was only then that Fleming realized the value of the discovery he had made years earlier. That discovery had been made at St. Mary's Hospital, Lon-don, where he worked. A governor of St. Mary's, the prominent physician

Lord Moran, was a close friend of the newspaper magnate Lord Beaverbrook. It would be in the hospital's interest to be associated with the antibiotics revolution, and so Beaverbrook focused the spotlight exclusively on Fleming—who, to his immense discredit, played along, treating Chain's and Florey's "contribution" dismissively.* This naturally infuriated Florey, who protested to the Royal Society and the Medical Research Council; but it was not in the interests of either organization to expose Fleming. Some restitution came in 1945 when the Nobel Physiology or Medicine Prize went to Chain, Florey and Fleming, but even then there was an injustice: anyone can make an accidental discovery; the real distinction for which the prize should be awarded is the realization of the discovery's importance and the experimental genius in making it exploitable as a life-saver.

There have been several celebrated instances in which Nobel Prizes have been awarded to one scientist for what has been mainly the work of another. Perhaps the best-known example is that of Antony Hewish, who received the 1974 Nobel Physics Prize for the discovery and early investigation of pulsars—work that had largely been done by his student Jocelyn Bell (now Jocelyn Bell Burnell). Hewish certainly supervised and guided her research, and played a significant part in the interpretation of her results, so there's no question of his not deserving at least a share in the Nobel accolade, and he himself has never sought to eclipse her glory; the fault in this instance lies with the awarders.

In two other cases of the mentor-student relationship, however, the behavior of the mentor seems more dubious. The US physicist Robert Millikan received the 1923 Physics Nobel for his work establishing the electrical charge of the electron. In essence, the relevant experiment involved measuring the charge on tiny oil droplets as they fell through an electric field, and calculating from the results to reach a figure one could deduce as being the charge an individual electron must bear. An experiment to attempt this using water droplets had been devised at Cambridge University but had given no good results (the water droplets tended to evaporate almost immediately); Millikan's laboratory at the University of Chicago was trying in 1909 to refine this in hopes of getting more usable results. After

---

* Decades later, Dr Christiaan Barnard behaved similarly towards the acclaim he received as the surgeon responsible for the world's first successful heart transplant. In fact, all of Barnard's heart-transplant operations were failures, as other surgeons had predicted they would be. They knew the science concerning tissue rejection was not ready for transplantation, so had held back from attempting the relatively simple operation.

months in which not much progress was made, a new arrival, graduate student Harvey Fletcher, had the idea of using oil droplets instead of water droplets. In Millikan's absence on other business, Fletcher set up an appropriate experiment and it worked beautifully. Thereafter the two men collaborated to achieve the goal of establishing the electron's charge, and they jointly prepared a paper to this effect that was published in 1910. However, to Fletcher's dismay, Millikan cited university protocol regulations to insist the paper be published as by Millikan alone. That paper brought Millikan a Nobel Prize; it is to Fletcher's great honor that he never expressed any ill will over the matter.

A similar controversy surrounded the award of the 1952 Physiology or Medicine Nobel to the US biochemist Selman Waksman for the discovery of streptomycin, the antibiotic that effectively conquered tuberculosis. In this instance the discovery had been made, albeit under Waksman's overall direction, by his doctoral student Albert Schatz, and the two not only worked together thereafter on streptomycin but were listed as coauthors of the pertinent paper. Even so, the Nobel went to Waksman alone.

So far, so much like the later experience of Hewish and Bell. However, Waksman then felt entitled to patent streptomycin in his name only, charging the pharmaceutical companies royalties for its manufacture. These royalties were very substantial indeed, and, even though much of the money was plowed by Waksman back into science—establishing the Waksman Institute of Microbiology at Rutgers University—Schatz not unnaturally felt he should have a share, and sued. The justice of his case was obvious: streptomycin had been a joint venture, and so the rewards likewise should be joint. This wasn't, however, how the US scientific community saw it: they were horrified any student should have the temerity to sue his mentor. All US scientific doors were firmly slammed in Schatz's face, and eventually he was forced to emigrate to South America to find work. To this day Waksman is almost always given sole credit for what is more correctly termed the Waksman–Schatz discovery of streptomycin.

Far more complicated in motivation was the public humiliation heaped by Sir Arthur Eddington—during the first one-third of the twentieth century the titan of astrophysics, a discipline he had almost singlehandedly created—on the young mathematical physicist Subrahmanyan Chandrasekhar. In 1935 Chandrasekhar, working under Eddington, presented a paper which proved—so far as any theoretical prediction *can* prove—that stars of mass above about 1.4 times that of our sun will at the end of their lives collapse infinitely. This was the first scientific prediction of the existence of black holes. What Chandrasekhar did not know was that for some years Eddington had been working on a (largely nonsensical) Grand Universal Theory, which theory would be obviated entirely if stars could collapse into singularities. At the presentation of Chandrasekhar's paper Eddington behaved abominably, making his young protege a laughingstock in the secure knowledge that none present would dare challenge the Father of Astrophysics.

Luckily Chandrasekhar refused to take this lying down, and persisted over subsequent years in maintaining the veracity of his calculations—which were verified independently, but covertly, by such pillars of the physics community as Niels Bohr and Paul Dirac. Meanwhile Eddington used his prominence to lambast Chandrasekhar at every opportunity, framing Chandrasekhar's math as incompetent while himself using numerous fudges—cheating, in other words—in his doomed attempt to prop up his own hypothesis. The result was that Chandrasekhar's prediction of complete stellar collapse was lost to astrophysics for something like four decades. Only in 1983 was he awarded his thoroughly merited Nobel Prize for this work. Leaving his personal affront aside, the disadvantage to astrophysics through the long delay in recognizing black holes was immense—and all because of Eddington's stubbornness and the sycophancy of the rest of the astrophysics community.*

OF DUBIOUS HEREDITY

In 1865, just six years after the publication of Darwin's *On the Origin of Species* (1859), the Austrian monk Gregor Mendel published the results of painstaking long-term experiments he had done with generations of garden peas. His concern was with the way in which characteristics were passed down from parent peas to offspring peas, and it soon became evident to him that these characteristics were embodied in the form of units that were essentially unchanging from one generation to the next: the offspring of (say) a tall and a short organism would not be all of medium height but, rather, some would be tall and some would be short. What made the offspring different from their parents was the shuffling of these various trait-determining units. What Mendel had discovered was the idea of the gene, and hence the whole foundation of the modern science of heredity.

---

* Arthur I. Miller's *Empire of the Stars* (2005) is a highly readable account of this whole affair.

The importance of Mendel's paper was not understood until 1900, when Carl Correns and Erich von Tschermak-Seysenegg realized that, if this was the way inheritance worked in peas, surely so must it work in all organisms, humans included. There was a revolution in the biological sciences. Even then, biologists didn't quite grasp the tiger they had by the tail: for decades it was assumed genes were somewhat abstract entities, codings rather than actual pieces of matter that could be examined and isolated. Only with the unraveling of the structure of DNA did it come to be appreciated what a gene actually is. Today we talk blithely of splicing and engineering genes; just a few decades ago such notions would have seemed science fiction.

Much earlier than this, in the mid-1930s, the UK statistical geneticist Sir Ronald Fisher analyzed Mendel's published data afresh. He found nothing wrong with Mendel's conclusions, and that many of his experiments could be easily replicated to give the expected results, but that in some instances Mendel's figures were so precisely aligned to the theoretical ideal that they represented a statistical near-impossibility. There were too many spot-on results in Mendel's experiments for credibility.

What exactly happened? Did Mendel, knowing he'd uncovered a truth about the way characteristics are inherited, fudge some of his data or even invent a few extra sets of experimental results to bolster his case? Was he unconsciously biased when recording his observations, seeing what he expected to see? Another possibility is that he did not in fact perform every part of every experiment himself, but had one or more assistants at work in the monastery gardens: it would be far from the first time an assistant skewed experimental results in order to make the boss happy.

This desire to please is frequently regarded, thanks to a book on the subject by Arthur Koestler, as the most probable explanation for the celebrated case of the midwife toad. The Austrian biologist Paul Kammerer performed a number of experiments in the first part of the twentieth century that appeared to confirm the Lamarckian idea of evolution through inheritance by offspring of characteristics acquired by the parents during their lifetime, rather than, or at the very least alongside, the theory of evolution through natural selection. The most dramatic of his results showed the apparent inheritance in normally land-breeding midwife toads, when forced to breed in water, of a callosity of the male palm that's found only in water-breeding toads. In water-breeding toads this callosity helps the male hold onto the female during mating; it hasn't developed in the land-breeding midwife toad because, basically, mating is easier on land. For it to appear as a congenital characteristic after just a few generations of enforced water-breeding implied that offspring were inheriting their parents' acquired characteristics.

In 1926 G.K. Noble examined Kammerer's specimen toad and found the dark patch which Kammerer had claimed as the relevant callosity was in fact due to an injection of India ink. Disgraced, Kammerer soon afterward committed suicide, and for a long while it was assumed he was a faker. It's now thought at least possible that his assistant "helped" the experiments along.

So, were the other monks "obligingly" serving up to Mendel the results he wanted? Or, feasibly, were they bored by the tedious pea-counting assignments he'd doled out to them? Did they—not realizing the importance of the experiment—invent results so they could bunk off? Whatever the truth, it's a remarkable instance of a breakthrough of paramount importance being made at least partly on the basis of faked experiments.

# # #

It's a lot harder to blame an anonymous assistant in a subsequent affair that has some echoes of the Kammerer fiasco. This time the scientist at the center of it all was William Summerlin, working in the early 1970s at, successively, Palo Alto Veterans Hospital, the University of Minnesota, and New York's Sloan–Kettering Institute. Summerlin's field was dermatology, specifically the problem of skin grafting. At the time a major problem in any attempt at grafting and transplantation was the natural inclination of the recipient's immune system to reject the grafted or implanted tissue. Summerlin was working on this using black mice and white mice as donors and hosts so the grafted skin would show up clearly. It was while he was at Sloan–Kettering that he was caught faking by an alert assistant, James Martin. Martin noticed something odd about the black "skin transplant" one of the white mice had received, and discovered it wasn't a skin transplant at all but had been drawn on with a felt-tipped pen! In the ensuing investigation it was discovered Summerlin's entire career of transplantation was full of dubious results, some seemingly the product of self-deception but most actually fraudulent.

In 1981–82 a similar tale emerged at the Department of Physiology at Harvard Medical School, where a young doctor called John Darsee had been researching into the efficacy of various drugs in the immediate aftermath of heart attacks. Caught faking one experiment, Darsee was permitted to continue his researches but now under close supervision. But not close enough. In the end almost all of the papers he'd published during a seven-year career had to be retracted, quite a few on the grounds that they were total fictions—complete with fictitious collaborators. The Darsee case became a *cause célèbre* in not just the scientific but the political world, primarily because of its poor—and slow—handling by Darsee's superiors and the scientific institutions concerned. This matter of poor, slow reaction to scientific fraud has been, alas, typical of many cases that have emerged over the past few decades.

PILTDOWN MAN

Probably the greatest paleontological embarrassment of all has concerned a fossil assemblage that appeared to be the remains of a Neanderthal-style hominid. The assemblage was "discovered" in early 1912 in a gravel pit near Piltdown Common, Fletching, Sussex, UK, and was named Piltdown Man, *Eanthropus dawsonii.*

The historical stage had been set for the "discovery." British paleontology seemed to have been in the doldrums for decades. While workers on the Continent were unearthing exciting fossil hominids, their UK counterparts were lagging sadly behind. Patriots were desperate to convince themselves that, even in the field of prehistoric hominids, British was best.

The solidity of this mindset among establishment scientists can be assessed by the reception given to Galley Hill Man. In 1888, at Galley Hill in Kent, a very modern-looking human skull was discovered by chalk workers and excavated by two local amateurs, Robert Elliott and Matthew Heys. After their initial examination of it, the skull found its way to Sir Arthur Keith, head of the Royal College of Surgeons. Basing his assessment on the stratum in which the skull had been found, Keith was swift to go out on a limb and say it was of Neanderthal vintage, if not older. This seemed to confirm to him the presapiens hypothesis of Pierre Marcellin Boule and others that humans indistinguishable from moderns had existed in eras of enormous antiquity. In fact, the skeleton in due course proved to be that of a Bronze Age individual shoved down unwittingly into lower strata by later chalk digging. Those who'd shyly suggested the fossil skeleton might be comparatively recent had been, at least at first, generally dismissed—if not ridiculed—as "overcautious."

So, when amateur geologist Charles Dawson made his "find" near Piltdown Common, he was assured of at least a tolerant reception. His further searches were assisted by some quite eminent figures of the day, notably the distinguished geologist Sir Arthur Smith Woodward, who examined the skull and proclaimed it that of a hominid, and the French polymath and Jesuit priest Pierre Teilhard de Chardin. Their haul included a fossil hominid jaw and various bits of animals otherwise unknown to the English south coast.

Despite occasional skeptical comment, Piltdown Man was for over forty years accepted as the UK's great paleontological discovery, although it was difficult to fit it into the accepted "map" of hominid evolution—indeed, that was the very reason Piltdown Man was so interesting.* Not until 1953 was the "fossil" shown—by J.S. Weiner, K.P. Oakley and W.E. Le Gros Clark—to be part of the skull of a relatively recent human being plus the jaw of a 500-years-dead orangutan, the remains having been stained to give the illusion of antiquity and to make them better match each other visually. The success of the deception owed less to the rudimentary nature of scientific dating techniques in the early twentieth century and more to the fact that no one really wanted to believe Piltdown Man wasn't genuine.

The identity of the hoaxer is still not known for certain. In *The Piltdown Men* (1972), Ronald Millar points to Sir Grafton Eliot Smith, the distinguished Australian anatomist and ethnologist, who was in the UK at the time and who was well known for his sense of fun. In a famous essay, "The Piltdown Conspiracy" (1980), Stephen Jay Gould turns the spotlight on Teilhard de Chardin. However, Charles Dawson—who on occasion had actually been seen experimenting on the staining of bones—has generally been regarded as the prime suspect.

There is, though, yet a further wrinkle to the story. In the late 1970s, a box was discovered in the British Museum's attic containing bones stained in the same way as those of Piltdown Man. The box had belonged to Martin A.C. Hinton, later the Museum's Keeper of Zoology, but at the time a volunteer worker there. In 1910, he had been denied a research grant application by Smith Woodward; just two years later the Piltdown remains were unearthed. Revenge? Today, Hinton is often cited as a likely culprit.

However, this may be a misjudgment. A more probable hypothesis, backed by anecdotal evidence, is that Hinton was initially just an observer of the hoax, and suspected Dawson as the perpetrator. As he watched Smith Woodward's claims for Piltdown Man escalate, though, Hinton decided to prick the bubble while at the same time making Smith Woodward

---

* At a less scholarly level, the US clairvoyant Edgar Cayce received the psychic insight that *Eanthropus dawsonii* was in fact an Atlantean, one of a number of refugees from the doomed continent who had fled to the British Isles.

look a fool; he therefore planted some extra remains that were all too obviously bogus: a reported example was a cricket bat hewn from an elephant's leg-bone. To Hinton's horror, however, when the cricket bat was unearthed Smith Woodward proudly announced it to the world as a prehistoric artifact of hitherto unknown type. At that point Hinton threw his hands in the air and gave up.

A decade after the Hinton discovery, yet another suspect emerged: William Sollas. He was fingered in a posthumously revealed tape recording made ten years earlier by J.A. Douglas. Both men had been professors of geology at Oxford University, Sollas at the time of the Piltdown Man sensation and Douglas succeeding him in the chair. Douglas could recall instances of Sollas purloining bones, teeth and similar items from the departmental stores for no specified purpose, and it's a matter of public record that Sollas (like many others in the field) thought Smith Woodward an overweening buffoon, such as it might be a pleasure to expose through a prank. At this late stage it's impossible for us to evaluate the importance of Douglas's circumstantial evidence.

A further possibly relevant fact is that we now know Dawson had a secret career as a plagiarist. At least half of his critically slammed but popularly well received two-volume *History of Hastings Castle* (1910) was discovered in 1952 to have been copied from an unpublished manuscript by one William Herbert, who had been in charge of some excavations done at the castle in 1824. And nearly half of a paper Dawson published in 1903 in *Sussex Archaeological Collections* is reputed to have been lifted verbatim from an earlier piece by a writer called Topley, although in this instance the details are vague. But how could any evidence defame him when he had the supportive testimony of one Margaret Morse Boycott?

> Mr. Dawson and I were members of the Piltdown Golf Club. Let me tell you this. He was an insignificant little fellow who wore spectacles and a bowler hat. Certainly not the sort who would put over a fast one.

ARCHAEORAPTOR

The November 1999 issue of the magazine *National Geographic* caused a sensation when it announced the discovery of a fossil that represented the missing link between dinosaurs and birds. In this article Christopher Sloan, the magazine's art editor, proposed the name *Archaeoraptor liaoningensis* for the fossil, whose stated history was that it had been discovered in China and smuggled thence to the USA, where it was bought at the annual gem and mineral show (the world's largest) in Tucson, Arizona, by Stephen Czerkas, owner of the Dinosaur Museum in Blanding, Utah. What Sloan didn't mention in his article was that his own scientific paper on the fossil had been rejected by *Nature* and *Science*.

Czerkas did the decent thing and sent the fossil back to China, where it was examined at the Institute of Vertebrate Paleontology and Paleo-athropology, Beijing. A researcher there, Xu Xing, soon spotted a strong resemblance between the rear half of the supposed fossil and that of an as yet unnamed dinosaur he'd studied; the front half, however, was quite different. Eventually the fraud was traced to a Chinese farmer, who had created the chimera by fixing bits of two different fossils together with very strong glue. The dinosaur whose rear Xu had recognized is now called *Microraptor zhaoianus*; the front part of *Archaeoraptor* was identified in 2002 as belonging to the ancestral bird species *Yanornis martini*.

Occasionally creationists pick on the *Archaeoraptor* fiasco as an example of how far "evolutionists" will go to "bolster their theory"; it is, so to speak, the "Piltdown bird." The creationist reasoning seems to be that, if *Archaeoraptor* were a convincing transitional fossil between dinosaurs and birds, its existence would undermine their own claims that evolution between "kinds" is impossible. What they conveniently ignore is that there are other fossils that fill the bill; in fact, *Yanornis martini*, used for the front part of the fake, is itself a good example of a transitional dinosaur/bird fossil.

THE IQ FRAUDS

In the field of human intelligence, two major arguments have been going on for decades. The first of these concerns IQ tests, and the extent to which they offer—or can offer—an accurate assessment of an individual's intelligence. The problem is that no one is certain exactly what it is that an IQ test is measuring, beyond the greater or lesser ability to do well in IQ tests. The first IQ tests were designed purely to determine whether or not certain borderline-retarded children would be able to cope with school. At this the tests seemed successful. Accordingly, enthusiasts seized upon them, expanded their scope far beyond the originators' conception, and declared they were intelligence tests applicable to anyone. What was forgotten during all this was that no one had—or has—been able to decide exactly what intelligence *is*: putative definitions are countless, and none enjoys universal support.

Moreover, any IQ test will inevitably be colored by the cultural perceptions of the person(s) who devised the test. IQ tests are superficially not tests of knowledge (or, at least, shouldn't be), but instead tests of reasoning facility; in fact, however, they rely upon elements that, while they may be absolutely basic knowledge in one culture, may not be at all so in another. The idea that the tests can be applied with geographic abandon has slowly withered; the idea of their applicability across cultural and/or social divides within the same geographical area has proved somewhat more resistant.

The second controversy concerns the matter of what governs the intelligence of an individual: is it a matter of heredity, or is it largely controlled by upbringing? The scientific debate over nature vs. nurture has overwhelmingly swung toward the "nurture" side of the argument, but controversy persists among media pundits and public. The issue is of importance because among those insistent that the major component of intelligence is inherited is a small but vociferous racist coterie intent on finding "scientific" support for their prejudices. The irony is that xenophobia, which is what many of them are displaying, is generally a trait of the least educated and least intelligent.

An extreme example of an argument in favor of the "nature" side of the nature/nurture equation is "The Fallacy of Environmentalism" (1979) by H.D. Purcell.[1] Purcell warns against the danger of treating all people as equal. Rather than dispassionately presenting the defensible case that heredity *as well as* environment plays a part in determining an individual's ability, he attacks everyone who, he thinks, disagrees with him: they are "liberals" or even, horror of horrors, "Marxists." The views of the latter aren't important because, of course, all Marxists believe in Lysenkoism! (see pp. 343–351)

His real concern is, apparently, that researchers who come up with statistics which show that, say, one race does better than another at IQ tests are subject to critical attack. What Purcell cheerily ignores is that, since IQ tests are themselves of debatable validity, the statistics he's so keen to defend are based on at least questionable data. As A.A. Abbie pointed out forcefully in *The Original Australians* (1964) on the subject of Australian Aborigines' poor results in such tests, *life itself* is an intelligence test for desert nomads: pass it or die.

The degree to which IQ tests could be put to the service of racist conclusions was exemplified by *The Bell Curve: Intelligence and Class Structure in American Life* (1994) by Richard Herrnstein and Charles Murray. The authors believed IQ was at least to a certain extent inherited, and pointed out a strong correlation between an individual's IQ and their eventual position in life: US society, they claimed, has to a large degree become stratified on the basis of IQ. A major flaw in their argument is again that IQ is certainly not a universally reliable indicator of intelligence. However, if we substitute "intelligence" for "IQ" into the authors' argument, then their point that the strata of the US's class-structured society correlate with intelligence has some limited truth (although Donald Trump, Jr. and Paris Hilton provide instructive counterexamples). But it is an inescapable fact that US society is also stratified along racial lines. The authors, ingenuously or disingenuously, pointed to certain results that showed African Americans on average doing worse at IQ tests than Asian Americans, with white Americans somewhere in between. There are very obvious cultural reasons for this (not least the matter of inherited wealth), which the authors failed to stress. The book created a *cause célèbre*, with people on one side attacking it as racist while those on the other seized on what they saw as "scientific" justification for their own racism: it was only right and proper that African Americans should be at the bottom of the heap. They were, oddly, less keen to infer that Asian Americans should be at the *top* of the heap.

As an aside, beauty is at least as significant a determinant as intelligence on where the individual ends up in the social hierarchy, yet there are few who would (publicly) promote the notion that the plain are inferior to the beautiful.

For decades the nature vs nurture debate was vastly complicated by the fraudulence of one man, a UK scientist called Sir Cyril Burt. Burt was a consultant psychologist to the Greater London Council (GLC) in the field of education for some two decades until 1932, thereafter being Professor of Psychology at University College, London, until his retirement in 1950. During his time with the GLC he was responsible for the devising and administering of intelligence tests, advising also on such matters as juvenile delinquency and the psychology of education. Not unreasonably, Burt collated the results of his IQ tests, and he became interested in the nature vs. nurture discussion.

After his retirement he returned to the matter, and soon produced some statistics that appeared to conclusively back his belief that heredity was the sole significant determinant of human intelligence. He had, he claimed, located over fifty sets of identical twins who had been separated at birth and raised in different families; despite the difference in upbringing, the pairs of twins showed a very close correlation of IQs.

The matter seemed to be more or less proven, and for a long time psychologists accepted Burt's findings at face value. A year after Burt's death, however, the Princeton psychologist Leon J. Kamin realized Burt's figures on the identical twins seemed too good to be true; Burt's earlier average correlation of IQs between the pairs, when he had been working with only about twenty pairs, matched to the third decimal place the average he claimed later, when supposedly working with over fifty pairs. The chances of this happening are statistically close to zero. Kamin also noted some vital logistic details missing from the experimental reports; for example, who had actually administered particular tests. On several occasions Burt referred to "fuller accounts" that could be found in "degree theses of the investigators mentioned in the text"—meaning, in effect, that they were unavailable, because tracking down a degree thesis is murderously difficult, even in the Google era. A clue appeared in a 1943 paper of Burt's, though: the degree essay cited was "filed in the Psychological Laboratory, University College, London." Kamin investigated and discovered that, not only was there no such essay filed at the university, none had ever been submitted—indeed, there was no evidence that even its supposed author had ever existed outside of Burt's imagination. In *Science and Politics of IQ* (1974) Kamin cast considerable doubt—to put it mildly—on Burt's findings.

Then, in 1976, the London *Sunday Times*'s medical correspondent, Oliver Gillie, conducted some further in-depth detective work. Among the many damning new items he uncovered was that the two women Burt had cited as his research assistants, Margaret Howard and J. Conway, seemed chimerical. It became evident Burt had very copiously faked his results—and not only in this particular study but also in others relating to the purported hereditary nature of intelligence.

This was embarrassing for people who had used Burt's findings as the basis for their own studies, among them the US psychologist Arthur Jensen. In 1969, Jensen had published in the *Harvard Educational Review* a controversial paper, "How Much Can We Boost IQ and Scholastic Achievement?", that purported to show African Americans scored less well in IQ tests than white Americans because they were inherently less intelligent, and not because black children were (as they still are) educationally deprived to a truly shocking extent. Jensen, along with his mentor, the prominent UK psychologist and author H.J. Eysenck—who'd studied under Burt—were two among an ever-dwindling few who continued to defend Burt after the revelations by Kamin and Gillie.

Little Albert

Burt is far from the only noted psychologist guilty of fakery. The US psychologist John Broadus Watson, a pioneer of behaviorism, made much of the conditioning experiments he had done with an 11-month-old infant, Little Albert. For decades these experiments were regarded as an important underpinning for behaviorism, even though other researchers had difficulty replicating them. Only in the 1980s were psychologists willing to come out into the open and say that, while Little Albert had certainly been a real child, either Watson had never performed the groundbreaking experiments with him or, if he had, he had misreported the results.

Subliminal Advertising

Also of psychological interest is the concept of subliminal advertising. This seems first to have surfaced in a 1957 report by US market researcher James Vicary that he had exposed over 40,000 New Jersey cinema goers to fleeting messages embedded in the movies they watched telling them to eat popcorn and drink Coca Cola; the result had been a 57 percent increase in popcorn sales and an 18 percent increase in soft-drink sales. Advertisers rushed to embed subliminal messages in their ads; contrariwise, consumer associations mounted protests. A popular and influential book on the subject was *Subliminal Seduction* (1973) by Wilson Bryan Key.

However, follow-up scientific research could find no effect one way or the other from subliminal messages, and in 1962 Vicary confessed he'd made up his figures; it's doubtful he even did the experiment. The notion of subliminal advertising was thus tossed into the litter bin where so many initially plausible scientific ideas go. Nonetheless, hopeful advertisers still apparently use the technique, if we're to judge by August Bullock's 2004 book *The Secret Sales Pitch: An Overview of Subliminal Advertising*.

Cures for Cancer

Of all the quacks, perhaps the most contemptible are those who peddle false cures for cancer. The deaths they have caused, through diverting their victims from therapies that might at the very least have offered a chance for survival, are a matter of record, and yet these individuals have shown the tenacity of a nasty yeast infection, using loopholes in the law to extend, often by decades, their lethal and avaricious careers.

It was a case of cancer that succeeded finally in bringing to an end the career of Ruth B. Drown, a quack who spent a lifetime using her remarkable Radio Therapeutic Instrument.* Mrs. Marguerite Rice of Illinois was told by her doctor in 1948 that a lump in her breast might signify cancer; he urged her to have a biopsy taken. But Mrs. Rice was frightened of the procedure, so she and her husband instead consulted Drown. Drown referred the Rices to a local practitioner, Dr. Findley D. John, and Mrs. Rice had several weeks' worth of expensive sessions with him before Mr. Rice decided it might be cheaper to invest in their own personal Radio Therapeutic Instrument. That bought, again not cheaply, Mrs. Rice spent hours each day "completing the circuit."

Some while later the Rices came across a newspaper article debunking Drown's therapy. They rapidly sought the medical attention they should have sought in the first place, and at the same time contacted the American Medical Association. The AMA in turn advised the Food and Drug Administration (FDA), who had for some time been looking for a means of prosecuting Drown but had had their hands tied by legal niceties. In the event it was not until 1951 that the FDA was finally able, bolstered by the Rices' evidence, to bring Drown to trial, in her home state of California—although by this time Mrs. Rice, her cancer inoperable, was too weak to attend. Drown was found guilty and fined $1,000 (nearly $10,000 in today's terms). Even after this, she could still continue to practice in California so long as she didn't sell her devices across state lines. So she carried on, with continued financial success. Only in 1963 did the California State Bureau of Food and Drug Inspection bring a case against her, and in the event Drown died before that case could be brought to trial. Mrs. Rice, of course, had died long since.*

Even more tortuous and even more alarming was the instance of Harry Mathias Hoxsey. Hoxsey claimed to have inherited his cancer cure from his father, who had died—of cancer!—in 1919; Hoxsey senior supposedly had observed how a horse had cured itself of a leg cancer by standing in a patch of certain shrubs, and had gone on from there to treat first animals and then human patients. He had passed the secret to young Harry on his deathbed.** However much truth there is in this story, it was in about 1922 that Hoxsey first used the compound, a paste, smearing it onto the lip cancer of a Civil War veteran and apparently curing him. It is indeed possible

---

* A far more detailed account than is possible here of both this case and the succeeding one can be found in James Harvey Young's *The Medical Messiahs* (2nd edition 1992)

** Hoxsey told various versions of the tale, but this is the gist of them all.

that the man was cured, because some years later, when the AMA was able to get hold of a sample of Hoxsey's paste, they discovered its active ingredient was arsenic. Arsenic is in this context an escharotic: it corrodes away nearby cells, cancer cells included. That it also destroys healthy cells may or may not be important, depending on the circumstances; more usually it is, and the dangers become even more extreme should the arsenic get into the bloodstream, as any mystery-fiction fan knows.

In 1924, Hoxsey opened a practice in Taylorville, Illinois, at the same time co-founding the National Cancer Research Institute—a grand-sounding name for a commercial operation designed to market Hoxsey's quackery. The other co-founders soon pulled out, so Hoxsey opened instead the Hoxide Institute, advertising for patients. Soon local newspapers were running articles about deaths at the Institute, while the *Journal of the American Medical Association* took up the cudgels concerning Hoxsey's fraudulence. Hoxsey responded by suing the AMA, a case that was dismissed. In a separate case, however, Hoxsey was convicted of practicing medicine without a license and fined $100, and in 1928 the Hoxide Institute closed.

At least in Taylorville. After a brief attempt to reopen for business in Jacksonville, Illinois, Hoxsey returned to his hometown of Girard, Illinois, with more success—although Arthur J. Cramp, a contemporary scourge of quacks, commented dourly: "Perhaps Girard will flourish briefly—especially the local undertaker and those individuals who have rooms to rent." However, as in Jacksonville, Hoxsey's Girard venture petered out swiftly. The story was repeated in several other states. In 1936 he reached Dallas, Texas, where he set up in grander style than ever before. But yet again he was convicted of practicing without a license, and this time the punishment—a five-month jail sentence and a $25,000 fine—was more than a slap on the wrist; unfortunately, the verdict was overturned by a higher court. Hoxsey then managed to obtain a Texas license to practice naturopathy, which made it difficult for the authorities to pursue him further.

By now he was going beyond the treatment of external cancers, prescribing "chemicals" for the treatment of internal tumors. The medical hypothesis he put forward to justify the use of his medicines was that cancer was a result of chemical imbalance within the body; his chemicals supposedly restored the balance, thereby checking the cancer and driving it out.

Visitors to his clinic were subjected to an impressive battery of diagnostic tests, the results of which indicated surprisingly often that the patient was indeed suffering from cancer. Later, when challenged in and out of court, Hoxsey and his "Medical Director," J.B. Durkee, could claim impressively high cure rates precisely because so many of their "successfully cured" patients proved, on proper examination, never to have suffered from cancer in the first place.

To the modern mind it seems completely incomprehensible that Hoxsey and his cronies could have been permitted to continue in their careers of destroying lives, but this is to underestimate the gravity of the matter. Hoxsey played politics astutely to become a significant figure, drawing prominent senators into his web. Later on he was allied for a while with the pro-Nazi Christian evangelist Gerald B. Winrod, upon whom Sinclair Lewis based his fascist demagogue Buzz Windrip in the classic dire-warning novel *It Can't Happen Here* (1935). Winrod accepted large sums from Hoxsey to plug the latter's fraudulent cures in Winrod's racist and antisemitic newspaper *The Defender*. Winrod's was not the only fascist organization Hoxsey flirted with.

Despite his connections in high places, legal proceedings dogged Hoxsey. Some were self-inflicted. In 1947, Morris Fishbein described Hoxsey in the *Journal of the American Medical Association* as a "cancer charlatan," and Hoxsey sued him for libel. He won the case, but was awarded not the million dollars he sought but a dollar apiece for himself and his dead father, whom he claimed had also been libeled.

The judge in that case, William Hawley Atwell, was rumored to have been one of Hoxsey's "satisfied customers," and so should have recused himself. He didn't. Neither did he recuse himself when the FDA brought their major case against Hoxsey, in 1950, accusing him of shipping fraudulent medicines across state lines; Atwell's conduct of the proceedings in court was so manifestly biased in Hoxsey's favor that it came as no surprise to the FDA that they lost the case. A prompt appeal to a higher court reversed the verdict, and Atwell was instructed to issue an injunction. This he did, but in such watered-down form that it had no effective worth. Only on a second attempt was Atwell compelled to issue the injunction the FDA had initially requested.*

But Hoxsey was still free to continue his practice in Dallas! He published a ghostwritten autobiography, *You Don't Have to Die* (1956), which sold well and attracted further customers. And, in James Harvey Young's memorable phrase, "Hoxsey's cancer treatment metastasized from Texas

---

* Atwell found further ignominy when in 1957 he again had to be forced by higher courts to abide by the law, this time the law concerning school desegregation.

into other states." All this time the FDA was powerless to do anything, because, incredibly, Hoxsey's activities were within the law.

The FDA was, however, entitled to issue public warnings about dangerous drugs, and in 1956 it began another campaign against him. Hoxsey brought a lawsuit against the FDA, but lost. Finally emboldened, the Texas legal authorities began to take action. Hoxsey did not give in without a struggle, but by about 1960 the cancer of the Hoxsey cure had been more or less stamped out.

But not entirely. As late as 1987 the feature-movie documentary *Hoxsey: How Healing Becomes a Crime* sought in part to rehabilitate his reputation.

While supercures for cancer have been almost the exclusive province of the quack, there have been a few touters of miracle cancer cures who have, at least for a while, seemed perfectly respectable—and sometimes even eminent. One such was Abdul Fatah al-Sayyab, a scientist who in the early 1970s rose to a position of considerable prominence in the Iraqi government thanks to the two drugs he had derived, tactfully called Bakrin and Saddamine in honor of the president and vice-president of Iraq at the time, Ahmed Hassan al-Bakr and Saddam Hussein. According to al-Sayyab, these drugs could cure certain types of cancer—hence his rise to high political status. In due course, alas, it was discovered the drugs had no effect on cancerous growth whatsoever, and al-Sayyab fell from grace.

The early 1980s saw a much more serious claim to have gotten a handle on cancer, thanks to Professor Efraim Racker of Cornell University and his graduate student Mark Spector. In consequence of a remarkable series of experiments performed by Spector, they put forward a new theory, the kinase-cascade theory, to explain the cause of at least some forms of cancer. In essence, Racker suspected that malfunctioning of a particular enzyme present in the walls of all cells, sodium–potassium ATPase, was symptomatic of cancer cells. Spector's experiments not only confirmed this but went further: the malfunctioning of the ATPase was due to the action of another enzyme present in all cells, a protein-kinase. Harmless in normal cells, the protein-kinase became actively harmful in cells that turned cancerous. More experiments showed that it, too, was being triggered by an enzyme—another kinase, in fact. All told, Spector reported, there were four kinases involved, acting in cascade fashion each to modify the next, with the last modifying the ATPase.

Spector then tied this model in with a recent discovery concerning the genes of viruses that cause cancer tumors in animals. The gene responsible, the so-called sarc gene, programs the virus to produce a protein-kinase. No one had been able to find a sarc gene in any organism other than a virus; at the time the only way to look for the gene was to search for its product, the protein-kinase, and this search had been in vain. Not when super-experimenter Spector turned to the task, however. He was soon able to show that some of the kinases he'd isolated were sarc-gene products.

A model for cancer's mechanism now seemed within reach. When a tumor virus invaded a cell, its sarc genes triggered a kinase cascade which ended in the adverse modification of the ATPase, so that in due course the cell turned malignant. If the mechanism of cancer could be understood, surely a cure or at least a preventative could not be far behind.

The trouble was that no one could replicate Spector's experiments— unless he happened to be there to help. Rather embarrassingly, not many tried; further, a couple of scientists who became suspicious of the necessity for Spector's presence kept their suspicions to themselves. The bubble was finally burst by Spector's Cornell colleague Volker Vogt, who decided to investigate for himself the gels Spector was using to provide radiometric proof of the presence of his kinases. Vogt discovered that a very clever, sophisticated trick was being used with the gels to produce the results Spector wanted—indeed, the whole model for cancer's mechanism proved to have been based on Spector's faked results.

It was about then that Cornell decided to check up on its star graduate student. It emerged he had neither of the two earlier degrees he claimed, but did have a couple of charges for forgery.

No one knows what Spector's motives were for faking the kinase experiments, although the quest for scientific glory must have been part of it— there was much talk of a Nobel Prize for himself and Racker. As for Racker, it seems he was so gung-ho for his ATPase hypothesis that he was in a state of denial as to the need to check his younger colleague's experimental results—or he may just have been bamboozled by a charismatic conman, as can happen to anyone. The main consequence of the whole hoax was, of course, far more important than the vast amount of time and energy spent on the kinase-cascade theory: cancer research was diverted down a blind alley for two long years or more.

THE VINLAND MAP

The curious saga of the Vinland Map began in 1957 . . . or perhaps around 1920. In 1957 the Spanish book dealer Enjo Ferrajoli and Joseph

Irving Davis of the London book dealer Davis & Orioli brought a volume to the British Museum to have it examined. Although the volume's binding was recent, its contents seemed anything but. Its main bulk was a medieval manuscript, *The Tartar Relation* by the Franciscan monk C. de Bridia Monachi, describing an expedition by John de Plano Carpini into Asia in 1245–47; but the real interest focused on the double-page world map, apparently of similar antiquity, bound into the front of the volume. This seemed to have been compiled around 1430–50 and based on earlier Viking and other sources; the coastlines of Denmark, Greenland and Iceland were depicted with a fair accuracy and the rest of Europe reasonably so, while elsewhere coastlines (and shapes) were considerably more fanciful or simply shown by wavy lines, the cartographer's indication that he was unsure of details—except for a coastline on the west of the Atlantic. On that side was shown an island whose western edge was again merely sketched but whose eastern, oceanward coast was more carefully done. It didn't take too much imagination to see in that coastline a representation of Baffin Land, Labrador and the Canadian Maritimes, with between them the Hudson Strait and the Gulf of St. Lawrence. Whatever the case, here was a map dating from before Columbus's 1492 voyage, and apparently based on sources dating from long before then, that showed the New World!

The experts at the British Museum, who included the head of the BM's Map Room, Peter Skelton, were unconvinced—not so much on cartographic grounds as because there were mistakes in the Medieval Latin of the map's annotations. A few months later, however, the US book- and manuscript-collector Laurence Witten bought the volume from its unknown owner for $3,500 (worth about $30,000 today). He showed it to his good friend Thomas Marston, Curator of Medieval and Renaissance Literature at Yale, and to Yale's Curator of Maps, Alexander O. Vietor. The two scholars arranged that Yale should have right of first refusal should Witten ever decide to sell it. At that time, however, it was little more than an absorbing curio, since no proper provenance could be established.

That there had initially been a third volume bound in with *Tartar Relation* and the map was obvious. Both items had suffered from bookworms, yet the holes that penetrated the manuscript did not match up directly with those of the map, implying that there'd once been a thickness of paper between the two sections. An inscription on the back of the map suggested the missing third item was likely to be a copy of the *Speculum Historiale*, a world history written by Vincent de Beauvais. The assumption was that an initial three-item volume had disintegrated with age, and that its contents had been rebound in two volumes—or perhaps the missing part had deteriorated so badly as not to be salvageable.

Some months after the meeting with Witten, Marston spotted part of the *Speculum Historiale* for sale in a book catalogue issued by Davis & Orioli, Joseph Irving Davis's company, and bought it. This time it was his turn to show off his new purchase to Witten. Witten borrowed the volume, compared the wormholes, and a few days later excitedly announced the missing third item had been found. No one seems to have remarked too much on the astonishing coincidence involved, nor about the fact that Marston's next action was to give his copy of the *Speculum Historiale* to Mrs. Witten as a gift, to go along with the two-part volume her husband had already given to her. The bizarreness of this latter event lies in the fact that the three items together hugely increased the historical plausibility—and value—of the map. Perhaps it was an attempt by Marston to entice the Wittens into making sure the map came Yale's way, because of course it was now something the university very much wanted.

In 1959, the Wittens put the items up for sale, but at a price Yale couldn't afford. To the rescue came a (still) anonymous donor, who contributed nearly a million dollars. Initially Yale kept the deal under wraps, not just to allow for further authentication but also to commission a lavish coffee-table book based on the map; contributors to the book included Peter Skelton and the BM's Assistant Keeper for fifteenth-century books, George Painter. Not until 1965, with the book's publication, did Yale announce its remarkable new acquisition to the world.

Publication day was October 11 of that year—a date chosen for maximum impact because it was the day before the US's annual Columbus Day celebration. Here, the publicists trumpeted, was proof Columbus had not been the discoverer of the New World, but its rediscoverer. Of course, it had been widely accepted before 1965 that the Vikings had preceded Columbus to the Americas, but that their efforts at colonization had been thwarted by a mixture of climate and native hostility. Nonetheless, the Italian American population was outraged at the "downgrading" of "their" hero, Columbus. This furor may have been one reason why, a few years later, the US declared Columbus Day an annual public holiday.

Skepticism about the authenticity of the map was not confined to gut reactions: plenty of cartographic scholars and medievalists were likewise dubious. That the matter of provenance was still hazy—Witten steadfastly refused to name the collector from whom he'd bought his volume—didn't help, but there were other concerns, mainly focusing on those errors the

BM experts had noted but also on some other oddities, such as historical annotations that conflicted with general belief at the time the map had supposedly been compiled.

In 1972 Yale sent the map to the research laboratory Walter McCrone Associates for chemical microanalysis, a technology that had been far more rudimentary back in the early 1960s. The results, released in early 1974, were depressing: the map's ink contained anatase, a synthetic that wasn't invented until the early twentieth century. A further analysis done in 1987 by the University of California at Davis's Crocker Nuclear Laboratory, this time using particle-beam techniques, cast some doubt on the anatase claim; however, ancillary results from Crocker supported the thesis that the map's ink was relatively recent.

There were of course two forgeries involved in the Vinland Map case: the forging of the map itself and the forging of the circumstances to convince Yale and Witten the document was genuine. The finger of suspicion for the first of these forgeries points to Yugoslav historian Luka Jelic. Jelic was convinced the Vinland colonies had embraced Catholicism, and maintained that one Eirik Gnupsson, known as Henricus, was their bishop; this latter conviction was exclusive to Jelic, and yet reference to Gnupsson's putative bishopric appears in the map's annotations. It is of course entirely plausible that Jelic made the map just for fun, with no fraudulent motive whatsoever. The second forgery, however, is a completely different matter. As yet no clear suspects have emerged, and perhaps none ever will.

There are question marks hanging over another celebrated documentary fraud. In 1883 William Benedict (born Moses Shapira) tried to get the British Museum to give him £1 million for 15 parchment strips on which were inscribed the original text of Deuteronomy, which he said he had bought from a Bedouin who'd found them in a cave near the Dead Sea. C.D. Ginsburg, the BM's advisor on Semitic scripts, examined them and was impressed. What Ginsburg seems not to have known is that Benedict's track record was far from clean: the Berlin Museum had already turned down his offer, having been stung by him a decade or so earlier when he'd sold them several hundred supposedly ancient clay figures that later proved to be modern.

In the interim before the BM's putative purchase, two of the strips were, with Benedict's agreement, put on public display, where they were examined by visiting French archaeologist Charles Clairemont-Ganneau. He proclaimed the strips false: someone—presumably Benedict—had cut them off a Bible scroll that Benedict had sold to the BM a few years before.

An alarmed Ginsburg, on re-examining the strips, agreed.

Benedict reclaimed his "obvious forgery" and soon afterward, in Rotterdam, killed himself—an act regarded as an admission of guilt until a few decades later when, between 1947 and 1956, the Dead Sea Scrolls came to light in caves around the Wadi Qumran. This wadi is on the Dead Sea's northwestern shore while Benedict's Bedouin supposedly found the Deuteronomy parchments on the sea's eastern shore; nonetheless, the geographical proximity has raised interest in the possibility that the Deuteronomy parchments could have been genuine. Unfortunately, the evidence was lost after Benedict's suicide.

A NEW ERA OF FRAUD

It seems that the rate of scientific fraud has picked up remarkably since about the third quarter of the twentieth century—*detected* scientific fraud, that is. The blame is almost certainly to be laid in large part on the insistence of universities and other scientific institutions—governmental science appropriations divisions not excluded—that scientists should publish prodigious numbers of papers in the learned journals as proof that they're making progress. In fact, surveys have shown that over half of such papers are never cited by other authors, and so have presumably made no contribution whatsoever to the advance of knowledge; one suspects that many papers are in fact never read *at all* after publication. Perfectly competent, worthwhile scientists are driven to publishing unimportant papers in the more obscure and occasionally less than wholly reputable journals simply in order to impress their paymasters. (We'll return to this topic later—see pp. 67–88.) The opportunities to publish fraudulent and/or plagiarized papers are, in this environment, endless.

There are other reasons for the upsurge. For example, there's very much more science being done currently than in previous decades; the increase in fraud cases implies nothing one way or the other about (a) whether the percentage of fraudulent scientists is rising or (b) whether science is slowly becoming better at policing itself so that more frauds are being caught. It is disturbing, though, that the scientific establishment is not, as it claims to be as a matter of fundamental principle, even-handed in its regard for senior and junior scientists: the work of senior figures tends to go largely unchallenged while any conflicting work by junior figures is far too easily dismissed. This imbalance encourages any tendencies senior figures might have toward dishonesty, if only through contributing to an understandable arrogance.

This was very evident in the case of the UK cardiologist Peter Nixon, who in 1994 sued the makers of a television documentary about his fabrication of data and his endangerment of patients; in court in 1997 he was forced to admit that the documentary was essentially accurate, something he must have known before bringing his legal action. Surely only an arrogance born out of the structure of the scientific establishment could have forced him to do something so foolish as to sue.

This weakness in the establishment's structure is even graver in the rare instances when two (or more) scientists collude to defraud. This happened during the 1990s in Germany when the Harvard-trained molecular biologists Friedhelm Herrmann and Marion Brach enjoyed linked meteoric careers. Because each covered the other's rear, with the pair jointly threatening any subordinate who might raise concerns about the authenticity of their work, and because no one could conceive of two charlatans being in cahoots, all might have remained undetected had the two not broken up as lovers, with the subsequent rancor ex-lovers can display. Once they began making acid remarks about each other's work, their juniors gained the courage to speak out. It appears the pair were fabricating data on computer to show predicted results, then writing up nonexistent experiments to support the falsified data.

Brach was first to break down, admitting fakery but claiming Herrmann had put her up to it. He claimed innocence: any fakery must have been Brach's doing. Other coauthors were dragged into the melee, with at least a couple of them seemingly having followed Brach's and Herrmann's example in their own supposed work. Something like a hundred papers, according to one investigative panel, were corrupt. But then the scandal snowballed, as yet further German scientists, completely unrelated to Brach and Herrmann, started to get caught in the investigators' net. Clearly this was not just a matter of an isolated pair of crooks: there was something more fundamentally rotten in Germany's scientific establishment—or, at least, in that part of it related to the biological sciences. Various structural changes came quickly, but it may still be a while before everything is sorted out.

The name of the microbiologist Hideyo Noguchi is obscure now but in his day he was something of a celebrity in his field. He was brought to the US in 1900 by Simon Flexner, renowned for having isolated the organism responsible for dysentery. Noguchi's research followed along similar lines . . . only even more so: he managed to isolate polio, rabies, syphilis, trachoma and yellow fever—an astonishing tally for a single scientist. After Noguchi's death—of yellow fever—pathologist Theobald Smith spoke for many when he remarked: "He will stand out more and more clearly as one of the greatest, if not the greatest, figure in microbiology since Pasteur and Koch." Unfortunately for Noguchi's reputation, research done over the next few decades slowly revealed the truth: he had isolated none of these diseases, and much of the rest of his work was likewise spurious. Whether he was deliberately engaging in fraud or whether he was simply guilty of self-deception, no one knows.

# # #

One consistent problem about exposing fraudulent scientists is that the people who blow the whistle on the fraud are often the first to be punished by the scientific establishment, and sometimes severely, while the perpetrator is commonly given the benefit of the doubt for years, or even forever. In short, the messenger is shot.

A typical example is that of University of Illinois psychologist Robert L. Sprague, who exposed the fabrication of data by Stephen Breuning. When Sprague first met Breuning he was much impressed by the young graduate student, who, working at a center for the educationally challenged, was in a position to help Sprague's researches into neuroleptic drugs, used in the treatment of such patients. The two men collaborated happily for a couple of years, until 1981, but then Sprague became concerned that Breuning's results were *too good*: there didn't seem to be enough hours in the day or days in the year for Breuning to have conducted all the studies he reported. The final straw was Breuning's claim that subjective assessments of the patients made by nurses showed 100 percent agreement; in practice, one's lucky to obtain as much as 80 percent agreement.

Sprague, rightly, blew the whistle on Breuning, by now working at the University of Pittsburgh's Western Psychiatric Unit. This was not an unimportant issue, because Breuning's "results" showed strongly that neuroleptics were often doing patients more harm than good; in extreme cases, taking a patient off a course of the drugs could result in an assessed doubling of the patient's IQ! In response to publication of Breuning's papers on the subject, physicians worldwide became reluctant to use neuroleptics. The University of Pittsburgh set up a committee to examine Sprague's claims but, although Breuning confessed an earlier fabrication and resigned, the committee declined to examine any of the equally fraudulent work he'd done at their own university: so far as it was concerned, Breuning was guilty of no more than a youthful indiscretion.

Appallingly, not only did the National Institute of Mental Health accept this coverup, it began to investigate *Sprague*. His federal grants were cut off. When in 1988 he testified about the case in front of Congress's Subcommittee on Oversight and Investigations, the first reaction of the University of Pittsburgh was to threaten him with a libel suit. Only when the subcommittee's chairman, John Dingell, stepped in did the university climb down and produce a grudging letter of apology. By then, of course, Sprague's career had suffered untold damage.

If this were an isolated example, matters would be bad enough.

It isn't.

# # #

A few decades ago, children's earaches were generally treated at home; they were regarded as something transient that would, if left alone, in due course clear up. Gradually, however, it became apparent that in at least a few instances earaches could be signs of something more serious: ear infections might lead on to deafness, meningitis or even, on rare occasion, death. With the advent of antibiotics it seemed at last there might be a quick, simple, easily available remedy. Research turned toward finding out if this was indeed the case, and, if so, to determine which antibiotics to use.

In the early 1980s, Charles D. Bluestone and Erdem I. Cantekin, working at the Otitis Media Research Center of the University of Pittsburgh, set about a large-scale double-blind clinical trial of an antibiotic called amoxicillin. They received a fair amount of government financial help via the National Institutes of Health, but seemingly not enough, because Bluestone—despite Cantekin's forebodings—persuaded a number of pharmaceutical companies to provide additional funding, not to mention over $250,000 in honoraria and travel expenses for himself; the companies involved included Ross Laboratories, Eli Lilly and Beechams. Further, again overriding Cantekin's objections, he redesigned the trial to include, alongside amoxicillin, two commercially produced and considerably more expensive variants: Ceclor, manufactured by Eli Lilly, and Pediazole, from Ross.

By this time Cantekin was significantly worried. Analysis of statistical data is complex at the best of times, yet it seemed clear to him that the interim results of the trials of amoxicillin showed no distinct advantage over the placebo that had been administered to other patients in the trial group. Bluestone's conclusion differed: he saw amoxicillin as having a definite edge. Part of the problem was that the two men had different ideas on the moment at which a cure could be said to have been effected: Bluestone considered that, if the ear was still clear of infection after four weeks, that could be checked off as a cure, while Cantekin, pointing out that ear infections often came back for a second round, insisted eight weeks was a better yardstick. Using Bluestone's rule, amoxicillin performed marginally better than the placebo, but not so using Cantekin's. And, if amoxicillin was ineffective, why bother comparing the performance of Ceclor and Pediazole against it?

In 1985, Bluestone and Cantekin considered the study complete, and set about producing their report for publication. Soon Cantekin found himself appalled by the conclusions reached by Bluestone and the paper's other coauthors, and withdrew from further proceedings. The finished paper was submitted to the *New England Journal of Medicine* (*NEJM*) . . . but so too was an alternative paper, written by Cantekin alone and, though based on exactly the same research, coming to an opposite conclusion: there was no clear evidence that antibiotics, specifically amoxicillin, were effective against ear infections. The journal's editors, understandably confused, approached the authorities at the University of Pittsburgh to ask which of the two was the "genuine" paper; those authorities responded that, since Bluestone was team leader, his paper was the official one—and so the journal accepted it and rejected Cantekin's.

Some while later the *NEJM* gave its reasoning: "The important question . . . is not whose interpretation is correct . . . but who has the right to publish." One reader retorted:

> Let me reword your statement just slightly to demonstrate its astonishing absurdity: "The important question . . . is not whether Galileo or the Pope is correct with respect to whether the sun revolves around the earth or vice versa, but rather who, as between the Pope and Galileo, has the right to publish his opinion."*

Cantekin did not take any of this lying down. He felt Bluestone's conclusions had been hopelessly contaminated by the commercial funding. Further, and more importantly, he was still unconvinced that antibiotics were helpful in ear infections. The significance of this goes beyond the matter of patients wasting money on—and companies reaping profits from—a drug

---

* Both the *NEJM*'s comment and the response are as cited in Michael O'Donnell's *Medicine's Strangest Cases* (2002).

that doesn't work. In a few instances patients would suffer grossly from their ear infections having been left, in effect, untreated.

For five years Cantekin waged his struggle unsuccessfully, accusing Bluestone of perpetrating a fraud. Various panels at the University of Pittsburgh and the National Institutes of Health rejected his claims. One NIH report did suggest Bluestone might have been—perhaps quite unconsciously—guilty of tilting his conclusions in favor of his funders, and recommended his further researches be subjected to a five-year period of "administrative oversight," but that was it.

At least, that was it until 1990. In that year Congress's Subcommittee on Human Resources and Intergovernmental Relations held hearings on scientific misconduct that could endanger the public health, and the Cantekin–Bluestone affair was one of the cases it considered. The subcommittee came down like a ton of bricks on the University of Pittsburgh and the NIH, especially for the fact that between them these two institutions had managed to suppress Cantekin's conclusions almost entirely: "Evidence of the ineffectiveness of antibiotics would have been available to physicians and the public several years ago if the medical school had not prevented Dr. Cantekin from publishing [his conclusions]."

The subcommittee's declaration may well have been colored by the fact that what could be considered the largest experimental trial of all—the administration by doctors of antibiotics to millions of patients suffering ear infections—pointed to conclusions similar to Cantekin's. As Cynthia Crossen reported in her account of the affair in the *Wall Street Journal* (January 3, 2001):

> . . . medical science has gradually come to its own conclusions about antibiotics and ear infections—and they are in line with Dr. Cantekin's. Although more antibiotics are prescribed today for children's ear infections—and for longer periods of time—in the US than anywhere in the world, several recent, independently financed studies have found that, for the vast majority of ear infections, antibiotics are little more effective than no treatment at all. Worse, physicians are now seeing in their own practices the potentially deadly consequences of too many children taking too many antibiotics—drug-resistant strains of bacteria.

In other words, Cantekin had been exactly right.

One might have thought the matter would be over after Congress's very public exoneration of Cantekin, but no. In 1991 Bluestone published a further paper based on the 1980s data, again concluding that amoxicillin "or an equivalent antimicrobial drug" was the treatment of choice for ear infections even though the data showed at best a borderline effect. His conclusions became part of a federal Clinical Practice Guideline. And so Cantekin finally turned to the law . . . and years passed. Not until 1999 did the legal tide begin to turn in Cantekin's favor. In the meantime he had been, ever since the whole affair blew up, completely ostracized by his colleagues at the University of Pittsburgh—not fired, presumably because that would have brought the matter even more emphatically into the public view, but with his salary frozen at its 1986 level. Yet again, the honest whistleblower was punished for acting in the interests of the public.

# # #

The Australian obstetrician and gynecologist William McBride was the man who in a 1961 letter to *Lancet* fingered the drug Thalidomide, often prescribed for pregnant women, as the cause of so many babies being born with shrunken limbs. With good reason, he became an international celebrity.

In 1981 history looked set to repeat itself when he and two junior colleagues, Phil Vardy and Jill French, published a paper raising the alarm about another drug commonly prescribed for pregnant women, Debendox (marketed in the US as Bendectin); the three scientists had administered the drug to pregnant rabbits, some of whose offspring had been born with, once more, deformed limbs.

The only trouble was that, when Vardy and French read the published version of the paper, they saw that their data had been significantly manipulated—aside from anything else, a couple of fictitious pregnant rabbits had been added to the roster. The two young scientists complained to McBride. They were promptly sacked, and locally blacklisted. They tried to take the matter to higher authority, but their petitions were rejected out of hand, as was the letter they wrote to the journal in which the falsified paper had appeared, the *Australian Journal of Biological Sciences*.

Meanwhile, the lawsuits began, about 1,300 of them: any parents whose children were born with abnormalities where the mother had taken Debendox not unnaturally went to court. The drug's makers, Marion Merrell Dow, found many of the cases going against them, not least because McBride was only too willing to appear as an expert witness. Fortunately for justice, Marion Merrell Dow appealed in all cases, the company's own clinical tests having found nothing wrong with the drug.

Even so, the lie might have remained undetected in perpetuity had the Australian medical journalist Norman Swan not managed, fully five years later, to worm the story out of a reluctant Vardy. Swan's subsequent broadcast, and a book by Bill Nicol, *McBride: Behind the Myth* (1989), clinched the case, yet still the Australian medical establishment obstructed any attempts to take action. Finally, though, the New South Wales Medical Tribunal was forced to hold a hearing, and at the end of lengthy deliberations found McBride guilty of fraud.

This, of course, did nothing to compensate Vardy and French for the destruction of their careers, nor did it do much for all those families who had invested everything in honestly mounted lawsuits against Marion Merrell Dow and now discovered themselves confronted by huge legal bills.

# # #

In 1995, the UK's General Medical Council struck the gynecologist Malcolm Pearce off its medical register. In the previous year he had published, in a single issue of the *British Journal of Obstetrics and Gynaecology*, a journal of which he was an editor, two papers concerning research which he had in fact never done. His coauthors—two on one paper, one on the other—had had nothing to do with the papers, having accepted their "coauthorships" as a "gift"; though two of the coauthors were relatively junior, the third, Geoffrey Chamberlain, was the journal's editor-in-chief as well as President of the Royal College of Obstetricians and Gynaecologists (RCOG) and Pearce's departmental head at St. George's Hospital, London. The two had also cowritten a book, *Lecture Notes in Obstetrics* (1992). In the paper which listed Chamberlain's name, Pearce claimed to have treated an ectopic pregnancy (one in which the embryo has implanted outside the uterus) with such success that both mother and baby were now doing fine. This would have been an historic feat, if true. The second paper concerned a large-scale trial on a rare ailment which would have involved immense logistical outlay, and yet none of Pearce's colleagues could remember him working on it.

Chamberlain and the two junior "coauthors" were reprimanded, and to be fair it was Chamberlain who initiated the probe once people had started muttering about the papers; further, Chamberlain stepped down from his editorial job and from his RCOG presidency. It seems that all concerned assumed the two papers were a lapse, and that all of Pearce's earlier work was still valid; yet experience of other fraudsters tells us this is unlikely.

Other disturbing aspects of the affair include the nepotism of editors publishing their own papers and academics accepting prestigious "coauthorships" of papers on work of which they in fact know nothing; it's a practice which, while common, is dishonest on numerous levels.

The creation of false data likewise lay at the heart of a US scandal of the 1990s, although the true scandal here was that the data had been fabricated not during the 1990s—that was merely when the crime came to light—but in the 1980s. The perpetrator was Roger Poisson, a researcher at St. Luc's Hospital, Montreal. He was a major contributor to clinical trials being conducted under the aegis of the US National Surgical Adjuvant Breast and Bowel Project, with Bernard Fisher of the University of Pittsburgh as the primary organizer. The point of the trials was to find out if, in many cases of breast cancer, lumpectomy—excision of the tumor, followed by other treatment—could offer just as great a chance of success as the then-prevalent option of mastectomy. To the great relief of many women, the conclusion of the trials, presented by Fisher and Poisson from 1985 onward, was that lumpectomy was indeed a perfectly viable procedure in many instances.

In 1990, Fisher's team grew suspicious of Poisson's data. In 1991 Fisher, having reevaluated much of Poisson's work, reported the matter to the National Cancer Institute. The NCI passed things on to the Office of Scientific Integrity, who took until 1993 to issue a definitive report that Poisson had indeed been doctoring his data. Not until March 1994, when the story broke in the *Chicago Tribune*, was the public made aware of what had been going on—yet surely this was a topic of crucial interest to said public. Further, in the period from 1990 to 1993 Fisher and the National Cancer Institute had done nothing to alert the scientific community to the fact that papers involving Poisson must be regarded as suspect. Fisher's defense was that, in re-evaluating the trials while excluding Poisson's contribution, there was no reason to change the conclusions. Those conclusions are still accepted today, but that does not excuse the cover up.

In 1978, Helena Wachslicht-Rodbard of the National Institute of Arthritis and Metabolic Diseases submitted a paper on anorexia nervosa (further details need not concern us) to the *New England Journal of Medicine*. The *NEJM* sent the paper to two specialists in the field for peer review; one was Philip Felig, vice-chairman of Yale's Department of Medicine. He gave the paper to a junior colleague, Vijay Soman, to do the actual reviewing. Soman's opinion, sent to the journal's editor as Felig's, was negative. The other review was positive, as was that of a third specialist to whom the editor then sent the paper as umpire. The whole process took three months, and in the interim Soman and Felig submitted to the *American Journal of Medicine* a paper, based on work by Soman, that bore an astonishing resemblance to Wachslicht-Rodbard's original. Indeed—as she discovered when, in a delightful symmetry, the *AJM*'s editor sent the paper to her boss Jesse Roth for peer review, and Roth passed it on to her to do the actual reviewing—bits of Soman's paper were actually copied verbatim from her own. In due course—a very long due course—it was proven that Soman, who'd been planning a similar study to Wachslicht-Rodbard's, had saved himself a bit of work by simply plagiarizing her material. Soman's career was, quite reasonably, truncated. Felig, who had been a powerful man in the field, was deemed to have behaved badly throughout the affair and suffered too, being forced to resign the position he'd just taken up as head of the Department of Medicine at Columbia University's College of Physicians and Surgeons.

Another plagiarism case that surfaced at about the same time was that of Elias A.K. Alsabti (if that was indeed his real name), who was born in Basra, Iraq, but was, by the time he came to the US in 1977, supposedly a citizen of Jordan, where he had, he said, gained an MB and a ChB (Bachelor of Surgery). He claimed a blood relationship with the Jordanian Royal Family, who had paid for him to come to the US to gain his PhD. Alsabti charmed his way into posts at various non-negligible institutions across the US and published prolifically: some sixty papers over about three years. He was able to publish so much because his technique was simple: he would take a paper from a little-known journal, copy it, alter the title, claim authorship (with or without invented coauthors), and send it off to another little-known journal in a country different from the first.

Since the vast majority of published scientific papers go largely unread, Alsabti might have continued this practice for very much longer than he did. He was just plain unlucky to get caught: a student of microbiologist E. Frederick Wheelock, of the Jefferson Medical College, Philadelphia, saw that a paper published in Czechoslovakia by Alsabti was almost identical to a paper, also "authored" by Alsabti, that had appeared obscurely in the US . . . and that both were just re-digested versions of papers Alsabti had stolen from Wheelock after being fired from Jefferson for incompetence. Wheelock blew the whistle, writing to the editors of all the major journals in the field; but, with the exception of *Lancet*, all shamefully declined to publish his letter on the grounds that this was just a personal quarrel between the two men. As a result, Alsabti's US career as a con man continued for several more years before so many examples of his plagiarism surfaced that the scientific establishment could no longer ignore them.

Alsabti was fortunate indeed not to have been caught earlier, in Iraq, where in 1975 he persuaded the government of Ahmed Hassan Al-Bakr that he had invented a brand-new cancer diagnostic technique. Soon he was running his own lab in Baghdad, while also visiting local factories and fraudulently charging the workers there a fee for screening them for cancer. Since the Iraqi government offered universal healthcare, Alsabti's fee-charging habits eventually came to the attention of the authorities, but by the time they moved in he'd fled the country, eventually ending up in Jordan, where he claimed political asylum and, in essence, pulled the cancer stunt all over again—to the extent that the Jordanian government did indeed finance his transition to the US. Ironically, the Jordanians were among the first to recognize the fraudulence he perpetrated once there, cutting off his funding when they spotted he was listing senior Jordanian scientists as coauthors without their knowledge. Meanwhile, even after Wheelock sounded the alarm, the US scientific establishment was letting Alsabti happily continue to loot . . .

Action was much swifter in the case of the Polish chemical engineer Andrzej Jendryczko, whose plagiarism was first revealed in 1997, the scope of it becoming more evident over the next few months. His method was to take articles from primarily English-language scientific journals, translate them into Polish, then publish them as his own. Since comparatively few people outside Poland read the lesser Polish journals, he must have thought he was safe enough. Perhaps as many as a hundred plagiarized papers, perhaps even more, were published under his name, the names of various coauthors being added with or without their knowledge. It seems there was a brief attempt at a cover up by the Medical University of Silesia, where Jendryczko had worked, but just weeks later the authorities there made a public admission and his career was over.

#

What has come to be known as the Baltimore Affair really concerns the immunologist Thereza Imanishi-Kari; the focus came to be on the role of David Baltimore, one of the world's foremost molecular biologists, because of his intransigence as he attempted to cover up Imanishi-Kari's misdeeds and destroy all dissenters—in particular the young researcher Margot O'Toole, who had blown the whistle on Imanishi-Kari.[2]

O'Toole had come across evidence that Imanishi-Kari had falsified results in at least one of the experiments she had conducted—an experiment that was fundamental to the work Imanishi-Kari's lab at MIT (it was just about to move to Tufts University Medical School, Boston) had been doing on behalf of David Baltimore's much larger lab at the Whitehead Institute. O'Toole was especially qualified to recognize the evidence because the experiment concerned was one that had led to a falling-out between herself and her boss, Imanishi-Kari: O'Toole had, working with Imanishi-Kari's colleague Moema Reis, obtained a result that seemed supportive of Imanishi-Kari's work, but had been unable to replicate the result when repeating the experiment on her own and so had, to Imanishi-Kari's fury, declined to publish the finding. Now she had uncovered proof that Imanishi-Kari herself had repeatedly gained the same negative result. If Imanishi-Kari was prepared to falsify one result in order to bolster the hypothesis Baltimore was currently proposing, in how many other instances might she have done the same?* Imanishi-Kari and Baltimore had collaborated on a whole series of research papers that all pointed in the same direction, the latest of which (Weaver et al.) was considered of great significance.* But if Imanishi-Kari had been doctoring the data . . . ?

O'Toole did not immediately blow the whistle; she did, however, say that Weaver et al. should be retracted pending further investigation of its subject matter. She did all the right things, taking her concerns privately to senior figures at MIT and Tufts; but here the familiar pattern emerged of the establishment putting every possible obstacle in the whistleblower's path. There was a very extensive cover up, during which administrators and senior scientists collaborated disgracefully to pretend there was nothing untoward and to drum O'Toole out of her research career.

O'Toole tried to put it all behind her, and behind her it would have remained had it not been for, once again, Imanishi-Kari's abrasive personality. A former graduate student in her lab, Charles Maplethorpe, had kept in touch with his old acquaintances there, O'Toole included. Disturbed by what was going on, and having himself had doubts during his time at the lab about the veracity of some of the research, he contacted Walter W. Stewart and Ned Feder, the two self-appointed "policemen of science" who

---

* The research concerned the behavior of transplanted genes and their associated proteins. It seemed very important at the time, since it would have been a clincher for one of the two sides in a major debate that was then much occupying the molecular-biology world. In fact, the side it was supporting has since been proven wrong.

some while earlier had exposed the Darsee imbroglio (see pp. 16 and 17). It was through Stewart and Feder that the case was brought before John Dingell's Subcommittee on Oversight and Investigations in 1988–89. Baltimore was one of the witnesses called, and he used the public opportunity to charge that Dingell and his colleagues were conducting not a reasonable investigation of a specific case but more generally a witchhunt against scientists and scientific research everywhere.

Baltimore had enormous standing and the motives of politicians when interacting with science are often corrupt, so it's hardly surprising that many scientists across the US took Baltimore's statement at face value without pausing to examine what was actually going on. A national movement swiftly grew . . . and as swiftly deflated a few months later when a report emerged from the Office of Scientific Integrity to the effect that Imanishi-Kari had indeed committed fraud, both at the time of Weaver et al. and later, when producing fabricated evidence to back up her claims of innocence. Further, the past she had constructed for herself was beginning to unravel; people who'd worked with her in earlier years began to come forward with their own doubts, and it seemed she had doctored her resume, adding at least one degree that she did not possess. As it was Baltimore who had turned the affair into a *cause célèbre*, it was he who took the brunt of the scientific community's anger.

And yet somehow his clarion call against "meddling politicians" lingered. There's every reason to be nervous of politicians meddling with science, imposing upon it their own ideologies (as we'll see repeatedly in this book); but this is beside the point. The function of Dingell's subcommittee was to ensure, with extensive scientific advice, that taxpayer money was not being squandered on foolhardy or fraudulent research. And it is at least arguable that, if science manifestly fails in its claimed ability to police itself, then other police must step in.

Even more alarming than that the false call to arms lingered is that so did the cover up. Imanishi-Kari took her case to the Departmental Appeals Board of the Department of Health and Human Services, which eventually, in 1996, found that, despite presenting complete forensic evidence, the Office of Scientific Integrity had not *proved* her guilty of misconduct. Thanks to this legalistic nicety she was once more able to claim federal grant money and was reinstated in her position at Tufts; it was and is widely assumed that her name has been cleared. Baltimore, who in 1991 had been pressured into standing down from his recently acquired position at Rockefeller University, became the Robert A. Millikan Professor of Biology at the California Institute of Technology and received the National Medal of Science in 1999. He is still rightly regarded as a great scientist—he won the Nobel Prize for Physiology or Medicine in 1975—and was a courageous and outspoken opponent of the G.W. Bush Administration's many genuine corruptions of US science.

It should be noted that in *The Baltimore Case: A Trial of Politics, Science, and Character* (1998) Daniel J. Kevles presents an exactly contrary view of events: Imanishi-Kari was the innocent victim, Baltimore a righteous defender, O'Toole a fantasist, Maplethorpe a slimy *agent provocateur*, and Stewart and Feder witchhunters.

The Imanishi-Kari case was not Baltimore's first brush with fraudulent scientists, although there's no question that he played any part in the frauds perpetrated by Mark Spector and John Long: he merely collaborated on research with them for relatively short periods of time prior to their exposure.

The Spector affair we've already noted. John Long was from 1970 a resident at Massachusetts General Hospital, training under Paul C. Zamecnik. Zamecnik was researching Hodgkin's disease, and thus trying to culture Hodgkin's disease tumors. Most of the cell lines were fairly short-lived, but Long managed to establish several permanent ones, a feat that brought him much renown—including the collaboration with Baltimore—as well as promotion and grant money. His problems began in 1978 when a research assistant, Steven Quay, gained an anomalous result while working with one of Long's cell lines. Long repeated the experiment in lightning-quick time, and got a more expected result. Baffled by the whole affair, Quay shrugged it off. A year later, however, his suspicions still clung, and in due course it emerged that Long had faked his repeat of the experiment.

Long resigned from Massachusetts General Hospital, and that was when the real trouble started. Without Long there to perform the experiments on his cell lines, other experimenters found that they, too, were getting odd results. Further analysis showed one of the cell lines came from a patient who had not been suffering from Hodgkin's disease while the other three were not human at all, but came from an owl monkey! It is not uncommon for cell cultures to be contaminated by cells from elsewhere with the latter taking over the culture; considerable precautions are normally taken to prevent just this. No one knows how Long's cell lines became contaminated, but it seems impossible he was not at least aware of something being awry. Otherwise, why fake his repeat of Quay's experiment?

# # #

Dr. Jan Hendrik Schön was working at the Bell Laboratories in New Jersey when he came to prominence in 2001. In a paper published in *Nature* in that year, "Self-Assembled Monolayer Organic Field-Effect Transistors," he and junior coauthors H. Meng and Zhenan Bao announced that he'd created a transistor on the molecular scale. This was an astonishing breakthrough: it wasn't just that electronic components could be made smaller but that organic-based chips could be made on a scale so small that silicon itself would break down. The promise was of electronics that were smaller, hugely cheaper, and far faster. It was headline news, and Bell Labs basked in the glory.

Schön received the Otto–Klung–Weberbank Prize for Physics and the Braunschweig Prize in 2001, and the Outstanding Young Investigator Award of the Materials Research Society in 2002, even though, almost immediately after the publication of that "breakthrough" paper, people began raising red flags: not only did some of his data seem too good to be true, some of it seemed to have been duplicated from one experiment to another. In May 2002 a heartily embarrassed Bell Labs set up a committee to examine Schön's work. The committee uncovered considerable evidence of manufactured data and the like (while also finding his many junior colleagues blameless). Bell Labs fired Schön in September 2002 when the committee delivered its report, while the editors of the more responsible scientific journals began the task of re-examining the papers they'd published; by March 2003 *Nature* and *Science* had between them retracted no fewer than 15 of his papers because of suspect or bogus content.

# # #

Paleontological fraud on a large scale came to light in 2004, with the figure at the center of the case being Professor Reiner Protsch von Zieten of the University of Frankfurt. He had built up an impressive record of sensational discoveries, notably his dating of a hominid skull found near Hamburg to 36,000 years old: so-called Hahnhöfersand Man was thus the oldest known German hominid and a possible missing link between the Neanderthal line and that of modern humans. Other ancient human remains Protsch von Zieten produced included Binshof–Speyer Woman, whom Protsch dated to 21,300 BP, and Paderborn–Sande Man, dated to 27,400 BP.

A flamboyant figure, Protsch seemed to be an archaeologist with a knack. However, Thomas Terberger of the University of Greifswald became suspicious, and sent the Hahnhöfersand Man skull to the radiocarbon dating unit at Oxford University for a second opinion. The verdict was that the skull was a mere 7,500 years old—still interesting but hardly sensational. Dating of the other two human fossils revealed that Binshof–Speyer Woman was about 3,300 years old, not 21,300, and, most embarrassing of all, that Paderborn–Sande Man had died c1750.

Terberger, in conjunction with UK archaeologist Martin Street, published a paper in 2004 reporting the awful truth, and the University of Frankfurt promptly investigated. More horrors soon turned up. The German police were especially interested to discover that Protsch had attempted to sell the university's entire collection of 278 chimpanzee skulls to a US dealer for $70,000. Under Protsch's orders, thousands of documents from the university's anthropology archives that detailed the vile human experiments of the Nazis in the 1930s had been shredded. There was nothing wrong with the dating of a "half-ape" skull called *Adapis*, already known to be of a species extant some 35 million years ago; however, Protsch reported that it had been found in Switzerland, far from its known territory, which made the finding significant—alas, it now proved to have been dug up in France, like all the other *Adapis* fossils. The newspaper *Der Spiegel* discovered that the professor, who claimed to be the descendant of a Prussian nobleman—hence the "von Zieten" part of his name—was in fact simply Reiner Protsch, son of a Nazi MP, Wilhelm Protsch, which might go some way toward explaining the illicit document-shredding.

Unsurprisingly, in February 2005 the University of Frankfurt forced Protsch into retirement. An initial reaction from the paleo-anthropological world was that large chunks of the accepted history of our ancestors might have to be rewritten, but later, when the dust settled, it was generally realized the damage done by Protsch's fraudulent dating was in fact only minor: his "finds" had been sensational but not terribly important.

#

A rather similar scandal had erupted in Japan a few years earlier, this time concerning the discoveries of Shinichi Fujimura, an amateur archaeologist and volunteer at scores of digs around Japan. Although he began his career in 1972, it was in 1981 that he made his first sensational "discovery," an artifact that he claimed to have found in strata up to 40,000 years old.

After that there was no stopping him. Among professional archaeologists he became known as "God's Hands" for the magical knack he had of turning up significant items, and despite his lack of paper qualifications he was made deputy director of the Tohoku Paleolithic Institute. Textbooks were rewritten as a result of his "discoveries." It seemed that at every site where he worked he came up with some significant artifact or another, and always the limits of Japan's known archaeological history were being pushed back farther and farther. This was important to many Japanese because it seemed the previously accepted hypothesis of the Japanese coming from the same stock as the Koreans would have to be discarded: the Japanese were of separate ethnic stock, just as the nationalists in Japan had always, in the teeth of existing evidence, claimed.

On October 23, 2000 Fujimura held a press conference to announce that his team's latest discoveries were a whopping 570,000 years old. However, a couple of weeks later, on November 5, the newspaper *Mainichi Shimbun* published some photographs of Fujimura digging a hole at the site and planting artifacts in it; this photo, the newspaper said, had been taken the day before the sensational find. Promptly Fujimura held another press conference, this time to confess gross archaeological fraud. Many of the artifacts he had unearthed over the years had come, he said, from his own personal collection, and he had planted them. He was careful to distinguish these from others he insisted were genuine; later researches have, however, indicated that these too were almost certainly fraudulent. Enormous shockwaves ran through Japanese archaeology, and one distinguished archaeologist, Mitsuo Kagawa, was so traumatized by accusations he might have been party to the fraud that he committed suicide. Meanwhile, of course, those textbooks had to be rewritten once more.

Fujimura, in his turn, was not without his precursors. Most notable among these was perhaps Viswat Jit Gupta, a professor at the Punjab University of Chandighar whose fraudulence was exposed in 1989 in *Nature* by the Australian paleontologist John Talent. For some quarter-century the paleontology and stratigraphy of the Himalayas had been, to say the least, puzzling: fossils were turning up that seemed to have no business there, and they were turning up in inexplicable strata—old fossils in higher strata than more recent ones, with no apparent geological disruption to explain how this could have come about. Talent pointed out that all the anomalous finds had been made by one man, Gupta, who had detailed them in over

300 papers. Quite how many further papers were corrupted through having built on Gupta's "work" is hard to contemplate. It seems he merely appropriated fossils whose disappearance would go unnoticed, planted them anywhere between Kashmir and Bhutan, and then, well, dug them back up again.

# # #

The People's Republic of China is notorious for its *laissez-faire* attitude toward technological piracy and intellectual property. However, in recent years there has been a significant push by the country's government to encourage R&D at the cutting edge of technology, especially in the computer and information science field. This, coupled with the relaxation of state communism, so that innovators can benefit substantially from their innovations, has put hitherto-unknown pressures on scientists to succeed: glory and riches are the carrot, while the threat of state disapproval is the stick. Observers within China reckon the corruption of science has in consequence become endemic there.

A particularly high profile case emerged in May 2006 with the downfall of Chen Jin, one of China's most acclaimed computer scientists. Chen had studied at Shanghai's Tongji University, then gained his PhD at the University of Texas at Austin. He worked for a while at Motorola's research center in Austin before returning to China to work for the same company's division near Shanghai. He next accepted a post at Jiaotong University, where in 2003 he created, he claimed, the Hanxin, a chip capable of processing 200 million instructions per second. Jiaotong University promptly established its own microelectronics school, with Chen as head. Eager to eliminate the need for China to import all its chips—an expenditure reputedly running into billions of US dollars—the Ministry for Science & Technology, the Ministry for Education and the National Reform & Development Commission financed Chen heavily. He was able, too, to set up a string of companies designed to exploit his achievements for his own profit. In 2004 the Hanxin II and the Hanxin III followed, both capable of even faster speeds.

Toward the end of 2005, however, disgruntled former colleagues of Chen's began posting to the internet, then writing to the government and to the authorities at Jiaotong University, reporting that much or all of Chen's research was faked: the supposed Hanxin chips were in fact Motorola's chips, carefully doctored to remove the Motorola identification.

Once Chen had determined the specifications of the Motorola chip, it was a simple matter to pass them on to manufacturers as if they were his own invention.

# # #

In 2008, orthopedic surgeon Evan F. Ekman, of Southern Orthopedic Sports Medicine in Columbia, South Carolina, was a worried man. He'd been asked by anesthesiologist Scott S. Reuben of Baystate Medical Center, Springfield, Massachusetts, with whom he'd collaborated on a number of studies (for example, a 2005 piece[3] in *The Journal of Bone and Joint Surgery* on the use of Pfizer's anti-inflammatory Celebrex for patients who'd undergone back surgery), to review a paper Reuben had written on the use of painkillers following a particular surgery of the knee. When he requested that Reuben tell him who was the orthopedic surgeon who'd been involved, he was met with a wall of silence.

Worse followed. Some months later Ekman discovered the paper he'd begun to review had now been published in the journal *Anesthesia & Analgesia*, and bore his name as coauthor.[4] According to the journal's editor, Steven Shafer, who had published a number of Reuben's papers but had now become suspicious of the anesthesiologist, Ekman's signature was on the paper's submission form as a coauthor. Closer examination revealed it had been forged.

Reuben's specialty, so far as his research reputation was concerned, was the short-term relief of pain after surgery, in particular using the anti-inflammatory drugs called COX-2 inhibitors; examples of such drugs are Vioxx, made by Merck, and Celebrex and Bextra, both made by Pfizer. Surgeons had begun to look askance at these drugs because animal studies were showing they could slow the healing of bones, and then in 2004 Bextra and Vioxx were taken off the market because of concerns that they increased the risk of heart attack.

Similar concerns were starting to emerge over Pfizer's Celebrex, but Reuben's published research reassuringly demonstrated the drug's effectiveness within his own specialist area. He also wrote directly to the FDA on occasion, enthusing over the effectiveness of the COX-2 inhibitors. And his studies had a profound effect on the field—the postoperative treatment of literally millions of patients was affected by the results he reported, with a corresponding sale of billions of dollars' worth of COX-2 inhibitors, notably Pfizer's Celebrex.

Which was just as well, so far as Pfizer was concerned, because the company had funded much of Reuben's work through at least five research grants, had selected him as a preferred speaker at its conferences, and had in general made sure he received his full quota of perks.

Patients may have been less delighted. While the COX-2 inhibitors are effective in relieving post-operative pain, there's evidence, as noted, that they slow the rate of healing and increase the risk of coronary events. It's not easy to evaluate just quite how much damage Reuben's advocacy of these drugs did over the years; it may well be that the field of anesthesiology suffered more disruption from his exposure as a fraud, with chunks of the accepted wisdom having to be rewritten, than it did as a direct consequence of his fraudulence.

In May 2008, a routine audit being done by Reuben's employer, Baystate Medical Center, revealed discrepancies concerning Reuben's research. By March 2009 he had admitted to Baystate that many of his published studies were fiction, based on data that had been faked to the extent that sometimes the patients from whose experiences the data supposedly derived didn't even exist. Ekman wasn't the only "coauthor" whose name had been taken in vain; others were startled to discover they were associated with published studies of which they knew nothing. (This wasn't lax of them, as it might seem at first glance. Keeping up with the latest medical literature in one's own specialty is a major chore. These "coauthors" were by and large orthopedic surgeons. There was no particular reason why they should have been scouring the anesthesiology literature.)

Brought to trial in early 2010, Reuben struck a deal, agreeing to plead guilty to a single count of healthcare fraud. In February of that year he was given a six-month jail sentence by Judge Michael Ponsor, fined $5,000, and ordered to pay $50,000 to the federal government and—take a deep breath here—refund $360,000 to Pfizer and others whom he'd swindled by not actually doing the research they'd paid for through their grants to him.

Up to 25 of Reuben's published studies have been retracted. Alarmingly, according to an August 2016 report in *Science and Engineering Ethics*,[5] almost half of Reuben's retracted studies are still being cited in published research, and in only about one-quarter of those citations is it being noted that the study in question has been retracted. Just as in any other sphere of human communication—the internet, the news media, the pub—it is far, far easier to put a lie into the common discourse than it is to get it out again once it's been exposed. It's an aspect of human behavior upon which Alex Jones and his ilk have built their careers.

It was in the wake of the Reuben case that *Retraction Watch* (http://retractionwatch.com) was set up in 2010. Alarmingly, Reuben today is far from the top of *Retraction Watch*'s (admittedly incomplete) Leaderboard[6] of scientists whose work has been so incompetent or dishonest that they've been responsible for multiple, usually career-spanning retractions. Top of the list when I last checked, in August 2017, was the Japanese anesthesiologist Yoshitaka Fujii (what is it about anesthesiology?), with a grand total of 183 retractions. Reuben languishes far down the list, at #15, with a "mere" 25 retractions. Jan Hendrik Schön (see page 47), by way of comparison, features at #9 with 36.

Still, as any fraudulent scientist will tell you, it's not the quantity that matters: it's the *quality*.

COLD FUSION

March 23, 1989 brought news that stunned the world. The international scientific community was thrown into a frenzy; more publicly, so was the international financial community, while the news media subjected everyone on the planet to a sustained blitz. Two University of Utah scientists, Stanley Pons and Martin Fleischmann, announced they had discovered a technique for cold fusion.

All other things being equal, a nuclear fusion reactor that doesn't require vast infrastructure and colossal investment need not be very efficient for it to make commercial sense. Hydroelectric generation, for example, is not efficient at all, yet makes plenty of sense because, once the equipment is in place, the running costs are minimal and the fuel costs zero. Similarly, because its setup and running costs would be trivial alongside those of any other form of nuclear reactor, to demonstrate the viability of cold fusion as a technology you don't have to show that it works *well*, just that it works *at all*. Little wonder, then, that there was such a furor in the wake of the Pons/Fleischmann announcement: not only did they claim to have cracked the nut of cold fusion, they maintained they had done so at room temperature using equipment most of which might be found in the average school chemistry lab. The amounts of energy they were talking about were small, but then so was the scale of the experiment.

Physicists and chemists worldwide dropped everything to try to replicate the experiment. Some were impatient with this confirmation process, including many financial speculators and a truly shameful number of politicians. The Utah State Legislature promptly hurled four and a half million dollars at Pons and Fleischmann. The US Office of Naval Research added an initial $400,000. It looked as if at least tens of millions of dollars might be on the way from the US government. When initial reports from other scientific researchers seemed favorable, further financing from industry looked like a foregone conclusion.

But trouble was brewing for the two chemists and their most ardent supporter, the University of Utah. Pons and Fleischmann had chosen to avoid the normal painstaking process of peer-reviewed publication and instead used newspapers and TV to announce their "breakthrough." Whatever spin the various interested parties tried to put on this decision, it was to say the least odd. Further, while some early efforts to replicate the results seemed to hint at confirmation, a disturbing number of others did the opposite—and soon the negative reports predominated. Matters weren't helped by some apparent belated fudging of figures by Pons and Fleischmann. There were also astonishingly crass attempts by the University of Utah (which promptly disclaimed responsibility when it became public) to use the threat of litigation to silence critics—this, more than anything else, destroyed Pons's and Fleischmann's credibility. With the onslaught of adverse experimental evidence came demolitions of the theoretical underpinning of Pons's and Fleischmann's research.

One aspect downplayed by Pons and Fleischmann and their supporters was that deuterium fusion is not quite the "clean" energy source the public imagines: their putative cold-fusion cell would in fact have been highly radioactive. Later, under pressure, they tried to explain this away; at the time, the trending physics joke was "Have you heard the bad news about Pons's lab assistant? He's perfectly healthy."

We shouldn't forget, however, that at least a few researchers believed they'd been able to replicate the Pons and Fleischmann results. The two Utah professors may not, as they thought, have discovered cold fusion, but they may have discovered *something*.

Similar doubts surround the claims made in 2002 by Rusi Taleyarkhan, then of the US Department of Energy's Oak Ridge National Laboratory, to have achieved cold fusion. His team bombarded a beaker of chemically altered acetone with neutrons and then with sound waves to create bubbles; when the bubbles burst, the team reported in *Science*, fusion energy could be detected. Other teams, however, have had difficulty replicating the results—including Taleyarkhan himself. He finally announced in 2004 that he had done so using the uranium salt uranyl nitrate. Brian Naranjo of the University of California, Los Angeles, who in 2005 reported that his team there had attained cold fusion by heating a lithium crystal soaked in deuterium gas, analyzed Taleyarkhan's results and concluded the scientist had detected not cold fusion energy but leakage from some other radiation source present in the laboratory; if so, it seems a rather elementary mistake for Taleyarkhan to have made.

ANDREW WAKEFIELD AND THE VACCINATION/AUTISM FIASCO

In February 1998, *Lancet* published one of the most influential medical papers of all time. Unfortunately it was bogus and, most medical scientists now believe, deliberately fraudulent on the part of at least some of its authors, who were led by Andrew Wakefield, a researcher in the medical school of the Royal Free Hospital, London, and widely regarded as an up-and-comer.

To the medically uninitiated, the title of that paper could have given no clue of the impact it was going to have: "Ileal-Lymphoid-Nodular Hyperplasia, Non-Specific Colitis, and Pervasive Developmental Disorder in Children." The hypothesis that it advanced, based upon what would later be shown to be a pitifully small experimental base, was that the residue of the MMR (measles, mumps, rubella) vaccine, the dead virus, could do damage to the infant intestine and thereby release malignant proteins into the bloodstream. These proteins, the hypothesis went, on reaching the brain might trigger the condition known as regressive autism.

Even as the paper was being published, some of Wakefield's coauthors were distancing themselves from it, stressing that its conclusions were speculative at best. Most forthright among them was Simon Murch, Senior Lecturer and Consultant in Paediatric Gastroenterology at the Royal Free and University College Medical School (cited by Paul A. Offit in his *Autism's False Prophets*, 2008):

> This link is unproven and measles is a killing infection. If this precipitates a scare and immunization rates go down, as sure as night follows day, measles will return and children will die.

Hindsight would demonstrate the prescience of Murch's remark, but already it was too late. Primed by strategic press conferences given by Wakefield even before the publication of the *Lancet* paper, the UK and Irish news media went wild. Responsible parents, acting in what they believed the best interests of their newborns, declined to subject them to the MMR vaccine, and it wasn't long before, in some areas of both countries, vaccination levels dropped below that required for herd immunity.*

The first major consequence was a measles outbreak in Dublin, Ireland, during the winter of 1999–2000, in which 111 children needed hospitalization and three died. In 2008, England and Wales had to declare that measles, once almost extinct, was again endemic. More recently, in the spring of 2017—thanks to campaigners from antivaxxer groups like the Minnesota Natural Health Coalition, the National Health Freedom Action, the Vaccine Safety Council of Minnesota, the Minnesota Vaccine Freedom Coalition and the Organic Consumers Association—the Somalian community in Minneapolis was stirred into a terror that vaccination of their infants could cause the children to develop autism; about six in ten infants went unvaccinated. The result was what Kris Ehresmann, the Minnesota Department of Health's Infectious Disease Division director, described as a "public health nightmare." Since the city's Somali residents often live in overcrowded conditions, and given that measles is highly contagious, a potentially lethal outbreak of the disease was a near certainty—and, sure enough, that's exactly what happened.

Overall, the effect of Wakefield's paper in the US can be judged by the fact that in 2000 the country was able to declare the disease extinct within its shores; in 2014 there were 667 cases. Mercifully, as the public became more educated about the claptrap being peddled by the antivaxxer scaremongers, in 2016 that figure had dropped back down to "only" 70. But in the following year, by early June 2017, the Minnesota outbreak alone had generated 73 confirmed cases, with 21 hospitalizations.

Of course, the MMR vaccine protects against not only measles but also mumps and rubella (whooping cough, pertussis). Another product of the antivaxxer hysteria has been an uptick in whooping cough cases. In late October 2010 California was reported to have suffered its worst outbreak of whooping cough in almost six decades, with about 6,000 confirmed cases and ten fatalities. Nationwide, in 2015 the CDC reported that 2014 had seen 32,971 cases of the disease, an uptick of 15 percent on 2013's figure of 28,639 but thankfully far below that for the peak year, 2012: 48,277. The big jump in US rubella cases came in 2003 and 2004, which roughly accords with the height of the antivaxxer panic that hit these shores from 2000,

* If enough people in a community—a "herd"—are immune to an infectious disease, the community as a whole is protected because the infection can't get a proper hold.

thanks to a widely publicized July 2000 congressional hearing organized by Dan Burton (R–IN), chair of Congress's Committee on Government Reform, with Wakefield as star witness.* Here are the CDC's figures for the relevant years:

2000	7,867
2001	7,580
2002	9,771
2003	11,647
2004	25,827
2005	25,616

Back in the UK, *Sunday Times* journalist Brian Deer had refused to join in the media hullabaloo over the conclusions of Wakefield's paper. As Deer delved into the paper's background, he discovered some facts that made his investigative reporter's antennae twitch—for example, that a lawyer who was trying to mount a class action against the MMR vaccine's manufacturers had hired Wakefield to try to find a scientific justification for his claim. Of the dozen autistic children Wakefield had used as his experimental subjects, several were the offspring of this lawyer's clients, while others had been referred to Wakefield by the antivaxxer organization JABS (Justice Awareness & Basic Support).

There was more, plenty more—more than enough to persuade Deer that this wasn't the dispassionate, untainted study that everyone, including the editors of *Lancet* and some of Wakefield's collaborators on the paper, had assumed it to be.

In 2004, aware of Deer's reports—summarized in that year's TV documentary "MMR: What They Didn't Tell You" (November 18, 2004), aired in Channel 4's *Dispatches* series—ten of the thirteen coauthors on the Wakefield paper withdrew their names and the General Medical Council, the UK's regulatory body for medical practitioners, initiated an investigation. That investigation was astonishingly slow-moving, especially bearing in mind all the damage that was being done to the UK's public health by the antivaxxer scare, but in May 2010 the GMC struck Wakefield and coauthor John Walker-Smith off the medical register—not for the reasons discussed above but primarily for ethical violations related to the tests carried out on the kids luckless enough to be Wakefield's experimental subjects.

---

* Another featured witness was pediatrician Mary Megson of the Medical College of Virginia (now the VCU Medical Center), who claimed her practice of treating autistic children with cod liver oil was so successful that one of them had enjoyed an IQ increase of 105 points. Really?

Deer's final *coup de grace* came in 2011 with his publication of three peer-reviewed articles in the *British Medical Journal*—"How the Case Against the MMR Vaccine Was Fixed" (January 5), "How the Vaccine Crisis Was Meant to Make Money" (January 11) and "The *Lancet*'s Two Days to Bury Bad News" (January 18)—presenting his evidence that Wakefield's paper had been not just sloppy or marred by poor and unethical experimentation but in fact fraudulent. Wakefield, he claimed, had hoped to make his fortune from the marketing of various vaccine-related schemes.

By now Wakefield had long ago left for the US, where he could be sure of a warmer welcome among a thriving antivaxxer movement that—thanks to the ever-increasing anti-science ideology espoused by powerful politicians like Dan Burton—extended into Congress. The Royal Free Hospital said its sad farewells to him in December 2001. He became Executive Director of the Thoughtful House Center for Children, in Austin, Texas, an establishment devoted to furthering antivaxxer ideas; the Dixie Chicks appeared at some of its benefits—well, no one can be expected to get *everything* right. Wakefield left Thoughtful House in 2010, after *Lancet*'s long-delayed retraction of his 1998 paper; now called the Johnson Center for Child Health and Development, the establishment appears to have washed its hands of him.

He's still active in antivaxxer circles in the US, though, having become something of a cult figure. His book *Callous Disregard: Autism and Vaccines: The Truth Behind a Tragedy* (2010) has a foreword by Jenny McCarthy, the archetypal celeb champion of vaccine denial, but little recognized for her medical expertise. He directed the documentary feature *Vaxxed: From Cover-Up to Catastrophe* (2016), best known for having been withdrawn by Robert De Niro from the Tribeca Film Festival. The *Wall Street Journal* review of the movie, by the epidemiologist W. Ian Lipkin—of Columbia University's Mailman School and College of Physicians and Surgeons—was headlined "Anti-Vaccination Lunacy Won't Stop," which tells you about as much as you need to know.

At the time of writing, no one knows the genuine cause of the "autism epidemic," although one of the things we can be certain about—thanks to some quite astonishingly extensive studies done around the world—is that vaccination isn't it. Rarely a month goes by without the newspapers carrying some fresh report of new research indicating a possible cause—genetic, environmental or otherwise—but as yet none has been confirmed.

The microbiologist Martin J. Blaser has proposed that autism might instead be caused by the depletion of the normal flora of bacteria in babies, either through its being massacred by the overuse of antibiotics (not vaccines!) in the earliest weeks of life or through its failure to be properly established in the first place as a consequence of the baby being born by caesarean section and thereby failing to inherit a microbial "standard toolkit" from the mother's vagina during birth. Blaser suspects depletion of the normal flora—the population of bacteria that forms a surprisingly high proportion of our physical selves—is at the heart of numerous conditions whose incidence has skyrocketed in recent decades, of which autism is just one.

THE CLONING OF A HUMAN BEING

The next great science-fraud scandal of the twenty-first century was the cloning scandal centered on the South Korean biomedical scientist Hwang Woo Suk. In his native land, Hwang was not just a prominent scientist but a star. His field was cloning, and in February 1999 he announced he'd created the world's fifth cloned cow, though he failed to publish a scientific paper on this. The same pattern of announcement sans publication was repeated a few months later when he said he'd cloned another cow. For the next few years he regularly claimed genetic-engineering breakthroughs, including the creation of a cow resistant to BSE. In 2004, one of his announcements concerned not food animals but humans: he had cloned a human cell, creating embryonic cells by inserting an adult cell's nucleus into a human egg. The following year, through a paper published in *Science*, he told the world he and his team had succeeded in creating 11 stem-cell lines from different individuals. He had improved the process of creating stem cells so radically that it seemed the great goal of obtaining stem cells that exactly match the individual's would soon be attained; this is important because, if a patient could be treated using his or her own cloned tissues, the risk of tissue rejection would (presumably) be entirely obviated. Whole cascades of cures seemed foreseeable.

Doubts soon arose. Some young South Korean scientists had been picking apart Hwang's published work and posting the mistakes they found in it on the internet; mainstream print and TV journalism in South Korea likewise became skeptical. However, the western scientific establishment, the western media, and most importantly of all the South Korean government took far longer to catch on. The crunch came when the young Korean scientists published on the internet two identical photographs taken from two separate papers written by Hwang, each purporting to show a different type of cell. Although Hwang was swift to claim the duplication was mere error—the wrong photo had been sent with one of the papers—his critics regarded it as evidence of fraud, and suggested he'd never cloned an adult human cell at all, instead deriving his stem-cell lines in the usual way, using spare embryos from fertility clinics. There were questions, too, about one of the illustrations to Hwang's 2005 *Science* paper, in which it seemed a supposed recording trace had been at least augmented by hand, if it wasn't completely an artifice. Soon members of Hwang's staff were going public about his methods and confessing that to their knowledge some at least of his claimed stem-cell lines did not exist.

On November 24, 2005, Hwang held a press conference to admit various ethical lapses in his research: he had been blinded, he said, by his zeal to aid humankind.

Despite its confessional tone, the press conference had the effect of drumming up huge popular support for Hwang among the South Korean population and, significantly, among the nation's lawmakers. The weekly TV program *PD Su-Cheop*, which two days earlier had broadcast a factually accurate segment severely critical of Hwang's work, was pulled off the air. (Its cancellation was reversed some weeks later.) On December 17, 2005, Seoul National University launched an urgent probe into Hwang's researches, seeking evidence of fraud; just twelve days later, on December 29, the university announced that all the stem-cell lines claimed by Hwang had been fabricated. On January 10, the university announced that the data in Hwang's two *Science* papers of 2004 and 2005 were likewise fabricated. On January 12, Hwang held another press conference, saying the results had indeed been falsified—but by others on his team, who he said had deceived him. By that time few were interested in his excuses.

The case led to the formation in early 2006 of the Hinxton Group, set up to try to insure that similar cloning frauds did not occur in the future. The scientific community's self-critical post mortem on how Hwang had been allowed to get away with so much for so long was extensive; in early 2007, for example, Donald Kennedy, editor-in-chief of *Science*, announced that in future the journal would no longer be content with the peer-review process alone for "high risk" papers, but would subject such papers to additional and intensive editorial scrutiny before publication.

Yet there is no doubt Hwang did do some authentic cloning. A March 2006 *Nature* paper provided independent confirmation, done through DNA analysis, that the Afghan hound Snuppy, which Hwang claimed to have cloned in April 2005, was indeed a genuine genetic copy.

The field of human cloning has attracted several other dubious claimants, some more overtly silly.

Raëlianism is a prominent UFO religion, founded in 1973 by sports journalist Claude Vorilhon, and now with tens of thousands of adherents worldwide. Raëlians believe very strongly in the merits of human cloning, especially as a way in which infertile and gay couples may have children that are genuinely their own. To this end, in 1997 the Raëlian Movement founded the company Valian Venture Ltd, which conducts research on human cloning and, through its project called Clonaid, a service to help couples who wish to have cloned offspring and can afford the necessary $200,000 fee.*

In December 2002, Clonaid's CEO Brigitte Boisselier claimed the company had cloned a human child, called Eve; in January 2003, Boisselier claimed a second cloned child had been born and that several others would soon follow in different parts of the world. These announcements naturally created international headlines, but further details (and evidence such as DNA samples) were not forthcoming, the excuse being offered that Boisselier might find herself in jail for having violated various countries' anti-human-cloning laws; it's therefore generally assumed the claims were false—or perhaps a "metaphysical truth."

Hwang's and Boisselier's claims were presaged more than two decades earlier by David Rorvik's book, *In His Image: The Cloning of a Man* (1978), which described how he had recruited some of the world's top geneticists on behalf of a billionaire ("Max") to clone a male heir, and had succeeded in doing so. Since Rorvik was no wild-eyed conspiracy theorist but a science journalist with a respectable track record, the press and public took the story seriously, even though specialists were skeptical. The claim began to unravel when the Oxford University developmental biologist J. Derek Bromhall—who served as technical consultant on the movie *The Boys from Brazil* (1978)—recognized his own doctoral thesis as the basis for the cloning details described in the book and took its author and publisher to court. Rorvik was unable to offer any evidence in support of his account, and so the defense case collapsed. It's thought Rorvik's motive in writing the book may have been to kick start the necessary public debate about the ethics of human cloning before its reality rather than after. Even so, Rorvik continued—as recently as 1997 in an article in *Omni*—to insist his story was genuine.

---

* The focus of the Raëlian Movement on cloning is not merely an opportunistic one: it is a religious duty, because according to their principles it is only through cloning that humans can attain eternal life and thus become one with the Elohim.

The Case of the Gay Canvassers

The December 12, 2014, issue of *Science* carried an article by Michael J. LaCour, a graduate student in the Department of Political Science at UCLA, and Donald P. Green, a professor in the Department of Political Science at Columbia, that immediately attracted mainstream attention worldwide. The paper was "When Contact Changes Minds: An Experiment on Transmission of Support for Gay Equality." The first few lines of its abstract read:

Can a single conversation change minds on divisive social issues, such as same-sex marriage? A randomized placebo-controlled trial [done among Los Angeles County voters] assessed whether gay (n = 22) or straight (n = 19) messengers were effective at encouraging voters (n = 972) to support same-sex marriage and whether attitude change persisted and spread to others in voters' social networks. The results, measured by an unrelated panel survey, show that both gay and straight canvassers produced large effects initially, but only gay canvassers' effects persisted in 3-week, 6-week, and 9-month follow-ups.

According to the study, it wasn't just the people with whom the gay canvassers spoke who became more favorably disposed toward gay marriage and other gay issues: there was a knock-on effect, with other members of the same household showing a similar albeit lesser shift of opinion. Overall, then, "Contact with minorities coupled with discussion of issues pertinent to them is capable of producing a cascade of opinion change."

We can easily see why the media were so interested. Passive bigots, so to speak, when confronted by a real-life gay person and discovering that person to be just an ordinary human being, were likely to rethink their preconceptions.

This seemed so in keeping with both commonsense and experience that it was little wonder many people took the study at face value. After all, it has long been known that racism—like other forms of prejudice, such as antisemitism—is generally most endemic in segregated areas. Although other factors, such as the presence or absence of gangs, can play a part, in regions where people of different colors and cultures can prattle over their back fences or while queuing at the post office, prejudices are generally very much less prevalent.

Uncertainty began to creep in when, in January 2015, two graduate students of the University of California, Berkeley, David Broockman and Joshua Kalla, decided to mount an extension of the LaCour/Green study, assuming their work would tend to confirm it. However, as they announced in their own paper, "Irregularities in LaCour (2014)" (done with Peter Aronow of Yale),[7] they soon became concerned about the integrity of the data being used:

> As we examined the study's data in planning our own studies, two features surprised us: voters' survey responses exhibit much higher test–retest reliabilities than we have observed in any other panel survey data, and the response and reinterview rates of the panel survey were significantly higher than we expected.

In other words, the results were too good.

As their doubts rose, the two researchers contacted Qualtrics, the survey company that, according to the original LaCour/Green paper, had been hired to do the data-gathering. Qualtrics responded that, not only had it never done the work in question, it didn't have an employee of the name given by LaCour.

By May 19, 2015, Green, contacted a week or so earlier by Broockman and Kalla, was convinced there was something seriously awry with his co-author's data, and wrote to *Science* requesting the paper be retracted. He was "deeply embarrassed by this turn of events and apologize[d] to the editors, reviewers, and readers of *Science*." The question obviously arose as to why he'd failed to consult the raw data that LaCour had sworn existed, instead confining his critique to the dataset LaCour provided. According to Green:

> Given that I did not have [institutional review board] approval for the study from my home institution, I took care not to analyze any primary data . . . Looking back, the failure to verify the original Qualtrics data was a serious mistake.

Well, yes.

In response to all these developments, LaCour admitted some of the data had been falsified, then claimed the raw data wasn't available because it had been accidentally deleted, then claimed it had been deleted in order to preserve the confidentiality of the survey respondents. He also gave varying accounts of how the study had been funded.

There was so much interest in the affair that, when the website *Retraction Watch* first announced the breaking news of the LaCour/Green study's debunking, the traffic became so intense it crashed the site's server.

Ironically, when Broockman and Kalla went on to repeat the LaCour/ Green experiment, they found that in large part the results were more or less as the original data had predicted. The only real difference was that it didn't make any difference if the canvasser was gay or straight.

The Plastics and the Perch

It was a story that could have been written with the environmentally conscious elements of the mainstream media—such of them as there are— in mind. A pair of Swedish marine scientists had discovered that the perch larvae they were studying would eat tiny fragments of plastic waste in the ocean in preference to their natural diet of brine shrimp—they would eat the plastic eagerly, in fact. The larvae as a result grew more slowly and also, according to the research, became less able to detect or recognize the chemical alarms that warned of predators. Inevitably this was leading to a decline in perch populations and thus a disruption of the food chain.

We know that plastic litter in the oceans is a major problem, and we know that too much of this garbage is getting eaten by marine creatures and thereby entering the food chain. Here was a dramatic example of the process in action, and it involved baby fish. No wonder it caught the attention of journalists.

The research concerned was summarized in a paper that appeared in the June 3, 2016, issue of *Science*, "Environmentally Relevant Concentrations of Microplastic Particles Influence Larval Fish Ecology" by Oona M. Lönnstedt and Peter Eklöv. The actual research was done at the Forskningsstationen Ar (Ar Research Station), on the island of Gotland, by Lönnstedt, a research fellow at Uppsala University; her coauthor, Eklöv, was her supervisor at the university. A few months after the paper's *Science* publication, Lönnstedt was awarded a $330,000 grant to continue the research by the Swedish funding agency Formas.

The paper puzzled Josefin Sundin, of Uppsala University's Department of Neuroscience, and her friend Fredrik Jutfelt, a biologist at the Norwegian University of Science and Technology, in Trondheim. Sundin had been at the Ar Research Station at the same time as Lönnstedt during the spring of 2015, when Lönnstedt claimed she was doing the experiments, but could recall nothing of any such research project. Jutfelt, who'd been at the station briefly during the relevant period, reported the same.

The two scientists and some of their colleagues scrutinized the paper in detail, and became concerned about its methodology. When they asked to see the raw data—as is customary in such instances—Lönnstedt told them it had been almost entirely lost when her laptop computer was stolen.

In June 2016 the group wrote to the authorities at Uppsala University expressing their doubts:

> We wish to report a strong suspicion of research misconduct in the following study by researchers at Uppsala University, published in the journal *Science* on June 3 2016 . . .
>
> We have identified a number of potentially critical flaws regarding the execution and reporting of this study, which include: (1) missing data (wrongly stated in the paper as being available in the supplementary materials and in the Uppsala institutional repository); (2) inconsistencies in the sample sizes reported and the microplastic exposure concentrations used; (3) issues with the statistical design and analyses; and, most worryingly, (4) large disparities in the way the experiment has been reported by the authors compared with the reports of eye witnesses. These issues, many of which should have been identified at the peer-review stage, bear directly on the validity and reproducibility of the results presented in the paper . . .

This was all pretty damning. However, in August 2016 the panel of experts convened by the university to examine the charges dismissed them as false. The whistleblowers persisted, and the case was passed along to Centrala Etikprövningsnämnden (CEPN; Central Ethical Review Board), whose officers wrote in April 2017 to the university to the effect that, yes, there was indeed cause for concern:

> In summary, and in accordance with what has been stated above, the Expert Group finds that Peter Eklöv and Oona Lönnstedt have been guilty of scientific dishonesty. . . . In view of the lack of ethical approval, the essential absence of original data for the experiments reported in the article, and the widespread lack of clarity concerning how the experiments were conducted, it is the opinion of the Expert Group that the article in Science should be recalled.

The CEPN report also criticized the university's review panel for having skimped in their task, and *Science* and its peer reviewers, for having published the paper in the first place:

> It may be considered particularly remarkable that the article was sent for publication without the presence of the necessary data. It is worth pointing out here that the journal *Science* was deficient in its checking in this respect.

Finally, in early May 2017, *Science* retracted the study—purportedly at the request of its authors. Lönnstedt and Eklöv continued to maintain that everything had been aboveboard, and that the research had been carried out exactly as claimed, but explained in a statement reported in *Nature* (May 3, 2017) that

> Science has to rest on solid ground and the results of this study, even if they are correct, will not be trusted as long as a suspicion of misconduct remains. Thus, we decided that we would not have the endurance to keep defending the paper, and hence decided to retract it.

Just as the process of peer review is imperfect, occasionally allowing bad or bogus studies to be published in even the best of scientific journals, there's always the possibility that later reviews can come to the wrong conclusion. We cannot be certain beyond all scintilla of doubt that Lönnstedt did indeed "enhance" her work. However, this was the conclusion of the scientific community, and there the matter rests.

2

SEEING WHAT THEY
WANTED TO SEE

The nineteenth-century UK scientist Lord Kelvin was one of the greatest physicists of all time, but he also had a habit of being comprehensively *wrong* about things. Two of his ringing predictions have achieved classic status: "Heavier-than-air flying machines are impossible" and "Radio has no future." He proved evolution was bunk through calculating that the sun wasn't old enough to allow the long time periods required by Darwinian evolution. Of course, Kelvin couldn't have had a clue that the sun "burns" through nuclear fusion, not by mere combustion; where he erred was in rejecting Darwinian evolution purely because he found the notion philosophically intolerable and then grabbing the first "proof" he could find to justify his revulsion.

Almost as celebrated was his accusation that Wilhelm Roentgen, who in 1895 announced the discovery of X-rays, was guilty of scientific fraud. Kelvin's argument was that the cathode-ray tube, the source of the rays, had been in use by physicists for a full decade before the purported discovery. If X-rays existed, someone would already have noticed them. Well, someone just *had* . . .

Irving Langmuir, winner of the 1932 Nobel Chemistry Prize, in 1934 coined the term "pathological science" in the context of the researches J.B. Rhine was carrying out on ESP. Langmuir's informal definition of the term was "the science of things that aren't so"—in other words, the application of science to a phenomenon that doesn't exist, or in pursuit of a nonsensical hypothesis. The pathology lies in the self-deception: the researchers or theorists are incapable of standing back and observing that they're engaged in a fool's errand.

In this chapter our primary focus will be on such science, although always with the thought in the back of our mind that sometimes it isn't obvious at the time just what science is pathological and what is merely new and unexpected.

First, though, we should look at a couple of areas where the science community has tended until relatively recently to look at its own activities through rose-tinted spectacles, to believe overconfidently in its own standards of objectivity. I'm referring to the problems of replicability and of sham and predatory scientific journals.

The Replicability Crisis

The problem of the lack of replicability (or reproducibility) of reported scientific results is not new. Back in the opening years of the seventeenth century, when Galileo Galilei was having his little squabble with the Vatican, one of the reasons so many of his contemporaries in the Italian scientific community were reluctant to step up and defend him—aside from the fact that doing so might draw the unwelcome attentions of the Inquisition, plus the insufferable arrogance of the man had pissed a lot of his fellow scientists off—was that there was a general aura of doubt about many of the experimental results he'd reported. These concerned not his astronomical observations but, for example, his experiments rolling balls down inclined planes.* Try as they might, his contemporaries had difficulty replicating Galileo's results. This was unsurprising: all the timings that Galileo presented with such confidence were, shall we say, tricky to achieve in an era when the clock hadn't yet been invented. Our best guess is that at least some of Galileo's experiments were never actually performed—that they were, rather, thought experiments. Luckily he was a genius, so he got the right answers!

But we're not all geniuses, and some of us are frauds. Although we tend to believe what we're told, especially if it's told in a paper written by accredited scientists and published in a revered science journal like *Nature* or the *New England Journal of Medicine*, we have no real way of telling if what we've been told is the truth unless we check it—or, at least, until someone else does the checking.**

---

* Those very same experiments, in fact, that would eventually lead to the data upon which Newton could base his theory of universal gravitation.

** Conspiracy theorists will tell you that *even then* you can't be sure. Professional scientists are all part of the same club, you see, and when they *say* they're checking each other's results all they're *really* doing is . . .

The matter of replication lies at the heart of the modern scientific method. The traditional scientific method, as derived from the philosophical reasoning of Francis Bacon and others, runs roughly as follows:

- study the relevant phenomena (gather evidence)
- think up a way those phenomena might be explained (form a hypothesis)
- imagine some as-yet-unobserved consequences of the hypothesis (make predictions)
- test your predictions (devise experiments)

A couple of other stages have been added recently:

- publication of your experimental methods, results and conclusions
- replication of those experiments by others

If other people get the same results, it can be assumed *anyone* would—i.e., that a genuine advance in our knowledge and understanding has been made. If other people try to repeat your experiments and find they can't, or get entirely different results, then obviously there's something amiss: perhaps you're a crook, or perhaps there's some confounding factor present in your or their laboratory that has gone unnoticed. Whatever the case, clearly more work needs to be done to establish the truth.

This is all a pretty idealized account. In reality, scientists follow varied routes in their investigations, although they'll generally touch most of these bases. The last two, publication and replication, have come to be regarded as the most important.

The starry-eyed version of the way that science works is that all published results are eagerly leaped upon by scientists around the world, and that attempts at replication follow publication more or less as a matter of course. In reality, the majority of published papers go essentially unread; of those that *are* read, in very few instances are there any attempts at replication.

The reasons are obvious. Most scientists would rather be doing their own work than checking the work of others; even if that weren't the case, their paylords generally want original results out of them, because it's the original work that brings the grant money—mere replication being, well, boring. Moreover, the journals themselves offer little incentive to anyone who might be interested in replicating the average experiment. Imagine you've repeated a published experiment and discovered that, sure enough, the original researchers were right: what they reported is exactly what happens. The chances of your finding a journal willing to publish this discovery are pretty small. Journal editors aren't especially interested in the research equivalent of Dog Bites Man stories.

The trouble is, they're not hugely interested in Man Bites Dog stories either. Even if your replication exercise has unearthed a problem in the original experiment, and unless that original result was something significant, you're going to find it difficult to persuade a major journal to do more than put a brief note from you in the letters column.

Since most scientists are evaluated by their paymasters in terms of the amount of work they've published, this again makes replication seem like a fruitless endeavor.

As a measure of the scale of the problem, a June 2015 *PLOS Biology* paper[8] estimated that, for the medical sciences alone,

> An analysis of past studies indicates that the cumulative (total) prevalence of irreproducible preclinical research exceeds 50%, resulting in approximately US$28,000,000,000 . . . [per] year spent on preclinical research that is not reproducible—in the United States alone.

Medical science faces a particular problem, as we'll see in Chapter 6, in that so much of its research is funded by the manufacturers of the products being evaluated.

There are also cultural issues that may affect reported results. A review cited by Jonah Lehrer[9] looked at 141 clinical trials performed on acupuncture done between 1966 and 1995; of these, 47 were done in eastern countries—China, Taiwan, Japan—and 94 in western ones—Sweden, the UK, the US. Of the 47 Asian studies, every single one found that acupuncture was a useful and effective therapy. For the occidental studies, that figure dropped to 56 percent.* Leaving aside the possibilities of deliberate deception, it seems the scientists involved were perceiving the evidence that their experiments produced through the filter of their cultural conditioning. We all tend to fall into this same trap in our more mundane attitudes, and indeed do so much of the time; it's just a surprise to find that scientists, who are supposedly behaving with objectivity as they experiment, can so easily fall into that trap.

In a 2016 poll of 1,576 scientific researchers reported in *Nature*, "More than 70 percent of researchers have tried and failed to reproduce another

---

* Which still seems a surprisingly high figure.

scientist's experiments, and more than half have failed to reproduce their own experiments."[10] Of the respondents, 52 percent agreed there was a significant replication crisis and a further 38 percent agreed there was a problem, even if they didn't regard it as major; just three percent declared there wasn't a crisis.

Very recently the scientific community has become far more aware of the problem, or at least readier to tackle a problem it has for the most part carefully ignored. In the field of psychology, perceived as one of the areas in which the lack of replicability was especially rife, the journal *Perspectives in Psychological Science* has introduced a relevant regular section, "Registered Replication Reports," and other journals are cleaning up their acts. In the 2016 *Nature* survey noted above, about half of the 24 percent of respondents who'd attempted to publish a successful replication had found a willing journal, and most of the 13 percent who'd offered an unsuccessful replication had found a publisher.

In 2011, the University of Virginia's Brian Nosek, a co-founder of the Center for Open Science, launched an initiative, the Reproducibility Project, dedicated to assessing just how bad the problem was in psychology. Nosek and a group of volunteer scientists selected at random a hundred studies that had been published in the year 2008 in three respected psychology journals—*Psychological Science, Journal of Experimental Psychology: Learning, Memory, and Cognition*, and *Journal of Personality and Social Psychology*—and tried to replicate their results, sometimes in consultation with the original authors.

Out of the chosen 100, Nosek and his team attempted to replicate the 97 studies that claimed significant results, and succeeded in replicating just 35 of them. Moreover, the results they obtained in the replications were less positive than those reported by the original authors: the reported correlations were there, all right, but they weren't as strong as the original authors had found them to be.

We should be clear here. What Nosek and his colleagues discovered was not a rampant lack of integrity on the part of experimental psychologists. Obviously, human beings being what they are, among those hundred studies there must have been a few examples of researchers over-egging their psychological pudding, and quite possibly an outright fraudulent study or two, but the vast majority of the original authors were reporting quite honestly what they perceived. In some instances they'll have seen stronger-than-replicated results because they saw what they hoped/expected to see; in others, they'll have been guilty, presumably unconsciously, of selective reporting to favor their preferred hypothesis; in yet others, their strong results will have been merely a matter of chance.

Obviously a hundred studies drawn from a single year and a single discipline (and, at that, from just three journals within that discipline) represent just a drop in the ocean. There's no way any group of scientists, no matter how large—and no matter how large their budget!—could check out every single seemingly noteworthy study that's ever been published.

There have been suggestions that governments should step in to solve the problem—by setting up a sort of Bureau of Replication, perhaps. That would be an expensive enterprise, and one that would be hard to sell to the taxpayer—because replication doesn't offer something as shiny and cutting-edge as the original product. It would on the other hand be money well spent, since the successful functioning of science (not just technology) is something we'll be relying on increasingly in the future. Maybe the Bureau of Replication could more practicably be considered by an international body like the United Nations. It's difficult to imagine it actually happening, though.

How big is the replication crisis? We're not quite to the point suggested by the title of a famous/notorious 2005 essay written by John P.A. Ioannidis of the University of Ioannina School of Medicine, "Why Most Published Research Findings are False,"[11] but certainly there's a case to be made that any published scientific findings of any importance, assuming they haven't been authenticated already through the normal workings of science (for example, every NASA space probe that arrives safely at its destination confirms all sorts of published physics), should be considered ripe for re-evaluation.

PREDATORY AND FAKE JOURNALS

In April 2017, the climate-science community was startled by a new paper, "The Refutation of the Climate Greenhouse Theory and a Proposal for a Hopeful Alternative" by a little-known Swiss scientist, Thomas Allmendinger. From the abstract:

> In view of the global acceptance and the political relevance of the climate greenhouse theory—or rather philosophy—it appeared necessary to deliver a synoptic presentation enabling a detailed exemplary refutation. It focuses [on] the foundations of the theory assuming that a theory cannot be correct when its foundations are not correct. Thus, above all, a critical historical re-

view is made. As a spin-off of this study, the Lambert–Beer law is questioned suggesting an alternative approach. Moreover, the Stefan–Boltzmann law is relativized revealing the different characters of the two temperature terms. But in particular, the author's recently published own work is quoted revealing novel measurement methods and yielding several crucial arguments, while finally an empiric proof is presented. . . .[12]

There's an uncanny echo here of the blurbs to all those millions of books self-published through Amazon CreateSpace whose eager authors believe they've discovered at last the secret of life, the universe and everything that orthodox scientists—the *fools!*—have been blithely overlooking. Yet this paper was published in a supposedly peer-reviewed journal, *Environment Pollution and Climate Change*, one of the OMICS Publishing Group stable.

After its first two issues, the journal had published no fewer than six studies purporting to overturn modern theories of global warming. Perhaps this is not surprising, since its editor-in-chief, Arthur Viterito, a geographer at the College of Southern Maryland, is a noted climate-change "contrarian" who has served as an advisor to a phony thinktank, the science-denying Heartland Institute. He has suggested global warming might not be an atmospheric effect at all, but instead due to undersea volcanoes. Needless to say, scientists have had difficulty following the physics of his argument.

Cited by DeSmog Blog,[13] climate scientist Michael E. Mann expressed his critical judgement with some pith:

This isn't science. It's politically motivated denialist garbage. . . . Such sham journals make a mockery of the scientific process and must be exposed for what they are. Associating in any way with this pseudo-journal would endanger one's scientific reputation. Keep your distance from this toxic mess.

Before the likes of *Environment Pollution and Climate Change* came along, climate-science deniers had understandable difficulties in finding journals willing to publish their offerings. Supposed titans of the climate-skeptic movement such as Sallie Baliunas and Willie Soon, both of whom have genuine scientific qualifications—but in astrophysics, which attentive readers will notice has nothing to do with climate science—have published scores of papers, according to their websites, but closer examination shows that almost none of these papers have appeared in conventional scientific journals: a frequent publisher for them has been the website of the George C. Marshall Institute, a now-defunct (2015) anti-science thinktank largely funded by ExxonMobil.

A coauthorship by this duo did make it into a scientific journal. The year was 2003, the journal was the lower-tier publication *Climate Research*, and their review study, "Proxy Climatic and Environmental Changes of the Past 1,000 Years," was partly funded by the American Petroleum Institute. In the ensuing furor it emerged that all four of the article's peer reviewers had declared it should be rejected but that one of *Climate Research*'s editors, Chris de Freitas, an avowed climate-science denier, had gone ahead and published it anyway. Half the journal's editorial board walked out in protest.

A revised version of the study, with added authors Craig Idso, Sherwood Idso and David Legates—all familiar names to students of climate-science denialism—appeared a few weeks later, in Spring 2003, in the journal *Energy & Environment*, whose editor, Sonja Boehmer-Christiansen, is, like *Environment Pollution and Climate Change*'s Arthur Viterito, a "climate skeptic." Neither journal is regarded as having any significant scientific impact, as measured by citations of their articles elsewhere, but Baliunas and Soon do have the heartwarming knowledge that their paper is frequently cited by James Inhofe.

#

The traditional business model of an academic journal relies on income generated from subscriptions. Advertising can make a major contribution as well, and there are also, since the rise of the internet, fees paid to view individual articles on a one-off basis, but the core of it all is the subscription list. Outside of a few journals such as *Nature* and *Science* and the *British Medical Journal*, vanishingly few of the subscriptions are bought by individuals: almost all are paid for by institutions of higher learning—university libraries and the like. This means that, assuming the journal has any worth, the subscriptions can be remarkably expensive—another reason why so few individuals subscribe. With a captive audience and relatively low expenses, publishing learned journals might seem an easy way to print money, and indeed this is exactly the complaint often made about the big journal publishers.

Their response is that the successful journals pay for their more specialized, lower-circulation siblings, and that part of what readers are paying for is quality control. The first point of that argument has some merit—at least until you check out the annual profits of the major journal publishers—but the latter part is more dubious. Time after time, well respected journal publishers like Elsevier and Taylor & Francis have harbored, among their extensive stables of journals, some pretty iffy horses.

Since the emergence of the internet, the number of supposedly learned journals has shot up because most of the costs of conventional journal publishing—printing, distribution, and so on—can be circumvented. All you need is a website, which doesn't cost much, a few academics to serve (usually gratis, because it looks good on their resumes) on your editorial board, some actual *editors* to do the nuts-and-bolts work, and then secretarial staff and an accountant. Mom, Dad and the kids could more or less do it all between them.

That's something of an oversimplification, but the fact remains that it has become very cheap indeed to run an academic journal, so long as you're not too finicky about academic standards (and even *then* your expenses needn't be too high). If you're an institution that already has a reasonable income, such as a pressure group, an astroturf organization, or a phony thinktank—of which there are now an inordinate number—you can even produce a printed version.

All of these factors came together in the early part of the twenty-first century to fuel the rise of the sham scientific journal—the journal that, while bearing the outward appearance of academic objectivity, is in fact presenting a political or crank agenda.

The *Journal of American Physicians and Surgeons* (*JAPS*), for example, is the voice of the Association of American Physicians and Surgeons, best regarded as a fringe group driven by assorted bigotries and a loathing for the concept of socialized medicine. Most of the relevant medical databases refuse to list the *JAPS*, on the grounds that its articles are primarily political, to the point of promoting pseudoscience and science denial on topics such as abortion, vaccines, homosexuality, global warming and even gun control (it's against). In an editorial called "Defending Science" in the legitimate science journal *Chemical & Engineering News*, Rudy Baum gave some of his readers what was almost certainly their first introduction to *JAPS*:

> It is not indexed by Chemical Abstracts Service, Pubmed, or ISI's Web of Science. It has published articles that question the link between HIV and AIDS and that link abortion to increased incidence of breast cancer and thimerosal-containing vaccines to autism. It is, in fact, the purveyor of utter nonsense.[14]

You could argue that Baum was pulling his punches a little. What he didn't state outright, but clearly felt, was that much of the content of *JAPS* was not just "utter nonsense" but ethically objectionable.

The Creation Science Association (CSA), founded in 1977, first published its supposedly scientific journal, *Ex Nihilo Technical Journal* (now known, after several name-changes, as *Journal of Creation*), in 1984. Ken Ham's creationist outfit, Answers in Genesis, has since 2008 published *Answers Research Journal*, which declares itself to be

> A professional peer-reviewed technical journal for the publication of interdisciplinary scientific and other relevant research from the perspective of the recent Creation and the global Flood within a biblical framework [that] will provide scientists and students the results of cutting-edge research that demonstrates the validity of the young-earth model, the global Flood, the non-evolutionary origin of "created kinds," and other evidences that are consistent with the biblical account of origins. . . . Papers can be in any relevant field of science, theology, history, or social science, but they must be from a young-earth and young-universe perspective. Rather than merely pointing out flaws in evolutionary theory, papers should aim to assist the development of the Creation and Flood model of origins.

The journal's first paper, "Microbes and the Days of Creation"[15] by Alan Gillen, a professor at the Christian fundamentalist Liberty University, tried to establish when during the Creation Week the microorganisms—bacteria, viruses and so on—were introduced into the scheme of things. God failed to include this information while dictating Genesis to Moses, and so modern thinkers have to try to work it out for themselves. Gillen seems incapable of understanding that microbes can exist quite happily without the presence of larger organisms, and so opts for them having been introduced piecemeal, at the same time as their "hosts." The really nasty ones, like HIV, didn't come along until the expulsion from Eden—the woman's fault, as usual.

Both these journals proudly describe themselves as peer-reviewed, which is undoubtedly true. However, that can't be taken as any mark of scientific validity if all the peers are people who believe the world is only a few thousand years old. Whatever the editors might *think* is the case, such a view is so divorced from established reality that there's no way it can be made consonant with science. The journals produced by the various creationist organizations are a manifestation of cargo-cult science, and their supposed peer review is simply part of the cult.

Autism Insights was a journal that ran for just six issues from 2009, being discontinued in 2013 by its publisher, Libertas Academica (slogan: "Freedom to Research," a fair clue that Here Be Crankery). The paper that brought this journal into the wider view appeared in its second issue, on January 27, 2010: "Clinical Presentation and Histologic Findings at Ileocolonoscopy in Children with Autistic Spectrum Disorder and Chronic Gastrointestinal Symptoms" by Arthur Krigsman, Marvin Boris, Alan Goldblatt and Carol Stott. This contained the long-promised details of lead author Krigsman's study replicating the results of Andrew Wakefield (see pp. 55–59) on the link between the MMR vaccine and autism. "Long-promised" is an appropriate term: it had been as long before as June 22, 2002 that Krigsman had told Congress's Government Reform Committee that he'd achieved this replication.

Krigsman was then working for the Thoughtful House Center for Children, in Austin, Texas, an establishment whose stated aim was to conduct research into autism and to treat sufferers; the center was apparently the brainchild of Wakefield himself, and it served as his employer for a few years. It was, of course, an antivaxxer outfit, although this is not to say that some of its other work might not have been of value.

The editorial board of *Autism Insights*, as listed at the time, offers us an insight into what could likely have been the journal's . . . *nuance*, shall we say:

BRYAN JEPSON, MD, Director of Medical Services, Medical Center, Thoughtful House Center for Children, Austin, TX

ARTHUR KRIGSMAN, MD, Director of Gastrointestinal Services, Pediatric Gastroenterology, Thoughtful House Center for Children, Austin, TX

CAROL MARY STOTT, PhD, Senior Research Associate, Research Department, Thoughtful House Center for Children, Austin, TX

ANDREW WAKEFIELD, MBBS, FRCS, FRCPath, Research Director, Thoughtful House Center for Children, Austin, TX, USA

Wakefield's position at Thoughtful House lasted just long enough for him to see the publication of Krigsman's favorable study; in February 2010 he and his employer parted ways.

Autism Insights was an example of the new pay-to-publish, or "open access," model of scientific journal. Whereas, in traditional science-journal publishing, readers have to pay through the nose to stay abreast of the latest developments in their field, in the open-access scheme it's the scientists doing the research who have to pay to see their studies published. *Autism Insights* was one of the more expensive of these ventures, charging its contributors a whopping $1,699 by way of "processing fee."

Not all open-access journals are duds. For example, the Public Library of Science (PLOS) first published its journal *PLOS Biology* in October 2003, and now has ten journals in its stable, all operating on the pay-to-publish basis. Even though PLOS is a non-profit, its fees are not cheap (between $1,495 and $2,900 per paper in March 2017) and it has amassed not inconsiderable assets over the years; nonetheless, there's no denying that its peer-review standards are high and that it has published a lot of valuable work. The big advantage to the authors is that going the PLOS route gets their work out there far faster than would be possible via conventional journals. At a more mercenary level, it can be useful to an academic's resume to have a publication *now* rather than a year or three down the line.

Even though the pay-to-publish model can offer advantages to individuals and the scientific community, it's obviously wide open to abuse—and abused it most comprehensively has been. Not only are there countless sham academic journals, there are also plenty of predatory journals—outfits that seduce young and naive (or, very often, crank and/or incompetent) researchers into paying fees to publish in journals whose only *raison d'être* is to extract cash from the gullible. Although these journals claim to employ peer review, that's almost universally an empty claim—as has been demonstrated time and time again by the often very amusing stings that prankster scientists have been able to perpetrate on these crooks.

Both sham and predatory journals tend to have imposing-sounding names—like *Environment Pollution and Climate Change*. It's very much the same phenomenon as the one that leads bogus thinktanks and astroturf organizations to bestow on themselves names that seem impressive even if, quite often, they're at complete odds with the actual aims and functions of the organization. Who would have thought, for example, that the Alliance Defending Freedom is a hate group seeking to restrict very significantly the freedom of anyone other than heterosexuals? (One of its particular charms is its enthusiastic backing of the forced sterilization of transsexuals.[16]) Who would have thought the organization Accuracy in Media seeks to impose not accuracy but a laughably far-right ideology on the mainstream news media? Who would have thought that the Air Quality Standards Coalition seeks to relax the regulations on air pollution?

A similarly Orwellian motivation lies behind the naming of far too many fake and predatory scientific journals. In fact, one of the criticisms leveled at the Directory of Open Access Journals (DOAJ), launched in 2003 at Lund University, Sweden, and maintained by volunteers, is that it's unwittingly assisting the predatory journals. Of the 9,642 journals that it currently (August 5, 2017) lists,[17] some are certainly predatory—and the unpaid DOAJ volunteers work hard to try to weed these out. In other instances reputable journals have had their titles near-cloned by predators, so DOAJ users might easily be deceived into trusting the wrong journal.

A very useful list of predatory journals was begun and put online in 2008 by Jeffrey Beall, scholarly communications librarian at the University of Colorado, Denver; it was Beall who coined the term "predatory publishing" to describe the activities of these journals. In 2013, Beall was threatened with a $1 billion lawsuit by the OMICS Group; based on dubious legality, the threat evaporated.

By January 2017, Beall's List, as it was universally known, ran to thousands of titles, but in that month it was suddenly taken offline. According to his university bosses, Beall's decision to discontinue the website was a "personal" one: "Professor Beall remains on the faculty at the university and will be pursuing new areas of research." According to Lacey E. Earle of Cabell's International, which had been working with Beall to develop a new blacklist of fake and predatory journals, the reason was more sinister. Earle tweeted that Beall "was forced to shut down blog due to threats & politics."[18]

But it's part of the nature of the internet that things that you think are dead aren't really so. At http://beallslist.weebly.com you can find this: "This website is a copy of Beall's list of predatory publishers. It was retrieved from cached copy on 15th January 2017. We hope you will find it useful."

And already the website owner has posted a few suggested additions to the list . . .

STINGS

In 2014 computer scientist Peter Vamplew of the Federation University, Victoria, Australia, got fed up of receiving spam trolling for submissions from him for the *International Journal of Advanced Computer Technology*, a free-online-access journal that he regarded as predatory. Repeated requests to be taken off the journal's mailing list had gone ignored.

Accordingly, Vamplew dug out a spoof paper that David Mazières of New York University and Eddie Kohle of the University of California at Los Angeles had produced in 2005. Called "Get Me Off Your Fucking

Mailing List," the "paper" consists of nothing more than those seven words repeated *ad nauseam*. For the benefit of the hard of comprehension, so to speak, there are also illustrative flow charts, whose message, when painstakingly unpicked, is . . . but you've guessed it by now.

Vamplew sent a copy of the paper to the *International Journal of Advanced Computer Technology* on the assumption that doing so might at last persuade the journal to take him off its . . . You've guessed right again.

Instead, the journal said it would like to publish the paper. "They told me to add some more recent references and do a bit of reformatting," Vamplew informed the *Guardian*.[19] "But otherwise they said its suitability for the journal was excellent." This was, you understand, after exhaustive peer review.

Oh, and the journal's editors also requested $150 from Vamplew for the privilege of publication—a request that the computer scientist sorrowfully denied.

In 2009, Cornell University graduate student Philip Davis, in conjunction with Kent Anderson of the *New England Journal of Medicine*, pseudonymously* submitted to the *Open Information Science Journal* a paper he'd created using freely available software that generates grammatically correct and sciencey-sounding but in fact meaningless text. The journal, part of the Bentham Science Publishing stable, accepted the spoof paper a few months later, requesting a fee of $800 for publication—at which point Davis pulled the submission and went public. The journal's editor-in-chief, who hadn't even seen the paper—it had been accepted by the Bentham bosses without his knowledge—resigned, saying Bentham's business practices were obviously dubious. Davis declared he'd been galvanized into perpetrating the hoax by Bentham Science Publishing's constant spam invitations to submit to the company's journals.

Christoph Bartneck, of the University of Canterbury's Human Interface Technology laboratory in New Zealand perpetrated a similar spoof in 2016. Invited by e-mail to submit a paper to the International Conference on Atomic and Nuclear Physics, even though that isn't his field, he used Apple's iOS autocomplete software to concoct . . . something:

> I started a sentence with "Atomic" or "Nuclear" and then randomly hit the auto-complete suggestions. The text really does not make any sense. After adding the first illustration on nuclear physics from Wikipedia, some refer-

---

* And supposedly from an academic based at the nonexistent Center for Research in Applied Phrenology, New York.

ences and creating a fake identity (Iris Pear, aka Siri Apple) I submitted the paper which was accepted only three hours later! I know that iOS is a pretty good software, but reaching tenure has never been this close.[20]

He wasn't asked to pay for the acceptance; on the other hand, registration to attend the conference to present his paper would set him back $1,099 (US dollars). Shortly afterward, he heard that the conference's organizer, the OMICS Group, was under investigation in the US by the Federal Trade Commission. According to the FTC's website:

> The Federal Trade Commission has charged the publisher of hundreds of purported online academic journals with deceiving academics and researchers about the nature of its publications and hiding publication fees ranging from hundreds to thousands of dollars.
>   The FTC's complaint alleges that OMICS Group, Inc., along with two affiliated companies and their president and director, Srinubabu Gedela, claim that their journals follow rigorous peer-review practices and have editorial boards made up of prominent academics. In reality, many articles are published with little to no peer review and numerous individuals represented to be editors have not agreed to be affiliated with the journals.[21]

In 2013, John Bohannon of *Science* concocted a bogus scientific paper claiming to have extracted a cancer-curing wonder drug from a common lichen and, under spurious identities such as biologist Ocorrafoo Cobange, of the Wassee Institute of Medicine, Asmara, submitted versions of the "study" to no fewer than 304 pay-to-publish journals. Over half of his submissions were accepted. Some of the journals concerned were at the lower end of the reputability scale, but others were associated with major companies such as Elsevier and Sage, or highly respected academic institutions such as Japan's Kobe University. According to Bohannon's account of the experiment,[22] "It was even accepted by journals for which the paper's topic was utterly inappropriate, such as the *Journal of Experimental & Clinical Assisted Reproduction.*"

In 2016, Alexandre Martin, an assistant professor at the University of Kentucky's College of Engineering, reformated for submission a piece of writing about bats that his seven-year-old son Tristan had done for his school science project. Clearly a plucky youngster, Tristan had produced an essay that ran a full 153 words and included such breakthrough scientific insights as "Bats are really cool animals!"

Martin *père* had assumed he'd have to make a few submissions of the "study" before finding a predatory journal dumb enough to accept it, but to his astonishment the very first to which he submitted the piece said yes: the magnificently titled *International Journal of Comprehensive Research in Biological Sciences*. A mere $60 toward "expenses" would be enough to ensure the paper's appearance.

Alexandre Martin was actually game to cough up the $60 in order to prove his point about the evils of predatory journals, but his wife put her foot down—not just on the grounds that it was a waste of family money but also because, if Tristan eventually decided to pursue a career in the sciences, he might come to rue the fact that his first publication was in a notoriously predatory journal.*

Even though the Martins weren't prepared to shell out the "token fee," the journal's editor sent them a proof of the article. The editor had made a few necessary changes and expanded the text very considerably—indeed, just about the only things Martin senior recognized from the original were the title and the author names. Briefly he felt some qualms about the amount of work the journal's editor had put into this piece.

But then a little bit of surfing on the internet revealed to him that the "revised" paper had been directly plagiarized from two previously published articles . . .

Tristan is listed as coauthor, with his dad, of the paper that finally did come out of this enterprise, appearing in the perfectly reputable journal *Learned Publishing*: "A Not-So-Harmless Experiment in Predatory Open Access Publishing."[23]

In March 2017 John H. McCool of Precision Scientific Editing saw his paper "Uromycitisis Poisoning Results in Lower Urinary Tract Infection and Acute Renal Failure: Case Report" accepted for publication, "after peer review," by *Urology & Nephrology Open Access Journal*, despite the fact that the medical condition uromycitisis doesn't exist outside a 1991 episode of *Seinfeld*; fittingly, the paper is full of *Seinfeld* references. In his submission McCool claimed that he and his nonexistent coauthors worked at the nonexistent Arthur Vandelay Urological Research Institute. All that the journal required from McCool was a $799 (plus tax) contribution.

Even though McCool never got around to sending the check, the journal went ahead and published anyway.

In his column dated July 22, 2017,[24] *Discovery* magazine's columnist Neuroskeptic confessed that he too had joined the stingers' stampede. In

---

* The journal soon afterward disappeared, so she needn't have worried.

this instance the paper submitted, "Mitochondria: Structure, Function and Clinical Relevance," was packed with references to the *Star Wars* universe; its "authors" were Dr. Lucas McGeorge and Dr. Annette Kin. The study's subject was the "midichlorians," the intracellular microorganisms that give the Jedi their psychic powers. Much of the paper was lifted directly from relevant entries in Wikipedia. Just in case anyone might be a bit slow on the uptake, Neuroskeptic included in the paper a chunk of monologue from *Star Wars Episode III: Revenge of the Sith* (2005).

Some of the targeted journals spotted the hoax, and at least one responded with appropriate wit: "The authors have neglected to add the following references: Lucas et al., 1977, Palpatine et al., 1980, and Calrissian et al., 1983." Alas, despite the acuity of their peer reviewer, that particular journal offered to publish the piece assuming a few changes were made. Among the other journals,

> The *American Journal of Medical and Biological Research* (SciEP) accepted the paper, but asked for a $360 fee, which I didn't pay. Amazingly, three other journals not only accepted but actually published the spoof . . . the *International Journal of Molecular Biology: Open Access* (MedCrave), *Austin Journal of Pharmacology and Therapeutics* (Austin) and *American Research Journal of Biosciences* (ARJ).

Meanwhile, in response to the submission, *Continuous Research Online Library*, a Zinianz journal, offered Dr. Lucas McGeorge an editorial post.

In March 2017, a group of Polish researchers reported in *Nature*[25] that they'd invented out of whole cloth a historian/philosopher of science called Anna O. Szust, who worked at the Adam Mickiewicz University in Poznan. Anyone with a passing knowledge of Polish would immediately have suspected foul play, because the middle initial and surname of the good doctor, run together, mean "fraudster"—in effect her name was "Dr. Fraud"! Among her qualifications she listed papers she'd published in peer-reviewed journals and chapters she'd published in academic books; unfortunately, none of the papers, chapters or even publishers existed. Even so, when "Anna O. Szust" sent out applications* for a job to 360 free-to-access online journals, no fewer than 48 made her an offer while four of those were ready to make her editor-in-chief.

---

* "Applications were identical, except that some contained an extra paragraph expressing Szust's enthusiasm for new open-access journals," the researchers admit in their *Nature* account.

This will doubtless annoy those of us who are accustomed to seeing applications for jobs for which we're eminently qualified go without an answer of any kind. All you really have to do, at least in the highly profitable world of dubious academic journals, is invent a few qualifications and append a mugshot showing yourself to be an attractive young woman: "My name is Lara Croft. I raid tombs."

The grandaddy of all successful hoax submissions to a scholarly journal was "Transgressing the Boundaries: Toward a Transformative Hermeneutics of Quantum Gravity" (1996) by the New York University physicist Alan Sokal.

Like much of the rest of the scientific community—at least, those members of it who'd even noticed—Sokal had become both amused by and infuriated by the school of academic sociologists calling themselves either postmodernists or social constructivists who had taken to regarding science as some kind of whimsical endeavor that was as prone to fads and fashions as, say, *haut cuisine*.

Yes, there was an argument to be made that the scientific establishment could find itself trapped by conventional thinking, as was the case in the first half of the twentieth century when geology refused to accept the overwhelming evidence in favor of continental drift. And, yes, it was also true that science often changed its collective mind—what was accepted as a "law" today might be recognized as bilge tomorrow. Yet these were very *old* problems within science. By the end of the twentieth century you'd have been hard pressed to find a scientist who didn't acknowledge that one of the great strengths of science is that it's in a constant state of revision. Unlike dogma, it's forever (at least in theory) open to new ideas and new criticisms, some of which might result in an overhaul of pre-existing ideas. Science isn't, as noted all through this book, about certainties (even though from time to time it does cough up entities that are as near to certainties as you're ever likely to find, entities it calls "theories"); what it's about is the best state of the art so far. For example, as of this writing the best theory to account for the large-scale nature of the universe is still Einstein's theory; even so, over the past few decades there have been numerous challenges to general relativity, some minor and some major, and *not a single well respected scientist has freaked out about this.* (Okay, it's possible there may have been one or two. I decline to do a tedious search for those hypothetical exceptions.)

This isn't the scenario the postmodernists saw.

The editors of the sociology journal *Social Text* in 1996 recognized that professional scientists were becoming increasingly pissed off not so much by the assaults being made on reality by the supposed "scientific sociologists"—claims such as that the uncertainty principle sprang from nothing more than Werner Heisenberg's yen to make himself more socially acceptable in the Nazi milieu—as by the deleterious effect this was having upon western culture at large. Those editors therefore decided the time was ripe for a special issue of their journal, to be called *Science Wars*. They sent out solicitations for submission and, oh dear, one of those landed on the desk of Sokal.

"Transgressing the Boundaries" is not an exemplar of skimpery: I have read published novels that are shorter than Sokal's paper. Plagiarizing joyfully, using the sociologists' bastardized version of the English language as if it were his native tongue, Sokal produced a piece that lay somewhere between Monty Python and, well, Monty Python:

> The pi of Euclid and the *G* of Newton, formerly thought to be constant and universal, are now perceived in their ineluctable historicity.

The editors of *Social Text*—which was not at all a predatory journal but a reputable magazine serving a supposedly reputable academic demographic—thought Sokal's piece was just fine, and published it.

It was a decision they've never lived down. Within days another Sokal essay, "A Physicist Experiments with Cultural Studies," appeared in the journal *Lingua Franca* and not just admitted the hoax but skewered the anti-scientific pretensions that underlay the *Social Text* editors' decision to publish his spoof. The follow-up book that Sokal did with Jean Bricmont, *Fashionable Nonsense* (1998), is well worth reading.

Sokal and Bricmont were in fact quite gentle in that book. In an essay called "In Praise of Intolerance to Charlatanism in Academia" (1996),[26] the philosopher of science Mario Bunge went a bit further:

> [The postmodernists] are not unorthodox original thinkers; they ignore or even scorn rigorous thinking and experimenting altogether. Nor are they misunderstood Galileos punished by the powers that be for proposing daring new truths or methods. On the contrary, nowadays many intellectual slobs and frauds have been given tenured jobs, are allowed to teach garbage in the name of academic freedom, and see their obnoxious writings published by scholarly journals and university presses. Moreover, many of them have acquired enough power to censor genuine scholarship. They have mounted a Trojan horse inside the academic citadel with the intention of destroying higher culture from within.

Sokal himself had doubts about the value of stings. He realized that, however silly "Transgressing the Boundaries" might seem to a physicist, or indeed to anyone who wasn't working within the professional postmodernist cage, to those whose daily work did exist within that framework it was actually quite difficult to make the distinction between his spoof and all the other crap academia they were expected to swallow:

> From the mere fact of publication of my parody I think that not much can be deduced. It doesn't prove that the whole field of cultural studies, or cultural studies of science—much less sociology of science—is nonsense. Nor does it prove that the intellectual standards in these fields are generally lax. . . . It proves only that the editors of one rather marginal journal were derelict in their intellectual duty.[27]

Very much later, in the spring of 2017, Sokal was to express reservations about the usefulness of another sociological sting, one that was widely reported as being in homage to his own effort.

This was "The Conceptual Penis as a Social Construct" by Peter Boghossian and James Lindsay (calling themselves Peter Boyle and Jamie Lindsay, supposedly of the "Southeast Independent Social Research Group"), published in the open-access journal *Cogent Social Sciences*.[28] Among its insights were such gems as:

> We conclude that penises are not best understood as the male sexual organ, or as a male reproductive organ, but instead as an enacted social construct that is both damaging and problematic for society and future generations. The conceptual penis presents significant problems for gender identity and reproductive identity within social and family dynamics . . . and is the conceptual driver behind much of climate change.[29]

Boghossian and Lindsay stated that the aim of their spoof was to point up the extent to which the peer-review system in scientific journalism had become broken. Any functional peer-review system, they observed, would not have let through such an asinine paper as theirs. They linked this to the problem of pay-to-publish predatory journals and the root of that problem, the publish-or-perish environment of modern academia, whereby university scientists are pressured by their bureaucratic overlords into publishing plenty of papers at the expense of actual research time, whether or not those papers have value.

All of these are more than reasonable points. At many of the low-end journals, of whatever kind, any peer review that takes place is a bit of a mockery, while everyone admits the pressure to overpublish is a real and destructive phenomenon. Yet the points were lost in the general atmosphere of crowing by the authors in their announcement of the sting in the magazine *eSkeptic*,[30] by the editor of that journal, Michael Shermer, and in the initial reporting of the hoax elsewhere in the media, including the blogosphere and Twitter. Authors, editor and journalists/bloggers/tweeters alike seemed to regard the success of the hoax as a comment on the ludicrousness of gender studies as an academic discipline. Richard Dawkins, for example, tweeted: "Son of Sokal? @PeterBoghossian brilliant hoax paper . . . satirising pretentious charlatans of Gender Studies."*

That's a far cry from Sokal's mild "It doesn't prove that the whole field of cultural studies, or cultural studies of science . . . is nonsense. Nor does it prove that the intellectual standards in these fields are generally lax."

In a follow-up piece,[31] the authors averred that the general depiction of *Cogent Social Sciences* as a bottom-of-the-barrel predatory journal that would publish *anything* so long as it came with a check attached was misleading. The journal is owned by the established academic publisher Taylor & Francis and publicly claims to reject 61 percent of the submissions sent to it. This rejection rate would not seem very high but, assuming the journal's claim is true, does give the lie to the notion that *Cogent Social Sciences* has no editorial filter beyond receipt of a check. Yet, if *Cogent Social Sciences* is not a predatory journal, then part of the authors' declared intentions in the hoax no longer makes sense.

They averred, too, that they had not meant to suggest that a main focus of the hoax was a demolition of the pretensions of gender studies: they'd deliberately put their criticism of the discipline fairly far down in their initial article about the sting to indicate that their other points were more important. Yet this overlooks the strength of their language in that section of their article:

> We intended to test the hypothesis that flattery of the academic Left's moral architecture in general, and of the moral orthodoxy in gender studies in particular, is the overwhelming determiner of publication in an academic journal in the field. That is, we sought to demonstrate that a desire for a certain moral view of the world to be validated could overcome the critical assessment required for legitimate scholarship. Particularly, we suspected that gender studies is crippled academically by an overriding almost-religious belief that maleness is the root of all evil. On the evidence, our suspicion was justified.

---

* To be fair, Dawkins does have a reputation for tweeting first and thinking later. To be even fairer, that's better than many prominent Twitter users, who tweet first and don't think at all.

Those are pretty sweeping statements.

As noted, Sokal himself wasn't overwhelmingly impressed. In an article published (yet again) by *eSkeptic*[32] he expressed his view that the paper was indeed very funny, but he raised a few objections to the hoax as a whole. While obviously something went very wrong with the peer review at *Cogent Social Sciences*, in general, in Sokal's estimation, the journal is of a standard that, while it probably wouldn't survive if it had to rely on subscriptions, is not entirely valueless. In other words, while you can argue that the pay-to-publish journal model as a whole means a lot of stuff gets into print that shouldn't, you can equally well argue that there's a benefit: some of the stuff that might otherwise not have gotten into print could very well prove to be of genuine value—it's just that journal editors didn't realize this at the time.

More to the immediate point, Sokal takes Boghossian and Lindsay to task for another claim: "Do the results of their experiment vindicate their conclusion that 'our suspicion was justified'? I would answer: yes and no, but mostly no."

He points out that *Cogent Social Sciences* is not in fact a gender-studies journal: its ambit is the social sciences in general. The gender-studies journal to which the authors initially submitted their hoax paper, *NORMA: International Journal for Masculinity Studies*, rejected it. And "it seems even less likely that this paper would have been accepted at a more prestigious gender-studies journal, such as *Gender & Society*, *Feminist Theory*, *Signs*, *Feminist Studies*, or *Men and Masculinities*."

Amanda Marcotte was blunter, tweeting to the authors: "If you're out to disprove maleness is the source of all evil, acting like a couple of dicks isn't the way to do it."[33]

WOO-WOO SCIENCE

John Taylor was a distinguished mathematician. In November 1973 he was asked to take part in a BBC television show with Uri Geller, who was making his first big international splash as a psychic spoon-bender. Taylor approached the encounter with the skeptical attitude of a scientist and left it as a gull, dazzled as much as any ordinary person might be by what he had seen Geller do. When the BBC's phone lines were clogged by callers reporting how, as a result of watching Geller perform, they'd discovered paranormal powers of their own, Taylor reckoned that here was a phenomenon meriting proper scientific investigation. So he set up experiments in which children apparently bent cutlery that was sealed in glass tubes that were firmly corked at each end. It was all a most impressive demonstration of psychokinetic power.

The big trouble with Taylor's investigation was that it seems never to have occurred to him that his experimental subjects might *cheat*; he even allowed some of the children to take home the tube-enclosed cutlery to give them the opportunity to focus their psychic energies on it outside the possibly alienating confines of the laboratory. When skeptics decided to do some elementary checking, by the simple means of spying on the children, they were able to film some of the little darlings removing the corks from the ends of the tubes and bending the spoons against the side of the table or underfoot . . . This the kids could do even when the experimenters were in the room by employing the elementary conjuring stratagem of distracting the adults' attention at crucial moments.

Eventually enough evidence of such chicanery emerged that Taylor himself recanted, publishing his conclusion—that there was nothing in the field of the paranormal worth investigating—in the book *Science and the Supernatural* (1980). His primary error lay in not realizing from the outset that a scientist who strays into an unfamiliar scientific (or in this instance pseudoscientific) field is little better equipped to evaluate it than anyone else; for example, a biologist's assessment of climate science is of little more value than yours or mine. In the case of the paranormal, the appropriate "science" is conjuring, not mathematics.

In a sense, Taylor was continuing in a long and distinguished tradition of scientists being fooled into giving charlatans credit through the misapplication of the scientific method. A weakness of the scientific method is that it assumes the integrity of all participants, whereas the whole field of the paranormal is infested with participants who have no integrity, whose purpose is to deceive. James Randi and, long before him, Harry Houdini have been two to point out forcefully that scientists should, when investigating anything pertaining to the supernatural, have as their vital colleagues professional conjurers, whose training in fakery better enables them to spot it.

Spiritualism famously ensnared the physicists Sir William Crookes and Oliver Lodge. Thomas Edison was so taken in by the Theosophy of Helena Blavatsky that he tried to invent a machine for communicating with the dead. The psychologists Gustav Fechner, Ernst Weber and Hans Eysenck fell into similar paranormal traps, while it's hard to know where to start with psychoanalysts like Carl Gustav Jung and Wilhelm Reich. Among

Nobel laureates we can mention the French physiologist Charles Robert Richet. At a more serious level of involvement we have of course J.B. Rhine, who devoted what might have been a sterling research career to ultimately fruitless investigations of ESP, and today Russell Targ and Harold Puthoff. And a 1997 survey found that 40 percent of US scientists believed they could communicate with God and that He was capable of answering their prayers.*

Letters from Distinguished Persons

The error of the respected French mathematician Michel Chasles, a member of the French Academy of Sciences and Professor of Geometry at the Imperial Polytechnic of Paris, was to stray not just into another scientific discipline but into the field of historical documents. In 1851 he met a plausible character called Denis Vrain-Lucas, who told him he'd come across some letters of historical interest. Their writers were prominent figures from French history and literature—Molière, Rabelais, Racine, Joan of Arc and Charlemagne, to name just a few. Chasles eagerly bought the letters, and urged Lucas to contact him should any more turn up. Naturally, they did—letters from Julius Caesar, Mary Magdalene, William Shakespeare and countless others, all of whom, strangely, had put their thoughts to paper in French. It was even more remarkable that some of them had put their thoughts to paper at all, because they were writing at a time before paper was invented . . . let alone the particular watermarked paper most of the correspondents had chosen to use.

Vrain-Lucas had served his apprenticeship, so to speak, in the lucrative field of false genealogy. Thwarted through lack of educational qualifications in his desire to make a career out of librarianship, he'd taken a job with a Paris genealogical company whose practice, when documents were lacking to confirm rich clients' relationships to famous historical figures, was to forge them. Vrain-Lucas proved quite good at this. When the head of the firm retired, he left his protege his collection of (genuine) autographs and documents—a handy toolkit for the aspiring literary forger. And then Vrain-Lucas found Chasles.

It took nearly twenty years before, in 1870, Chasles finally admitted he'd been suckered and took Vrain-Lucas to court. Even then, the charge wasn't

---

* To be fair, the same survey showed that a whopping 87 percent of the US adult population believed the same, so a scientific education is obviously not entirely wasted.

one of selling him faked documents; the case was brought because Vrain-Lucas had failed to deliver 3,000 autographs Chasles had paid for. At the trial it emerged that the next manuscript Vrain-Lucas had been planning to sell him was Jesus's own copy of the Sermon on the Mount.

Yes, it too was in French.

N-Rays

In 1903 the prominent French physicist René-Prosper Blondlot discovered N-rays, produced naturally by various materials, including many of the metals, but also by the human nervous system—notably, when people talked, by the part of the brain that controls speech, Broca's area. (He called them N-rays in honor of his employer, the University of Nancy.) His findings were confirmed by other French scientists, although outside France experimenters had difficulty in reproducing his results.

Using adapted spectroscopic equipment—in which the lenses and prisms were made of aluminum—Blondlot could project N-ray spectra. It was exceedingly difficult to discern the lines of the projected spectra: only those with finely attuned eyesight could do so, and the average, untrained observer was likely to see, well, nothing.

The US physicist Robert W. Wood attended several of Blondlot's demonstrations and concluded that he, Wood, must be an average, untrained observer—or perhaps not. Wood performed various tests and soon decided N-rays were a figment of the French scientist's imagination. At one point, while Blondlot was describing a projected N-ray spectrum, concentrating on the lines in front of him, Wood surreptitiously removed the aluminum prism from the N-ray "spectroscope." Blondlot continued his description, unperturbed.

In 1904 the French Academy of Sciences bestowed on Blondlot the prestigious Leconte Prize. In that same year Wood published an article recounting his experiences, and non-French scientists promptly stopped looking for the rays. Within France, however, physicists not only kept looking for them but also in many cases found them. For some decades thereafter it was a source of nationalistic pride that French scientists could detect and study "their" rays while counterparts abroad neglected this important field entirely. In the end, of course, even Gallic stubbornness could withstand reality no longer.

Blondlot and his compatriots sincerely believed they could see the N-ray spectra, and that Blondlot had made an important discovery. Part of what made them see what they wanted to see was undoubtedly nationalism: physicists in other countries had been discovering exciting new rays, making French physics look shoddy by comparison. Another large part of it may simply have been ambience: the climate in French physics became one in which the existence of N-rays was accepted as a given, and so naturally any physicist worthy of the name was able to see the spectra and conduct successful experiments to investigate the rays' properties.

ASTRONOMY DOMINOES

There are plenty of examples of astronomers managing to fool themselves by seeing what they want to see rather than what's actually there—hardly surprising since, throughout much of astronomy's history, observation was a matter of peering for long periods at objects that, even through telescopes, were incredibly faint. Exactly the same sort of thing happened in the early days of microscopy: as they squinted through their imperfect lenses at images that lurched in and out of focus, some seventeenth-century microscopists were even able to observe homunculi—tiny, fully formed human beings that supposedly inhabited what we would now call sperm cells.

The classic example of this phenomenon in astronomy is of course that of the "canals" of Mars. In light of observations he'd made during 1877–81, the Italian astronomer Giovanni Virginio Schiaparelli declared he could see straight markings criss-crossing the planet's surface. He called these markings "*canali*," meaning "channels"; the word was translated into English as "canals," with all the connotations the latter word brought with it. The assumption spread rapidly that Mars was populated. Not just canals were seen: there were also fluctuating markings that could surely be nothing else but seasonal variations in the Martian vegetation cover. It didn't take long before the two concepts were put together, so that the obvious purpose, aside from transport, for which the Martians built their canals—a huge endeavor, bearing in mind that no human enterprise at all would be visible to astronomers on Mars—was irrigation of their crops. Right up to the time that the first space probes began sending back photographs of the Martian surface it was not unrespectable to speculate about the nature of Martian vegetation. By then, however, the notion of the canals had generally been abandoned.

The two chief champions of the canals hypothesis were Camille Flammarion, who believed all worlds were inhabited and who had reported observing vegetation on the moon; and particularly the US astronomer Percival Lowell. For fifteen years Lowell studied Mars, mapping the canals incessantly; his books *Mars and Its Canals* (1906) and *Mars as the Abode of Life* (1908) were enormously influential, not only among some astronomers but also, more significantly, among the general public. Both Nikola Tesla and Guglielmo Marconi claimed to have received radio signals from Mars. All of this was fantasy based on a mistranslation, on the limitations of earthbound telescopes, and on human psychology.

Less well known than the Martian "canals" fiasco is an embarrassment of the great astronomer (and surprisingly fine composer) Sir William Herschel. Herschel's strongest claim to fame is that he discovered (1781) the planet Uranus, the first "new" planet to be identified since classical times. When he began studying the celestial object he'd identified he thought it was a comet approaching the sun; accordingly, his observational records show a slow increase in the object's size. In due course he realized the truth, but still felt no need to reconsider his observations: clearly the planet was at a stage of its orbit where it was approaching the earth. Only once Uranus's orbit had been properly calculated did it became clear that, during Herschel's observational period, the distance between the "new" planet and the earth was actually *increasing*; an accurate record, which would almost certainly have been beyond the limits of Herschel's instrumentation, would have shown the mystery object getting, if anything, smaller.

Another great UK astronomer, Sir John Flamsteed, a contemporary of Newton's, had a theory whose proof would lie in Polaris being further from the celestial north pole in summer than in winter. He made the appropriate observations, and sure enough found this to be the case. Other astronomers repeated the observations and found no such effect (there is none). When confronted, Flamsteed blamed his instruments, which he thought must have introduced the error. It seems never to have occurred to him that he'd simply deceived himself.

Closer to our own time, in the 1910s and 1920s astronomers at the Mount Wilson Observatory, led by Adrian van Maanen, reported that the spiral nebulae could be seen to be slowly unraveling. At the time it was thought all nebulae were clouds of gas and dust within our own, unique galaxy. That some of them should display structure was a puzzle, resolved by the theory that the nebulae formed as ellipsoidal balls and then slowly disintegrated to become irregularly shaped clouds; van Maanen's discovery, based on close comparison of photographs made at different times, appeared to confirm this notion. Such was the prestige of Mount Wilson that none dared contradict the conclusion, and the astronomy establishment grew quite sharply defensive when questions were raised. Any dissent remained stifled until Edwin Hubble showed in 1924 that certain nebulae—including those of ellipsoidal ("elliptical") and spiral form—are hugely distant galaxies; the data that van Maanen had recorded were quite simply impossible.

While there's no guilt in van Maanen and the others having deceived themselves when believing they could see incredibly small differences in the photographs, a strong finger of condemnation should be pointed at those in the astronomy establishment who bullied everyone else into going along with the false data even when these grew increasingly dubious. Among them was Harlow Shapley, who most certainly should have known better.

GRAVITATIONAL WAVES FROM THE EARLY UNIVERSE

Most physicists accept that in the very earliest moments after the Big Bang—between about 10^{-36} seconds and about 10^{-32} seconds after it—the universe expanded at an extraordinary speed; after this period of very rapid inflation, the rate of expansion slowed down considerably. Although no one really has a clue why the early universe should have behaved like this, the assumption that this very brief phase occurred rather neatly explains quite a few aspects of the universe we see today that would otherwise be puzzling. However, critics of inflationary theory—and there are quite a few—point out that, while inflation may neatly explain some of the phenomena we observe, this doesn't preclude *other* explanations. Moreover, the theory of inflation creates problems of its own. What is needed is a genuine prediction that inflation theory makes of something we *haven't* yet observed—followed, of course, by confirmatory observation.

One such prediction is that the cosmic microwave background* should show traces of primordial gravitational waves, and that we should be able to detect those traces in the form of polarization effects. The cosmic microwave background does indeed show polarization effects, known as E-modes, which are a product of electromagnetic scattering; these were predicted before first being observed in 2002.

What cosmologists seek as confirmation of inflation theory, however, are something rather different: the so-called B-modes. There are two ways

---

* The low-level radiation that fills all space and that is, in effect, the relic of the vast energies released during the Big Bang.

in which B-modes could be produced: by the gravitational lensing of E-modes (such "virtual" B-modes would be fascinating in themselves, but aren't of relevance here) and by primordial gravitational waves. Detecting B-modes of this latter type would be a strong confirmation of inflationary theory in general, while the details of their nature and strength would offer evidence in favor of one or other of the various rival models of inflation that cosmologists have devised.

Based in Antarctica, the Background Imaging of Cosmic Extragalactic Polarization (BICEP) instruments are designed to examine polarization in the cosmic microwave background. In March 2014 the leaders of the project associated with the second of these instruments, appropriately enough named BICEP2, held a press conference to announce they'd detected what appeared to be relevant B-modes. Here at last was direct evidence of the elusive primordial gravitational waves, as predicted by inflation theory.

Needless to say, there was considerable excitement in the physics world. In the early days after the first announcement there was a common assumption that the BICEP2 team would be picking up a Nobel pretty soon.

It wasn't too long, however, before voices of doubt arose, some suggesting the researchers had rushed to judgment. Other groups offered different explanations for the observations the BICEP2 team had made—in particular that the results might be due not to anything as exotic as primordial gravitational waves but to intervening dust within our galaxy. By the time the scientists of the BICEP2 Collaboration, as it was more formally called, published their results in a scientific journal,[34] on June 19, they were expressing considerably more caution about their conclusions. A particular factor in this reconsideration was that the European Space Agency's orbiting Planck Telescope had begun to release new and far more accurate data about the distribution of galactic dust and the polarization effects it can create, with yet more and yet better data promised by the end of the year.

Speaking in London on the day the BICEP2 study was published, Professor Clem Pryke of the University of Minnesota, a central member of the BICEP2 crew, spelled it out: "Look, the scientific debate has come down to this—we need more data."[35]

Sure enough, the fuller Planck results, announced in September 2014, indicated that the signals the BICEP2 Collaboration had attributed to primordial gravitational waves were in fact completely explainable in terms of cosmic dust.

This can easily be seen as a classic example of scientists seeing what they wanted to see—the confirmation of inflation theory would have been a tremendous breakthrough—but it was also a classic case of science working how it's supposed to work: an initial erroneous claim on the part of some workers was very quickly scaled back in response to the sensible criticisms of others, and finally abandoned in the light of new and improved data.

Polywater

In 1962, the Russian chemist N.N. Fedyakin, of whom little else is known, announced the discovery of what appeared to be a new form of water. Experiments showed it had a lower freezing temperature (about −40°C) than water yet tended to solidify at a much higher temperature than this (about 30°C), boiled at about 250°C rather than the customary 100°C, was at least several times more viscous, and, depending upon circumstances, appeared to be about 40 percent denser or 20 percent less dense. This kind of water could be found when ordinary water vapor condensed in very narrow glass or fused-quartz tubes; it was christened "anomalous water," "water II," "orthowater" and much later, by UK and US researchers, "polywater"—the last because it was assumed to be a hitherto-unknown polymer of water.

Fedyakin and colleagues at Moscow's Institute of Physical Chemistry did much research on polywater. One colleague, Boris Derjaguin or Deryagin, came to the UK in 1966 and lectured on polywater to the Faraday Society, prompting UK scientists such as J.D. Bernal to repeat the Russian experiments successfully. A couple of years later, the main experimentation locale shifted to the US, where copious papers on polywater and its deduced properties were published until the early 1970s, when it was discovered, as spectroscopy became more sophisticated, that polywater was merely water contaminated by impurities from the tube walls.

In effect, there were two forms of polywater, depending upon whether the capillary tube used was of quartz or of glass. In the former case, tiny pieces of quartz from the tube walls detached to enter the water and form an impure silica gel (quartz is a form of silica). In the case of a glass capillary, various contaminants of the glass—produced, for example, during the heat-process of pulling the tube—became dissolved in and/or combined with the water. The frequently repeated calumny that among the fatty compounds found in polywater was sweat from the experimenters' fingers has its origins in the fact that the spectra of "glass-based" polywater were not dissimilar to those of human sweat.

In a tightly restricted sense, polywater did exist: the substance whose properties and behavior the various chemists investigated was not an illusion. Their shared error was, rather, in being insufficiently critical of explanations—notably by Derjaguin, eager to capitalize on a phenomenal discovery—as to what the substance actually *was*. As soon as more rigorous analysis was brought to focus, the truth emerged. This should have happened some years earlier.

At the time, polywater was not merely an arcane chemical concern. There were fears among some of the scientists involved that, should polywater accidentally be let loose in the wider world, all of the world's ordinary water could "flip" into this form—a terminal catastrophe for life on earth. But there was a bright side, too. As an excited *New York Times* writer commented:

> A few years from now, living-room furniture may be made out of water. The antifreeze in cars may be water. And overcoats may be rainproofed with water.

MITOGENETIC RAYS AND KIRLIAN PHOTOGRAPHY

In 1923, the Ukrainian physicist Alexander Gurwitch (Gurvich or Gurwitsch) reported observing the behavior of curious short-wavelength (ultraviolet) electromagnetic rays emitted by living matter: they transmitted happily through quartz but were blocked by glass. The effect of these rays seemed to be to communicate some life-promoting energy from one organism to the next. Since Gurwitch's original experiment appeared—and still appears—to be sound, and since other researchers conducted experiments which likewise seemed explicable only in terms of his mitogenetic (or mitogenic) rays, it was no wonder the rays' existence was generally accepted throughout most of the scientific community for much of the 1920s.

The notion of mitogenetic rays did not spring from nowhere. It was widely thought at the time that the processes of an organism's development, such as that of the embryo in the womb, would forever be beyond the capacity of physics or chemistry to comprehend. Thanks to a period spent in Germany with pioneering developmental biologist Wilhelm Roux, Gurwitch begged to differ—and quite rightly. He came to believe organic development must be under the control of a "supercellular ordering factor," a sort of energy field that instructed each newly forming cell as to its correct place in the developing organism. It should be possible, he reasoned, to detect this field. The energy for the field must come as a sort of by-product of ordinary metabolism, and the best source for it would therefore be a mass of growing cells. While the main purpose of the energy they produced would be to guide the growth of further cells within their own organism, it ought to be possible to use it to stimulate growth in a different organism, should the second be placed very close to the first.

In Gurwitch's initial experiment, putting the tip of a growing onion root alongside the tip of another onion root seemed to promote cell division in the target root. Other researchers did equivalent experiments with yeast, bacteria, etc., with similar results.

Soon studies extended to the larger scale. Injecting cancer cells into a healthy animal, it was found, would reduce the animal's effusion of mitogenetic rays, although the tumors themselves would continue to be strong emitters even if the animal were killed. A rabbit's eyes were a good source of mitogenetic rays but, if the animal were starved, the intensity declined; after a few days' further starvation, however, the emission of mitogenetic rays would start up again, although now with a different wavelength, reflecting the different metabolic processes at work. The blood of children suffering from Vitamin D deficiency displayed lesser ray-emission than other children. And so on.

Nevertheless, attempts to measure or even detect the rays by purely physical means—such as photoelectric cells—came to nothing. By the early 1930s, more and more researchers were beginning to call into question the very existence of the rays. In 1935, reviewing the 500 or so papers published on the subject, A. Hollaender and W.D. Claus concluded in the *Journal of the Optical Society of America* that the whole mitogenetic-ray scenario was the product of badly executed science at every level: researchers were looking for results so minuscule as to be statistically insignificant, and of course were finding what they expected to find.

Mitogenetic rays faded from the journal pages soon thereafter, though with the occasional scattered exception right up until the 1960s. Just to confuse the picture, chemical reactions in living cells *can* emit tiny amounts of electromagnetic radiation—but as visible light, not ultraviolet. It's feasible that some of the original experiments detected this effect; later claims were, however, products of mob delusion.

It's not too much of a stretch to relate Gurwitch's mitogenetic rays to the much later fad for Kirlian photography, whereby people deceived themselves into believing they could take photographs of an organism's energy field, or aura. In 1939 the Soviet hospital technician Semyon Kirlian noticed, while observing a patient receiving treatment from a high-frequency generator, tiny flashes in glass electrodes as they were brought close to the patient's skin. Fascinated, he and his wife Valentina experimented by taking photographs of human skin placed in a powerful electric field. Their first target was Kirlian's hand. When photographed this way, it seemed to have a glowing aura.

The craze for Kirlian photography started some decades later when the experiment was recounted in the book *Psychic Discoveries Behind the Iron Curtain* (1970) by Sheila Ostrander and Lynn Schroeder.* Even though many researchers were unable to reproduce the results, claims for Kirlian photography swiftly grew: the aura of a leaf would retain its shape even if you tore off part of the leaf, the aura of someone who'd just drunk a glass of vodka was brighter than before, and so forth.

While there was no disputing that living organisms photographed in such circumstances showed a glow, this was soon explained as merely the product of a familiar process called corona discharge. As for the claims that the aura was an "etheric body" whose brightness varied with the well-being of the organism, John Taylor reported in his confessional *Science and the Supernatural* (1980):

> When all of these factors are carefully controlled, there is no change of the Kirlian photograph [of a fingertip] with the psychological state of the subject. It seems that the most important variable is the moisture content of the fingertip, and the Drexel group even concluded that "corona discharge (Kirlian) photography may be useful in detection and quantification of moisture in animate and inanimate specimens through the orderly modulation of the image due to various levels of moisture."

Such a mundane conclusion was of course ignored by the media and the public, who preferred to think the glow around the vodka drinker's hand brightened not because he was sweating a little more but because he was, well, a bit lit up.

MENSTRUAL RAYS

Amid all the accounts of various rays that have later proved illusory, one particular ray-related phenomenon has gone under-reported.

---

* Quite why a perfectly ordinary scientific experiment should be regarded as a "psychic discovery" is anyone's guess.

A considerable body of popular fallacy surrounds menstruation. In some parts of Australia the Aborigines maintain that if a man goes near a menstruating woman he will lose his strength and grow prematurely old. Similar myths are alive in western society. Eating ice cream during one's period is risky—but not as bad as washing one's hair. Bathing during a period can lead to tuberculosis. Flowers wilt and crops are blighted as menstruating women walk by.

That there might be some truth to this latter superstition occurred to a German bacteriologist called Christiansen in 1929. He noticed that the fermentation starter cultures in his lab failed periodically, and linked these failures to the menstrual cycle of the female technician who tended them. Clearly either menstrual blood or menstruation itself was a source of radiation, and like a good scientist Christiansen investigated. The properties of menstrual rays proved to be not unlike those of Gurwitch's mitogenetic rays—they passed through quartz yet were blocked by glass, etc.—but their effects on developing organisms were negative where those of mitogenetic rays were positive.

Similar effects were noticed in the US, at Cornell University, by the bacteriologist and mitogenetic-ray enthusiast Otto Rahn, in whose laboratory yeast cultures had monthly difficulties when tended by a particular female student. As had been observed by Christiansen, the effect was seasonal: in winter it was barely noticeable while in summer it was at its strongest. Rahn, too, conducted investigations, and concluded that menstrual radiation was just the most obvious case of a more generalized radiation human bodies could give off in certain circumstances, such as hypothyroidism, and herpes and sinus infections. Rahn went on to publish a book on the subject, *Invisible Radiation of Organisms* (1936). One hardly need add that the topic isn't much discussed by scientists today.

A Microbial Muddle

We noted briefly how the early microscopists were deceived by poor optical equipment and wishful thinking into seeing things that weren't there. They haven't been the only ones. According to G.M. Boshyan in his *On the Nature of Viruses and Microbes* (1949), something of a bestseller in Moscow, viruses and microbes are really one and the same thing, transmutable into each other. Boshyan was a student at the Moscow Veterinary Institute, and he'd made a most remarkable discovery while peering through his microscope. When he treated microbes by boiling or chemicals they did not die: instead they were transformed, depending upon circumstances, into either viruses or bacteria. With further treatment, clusters of viruses could be induced into massing together to form a bacterium, or a bacterium could be persuaded to break down into its constituent viruses.

Anywhere else in the world, Boshyan might have been advised to take his researches in other directions, but this was the USSR under the sway of Lysenkoism (see pp. 343–351), one of whose pillars was that environmental factors could persuade species to transform one into another. Boshyan was hailed as having made a major breakthrough in molecular biology and given a large laboratory of his own; further, the discoveries were of such importance as to be classified by the security services in case of theft by foreign powers.

Despite the dangers of disputing the official line on the biological sciences during Lysenko's sway, the Soviet Academy of Medical Sciences decided to investigate. For a long while their panel was stalled by the classified nature of Boshyan's work. When, finally, they were permitted a demonstration, their unequivocal conclusion was that Boshyan had failed to clean his microscope slides properly. For once, reason prevailed: Boshyan's entire theory swiftly unraveled, and he was thrown out of the Institute. The case seems to have been one of synergy: indoctrinated by both Stalinism and Lysenkoism, Boshyan saw what he wanted to see, which was also what his masters wanted him to see, a situation that reinforced his certainty that he saw it . . .

RECOVERED MEMORIES

No one could doubt child sexual abuse is an abhorrent crime, and that its incidence is far more widespread than any of us realized until the last few decades. At the same time, it seems far *less* prevalent than was generally reported during the height of the hysteria in the 1980s and 1990s, when claims were based on supposed science involving the recovery of repressed memories, which often linked sexual abuse to parents' participation in Satanic rites. The notion was that many children found the experience of sexual abuse sufficiently traumatic that they dissociated themselves from the act so comprehensively as effectively to become multiple personalities; later, they had no memory of the abuse because it happened to "someone else." They did, however, display a complex of symptoms—lackluster eyes, inappropriate responses to certain triggers (especially sexual ones), and so forth—and these could be read by the wise therapist. With encouragement, the patient might be able to retrieve the memories that had been "lost," and from there it was but a short step to the criminal conviction of the accused abuser and the recovery of the patient through "closure" (whatever that is).

Hypnosis was a popular tool for the recovery of memories, and many an innocent person was jailed before people realized that all the hypnotic subjects were doing, like hypnotic subjects in any other context, was telling the hypnotist what they thought s/he wanted to hear—reflecting back the therapist's own preconceptions, in other words. What those preconceptions were was only too easy for the hypnotized patient to deduce from the questions themselves. Even after hypnosis had been abandoned as a tool, and a fraction of the falsely imprisoned had been released,* the recovered-memory movement continued unabated, a political rollercoaster powered by unlikely allies such as militant feminists and the religious right (thrilled by the Satanism angle).

Yet from the very outset there was little scientific evidence that repressed memories exist at all. To be sure, victims of concussion typically have a "blank period" prior to the trauma: it may be a long while, if ever, before fragmentary memories return of the minutes or hours leading up to the blow. Abrupt psychological trauma can similarly engender a "blank period." But these are very, very different processes from a pattern of repeated sexual abuse and its usual accompaniments of threats, false complicity, shame, secrecy, etc.; even if the model of memory-loss through dissociation during the act itself could be accepted, it's hard to see how the rest could be forgotten. It's today the overwhelming consensus of scientific opinion that repressed-memory syndrome is a fiction, that the "memories" are artifacts created by leading and/or aggressive interrogation, yet this only slowed the political juggernaut; it didn't stop it.

It's important here to recognize that the falsely prosecuted and imprisoned were not the only victims of the recovered-memory cult. Countless people, usually women, came to have their lives haunted by "memories" of appalling acts that never took place. No one can assess the psychological damage inflicted on them. Likewise, no can assess the damage done by the failure, because of the focus on fallacious "memories" of abuse, to address whatever was the real problem that brought the patients into therapy in the first place.

Nor should we forget that the techniques used by repressed-memory therapists have elicited similarly convincing accounts of UFO abductions.

---

* Some still rot in jail, pariahs even there because of a crime they did not commit.

PRAYER POWER

The prayers of others can make a woman pregnant! No, no, this is not your daughter's despairing excuse; it's according to a paper published in 2001 by the *Journal of Reproductive Medicine*: "Does Prayer Influence the Success of In Vitro Fertilization–Embryo Transfer?" by Kwang Y. Cha of the Cha Hospital, Seoul, and Rogerio A. Lobo and Daniel P. Wirth of Columbia University, New York. They reported an experiment in which 199 women who required in vitro fertilization were separated into two experimental groups. Without the subjects' knowledge, volunteer Christians in the US, Canada and Australia prayed over photographs of the women in one group for a successful fertilization. The women in the other group were left unprayed-for as a control.

The results reported in the *Journal* were impressive. Of the prayed-for women, 50 percent attained a successful pregnancy, while of the others only 26 percent did so (which latter percentage is about the standard rate for in vitro fertilization attempts). According to Lobo, given as the paper's senior author, the three scientists spent some time considering whether or not they should publish their results, because the difference in pregnancy rates was so improbably dramatic—too good to be true.

Others reached that same conclusion. Even while newspaper headlines and evangelists were bellowing worldwide about proof of the power of prayer, researchers began investigating the paper's claims. One such, Bruce L. Flamm, Clinical Professor at the Department of Obstetrics and Gynecology, University of California Irvine Medical Center, published an extensive three-part analysis in 2003–2004 in the *Scientific Review of Alternative Medicine*. He found much that was troubling in the experiment's methodology and conclusions—and indeed in the modern trend for peer-reviewed journals to publish, without seemingly doing much in the way of peer-reviewing, papers that invoke the spiritual or supernatural: "It is one thing to tell an audience at a tent revival that prayers yield miracle cures, but it is quite another to make the same claim in a scientific journal." A further worrying fact was that one of the paper's three authors, Wirth, had published extensively in support of the supernatural; was it reasonable to expect his contribution to be unbiased?

Drawn into the spotlight, Lobo began wilting. Columbia University issued a statement on his behalf saying that the first he'd known of the study was some six to twelve months after its completion; he had merely provided "editorial assistance" for the paper . . . and allowed himself to be listed as senior author. And then in 2002 coauthor Daniel P. Wirth—who had given his affiliation here and in other paranormal-supporting papers as "Healing Sciences Research International," an institution in California that he headed and that seems to have consisted primarily of a PO box—was indicted for large-scale fraud. In April 2004 he and an accomplice pleaded guilty to conspiracy to commit mail and bank fraud, and agreed to forfeit assets totaling over $1 million gained by various schemes. In this context, with Kwang Cha and (by now) Lobo unwilling to return phone calls or e-mails, it seems legitimate to speculate that they were as much taken in as anyone else by Wirth, and that the results of the experiment were "manipu-lated"—if indeed the experiment ever took place at all.

The Lobo et al. study has not been the only controversial paper on the purported therapeutic effects of prayer power to be published in a peer-reviewed medical journal. Mitchell Krucoff, a cardiologist at Duke University Medical Center in Durham, NC, and his team focused on the thera-peutic effects on patients suffering congested coronary arteries of prayers offered up on their behalf by people unknown to them. A pilot study was published in 2001 in *American Heart Journal*, with a fuller study in *Lancet* in 2005. In the wake of the 2001 paper Krucoff, interviewed about his excit-ing results on the Discovery TV channel, summarized:

> We saw impressive reductions in all of the negative outcomes—the bad outcomes that were measured in the study. What we look for routinely in cardiology trials are outcomes such as death, a heart attack, or the lungs fill-ing with water—what we call congestive heart failure—in patients who are treated in the course of these problems. In the group randomly assigned to prayer therapy, there was a 50% reduction in all complications and a 100% reduction in major complications.

He made similar startling claims elsewhere. However, Andrew Skolnick, Executive Director of the Commission for Scientific Medicine and Mental Health, compared Krucoff's public statements with the results recorded in the *American Heart Journal* paper itself, and discovered that to say "In the group randomly assigned to prayer therapy, there was a 50% reduction in all complications and a 100% reduction in major complications" was not entirely accurate. Just for a start, while all of the control group had survived until the end of the experiment, one of the prayed-for group had died— definitely a "major complication." Pressing Krucoff on this point in a 2004 interview, Skolnick received the reply: "Well, the difference between zero and one in a cohort of 30 people is no difference." Even if one could accept that, what it most certainly isn't is a reduction.

Anticlimactically, the more complete study published in *Lancet* showed absolutely no effect one way or the other of prayer on the patients. However, this is not the impression you'd get from the press release issued on July 14, 2005, by Duke University or even *Lancet's* editorial, both of which imply that the results are, for some unexplained reason, hopeful. In the press release Krucoff says:

> While it's clear there was no measurable impact on the primary composite endpoints of this study, the trends and behavior of pre-specified secondary outcome measures suggest treatment effects that can be taken pretty seriously when considering future study directions.

Translated into English, this means: "We found no effects but we'll try again." As presumably Krucoff will. Inevitably, if a number of studies are done, there'll finally be one which, purely by chance, appears to show the prayed-for patients doing better than the control group.

In March/April 2006, the *American Heart Journal* published another paper, done by Herbert Benson (a cardiologist from the Mind/Body Medical Institute in Massachusetts) *et al.*, reporting on the effects of prayer on 1,800+ patients recovering from heart surgery, an experiment conducted over more than a decade. Heralded as the most scientifically rigorous study in the field to date, this found that the prayers of strangers had zero effect on the patients' recovery. Perhaps counterintuitively, when patients knew they were being prayed for they had higher rather than lower rates of post-operative complications. Benson and his colleagues suspected that knowledge of the prayers inappropriately raised the patients' expectations for a rapid recovery. In other words, prayer power's bad for you!

Despite the very public discrediting of the earlier reports, the belief in the curative power of prayer seems to be deeply ingrained. In 2006 it was reported that the Church of Christ, Scientist, was gearing itself up to resist the imminent threat of a bird flu pandemic by use of prayer—and not necessarily only with other Christian Scientists: apparently, *in extremis*, anyone would do. This was in accordance with the Christian Science doctrine that disease is a spiritual rather than a material phenomenon, its undeniable physiological effects being due to the fear of the sufferer. According to Mary Baker Eddy's *Science and Health* (1875), "Disease is an experience of so-called mortal mind. It is fear made manifest on the body."

In late 2005, the televangelist Darlene Bishop was sued by the four children of her late brother, the songwriter Darrell "Wayne" Perry, for having persuaded him to give up the course of chemotherapy he was undergoing for his throat cancer and to rely instead on prayer and God's healing powers.

CORRUPT SCIENCE IN THE COURTROOM

Clinical ecology is a "science" that's taken seriously in courtrooms but less so outside them. The initial premise of clinical ecology is not absurd. In *Dirty Medicine* (1993), Martin J. Walker sums it up thus: "The philosophy and practice of what is generically called clinical ecology assumes that the mechanical and chemical processes of the Industrial Revolution, the electrical and the nuclear age, have all had a deleterious effect upon the health of individuals and societies." There is much to applaud here: it is thanks in part to the efforts of clinical ecologists that we have become widely aware of things like the dangers of lead in our gasoline and mercury in our fish. We can also sympathize with Walker's general sentiment when he says, "The basic demand of clinical ecology is . . . a radical one: that the industrial means of production be reorganized to suit the health of the whole of society."

Where clinical ecology falls down is that it soon becomes possible to attribute just about any ailment to just about any element of the environment, which opens up a vast ballroom in which litigation lawyers and sham "expert witnesses" can dance. And the dubious science purveyed in the courtroom for reasons of financial gain inevitably diffuses into the public consciousness, where it all too easily becomes "accepted scientific fact." Thus we've had scares that proximity to high-voltage electrical cables and junction boxes causes cancer. We've had scares about atmospheric levels of radon so low as to be positively homeopathic. We've had scares about "chemical AIDS," which supposedly involves chemicals in the environment suppressing the immune system. We've had scares about fluoridation of drinking water. All nonsense.

The big legal triumph of clinical ecology, according to Peter Huber in *Galileo's Revenge* (1991), came in 1985 in the town of Sedalia, Missouri, where the local chemical plant, Alcolac, was blamed for pollution that caused a number of the residents to suffer "chemical AIDS." Appearing for the plaintiff as medical experts were Bertram W. Carnow, who "registered for the board certification exam in internal medicine in 1957, 1958, 1960, 1961, 1962, 1963, and 1964, but withdrew twice and failed five times," and

Arthur Zahalsky, an immunologist who "never actually studied immunology in graduate school; but he does claim to have audited immunology classes at Washington University in St. Louis."

The field of litigation is in general rife for scientific corruption. As Huber points out from a lawyer's perspective, "[A]s you labor to assemble your case, the strength of the scientific support for an expert's position is quite secondary. It is the strength of the expert's support for your position that comes first." In other words, who cares if the testimony of your "expert witness" is junk science or plain lies so long as it wins *your* case?

And junk science such testimony often is. It is surprising that the laws to curtail perjury are not more often invoked. Any other witness who presented an extreme implausibility as if it were an established fact would at the very least be cautioned, but the notion has crept into the court system, stated or unstated, that scientific fact—or scientific consensus, which is almost synonymous—is somehow open to debate, or even democratic vote.*

There are further factors involved. Science in the real world is assessed by the community of other, relevantly qualified scientists; in the courtroom it is only by chance if a member of a jury has the knowledge to dissect "expert" testimony. Generally it's lay people who're trying to cope.** Plausibility becomes everything; truth becomes immaterial. Similarly, whereas any competent scientist will hedge any statement of presumed fact with qualifications—"so far as is understood," "the general consensus of opinion," and so on—this tentativeness, which in science is a sign that someone knows what they're talking about, is all too often regarded by the lay person as the opposite: the purveyor of the definitive statement is regarded as the more authoritative witness, the more plausible "expert."

The biggest scientific corruption of all in connection with the legal system arises because "expert witnesses" are paid if they agree with the lawyer who wants to hire them, and not paid if they don't. The pay is often quite handsome—to the extent that many second-rate scientists find it all too tempting to abandon their academic careers for the riches offered by a career as a serial "expert."

Corrupt science in court affects us all. A couple of decades ago successful lawsuits against the manufacturers of whooping-cough vaccine drove

---

* This is a corruption that plagues modern western society in general.

** And sometimes they cope very badly indeed. In the 1943 paternity suit brought by Joan Berry against Charlie Chaplin, Chaplin's lawyers presented definitive scientific evidence that the actor could not have been the father of Berry's child: blood-typing proved it impossible. Nonetheless, the jury decided in Berry's favor.

millions of parents, quite reasonably, to reject it for their offspring on the grounds that it could cause all sorts of disabilities, up to and including brain damage and death. The truth was that, since vaccines work by inducing mild forms of the disease against which they protect, there's always a remote chance that some child, somewhere, might indeed be affected by whooping-cough symptoms as a result of being given the vaccine. There was no such instance on record, and numerous large-scale clinical trials had failed to turn up even a trace of such an effect, but it was not completely outside the bounds of possibility. That was enough for certain courts, prompted by "expert witnesses," to rule that the vaccine was at fault in a few tragic cases . . . and the panic began.

Of course, the full-scale disease itself *does* produce those symptoms, and infants suffer life-wrecking damage and death because of them; no one has yet assessed how many deaths were caused through the courts' acceptance of corrupt science deterring parents from having their kids vaccinated, but a conservative guess is thousands.

Similarly, the morning-sickness drug Debendox, also known as Bendectin, was driven off the market in the US because of a few high-profile judgments based on fake evidence. Morning sickness, although mothers tend later to joke about it, is in its worst cases seriously debilitating, extremely unpleasant, and on rare occasion life-threatening, and can cause miscarriage. Again, no one has attempted to calculate how many mothers and their unborn offspring died because of the court decisions, but it's a statistical certainty there were quite a number, not to mention the countless pregnant women who've suffered unnecessary misery.

One legal field in which science has been extensively corrupted through an over-willingness by prosecutors and investigators alike to rely on the lessons of experience, even where the lessons rest on merest guesswork, is arson investigation. In a number of instances, unfortunates have, based on evidence presented as scientific but in fact little more than folklore, been executed as murderers for setting fatal fires that have since been identified as accidental.

One such was Cameron Todd Willingham, executed in Texas in 2004 for setting the fire that killed his three children in Corsicana, near Dallas, in 1991. At Willingham's trial the Deputy State Fire Marshal, Manuel Vasquez, presented damning evidence of the fire having been artificially set. A US group called the Innocence Project, founded in 1992 to re-investigate dubious convictions via proper scientific analysis, commissioned a panel of five arson experts to sift through the evidence in Willingham's case; even before Willingham's execution, some of this evidence was presented to Texas Governor Rick Perry who, for short-term political reasons, ignored it. The conclusion of the panel was that the fire had been accidental; there was no reason whatsoever to doubt Willingham's own account. The "scientific" evidence that had convicted him represented Vasquez's no doubt sincerely held opinion, but that opinion was not science-based.

One of the items of evidence Vasquez presented was the way that some of the glass in the Willingham home had been crazed; this, he told the court, was a clear indication of the high temperatures generated by fire accelerants, and could have been caused in no other way. This indeed was a traditional belief among arson investigators at the time, but it relied on hunch rather than empirical evidence. Two years later, in 1993, using the aftermath of the 1991 wildfire in Oakland, California, as their testing ground,* a team of investigators including John Lentini, later one of the Innocence Project's five experts in the Willingham case, put much of arson investigation's conventional wisdom to the test. They found crazed glass aplenty, and laboratory follow-up work showed that no amount of heating would produce the "telltale" crazing; what produced the crazing was rapid *cooling* as might occur when the glass was hit by the water from a fireman's hose. Decades of conventional wisdom on the matter had been demolished. Vasquez could not have known this; the results of the work by Lentini *et al.* were, however, publicly available a full decade before Willingham's execution.

Eight months after Willingham's judicial murder, Texas exonerated another man, Ernest R. Willis, who on similarly fallacious evidence had been convicted and sentenced to death for killing two women in a house fire in 1986 in Iraan, Texas. For his 17 years of wrongful imprisonment Willis collected $430,000 in compensation from the state. The same "new" (in fact, decade-old) science that exonerated Willis had been rejected in Willingham's case.

The 1993 team at Oakland made various further groundbreaking scientific discoveries. Another indicator that had been used for years as evidence of high temperatures, and hence of the use of accelerants, was melted steel—typically melted bed springs. The team found numerous examples of bed springs that appeared to have melted but which, when examined in the laboratory, proved instead to have suffered extensive oxidation, which

---

* The advantage of the Oakland site, which contained some 3,000 destroyed homes, was that, whatever the cause of the original fire (in fact, natural), there was no question but that homes away from the epicenter had been ignited naturally.

happens at far lower temperatures. A more reliable indicator of high (albeit somewhat lower) temperatures remained melted copper; the patterns within the home of the melting, something into which investigators had for a long time read much, was, however, random, as were the patterns with which steel objects like bed springs oxidized.[36]

Some while later, Lentini got involved after the conviction of Han Tak Lee for having murdered his mentally ill daughter by setting a fire at a religious camp in Stroud, Pennsylvania. In this instance the "scientific" evidence presented by the state was a mixture of disproven conventional wisdom and straightforward hokum; the principal "expert," Daniel Aston, a part-time suspicious-fire investigator, claimed to have examined some 15,000 fires, whereas the busiest full-time investigator might be expected to examine something under 5,000 fires in the space of an entire career. Nonetheless, Aston's claim went unchallenged by the defense, the same going for the astonishingly detailed conclusions he presented. Lentini's own conclusions, given in his paper "A Calculated Arson,"[37] are worth quoting:

> The quality of the evidence presented by the Commonwealth [of Pennsylvania] speaks for itself. Fuel loads calculated to six significant figures, hydrocarbon "ranges" being interpreted as evidence of a mixture, furnace operating instructions being touted as normal fire behavior, and a host of other "old wives' tales" were used to convict Han Tak Lee.
>
> There are over 500,000 structure fires every year in the United States. (Approximately 15% are labeled suspicious or incendiary.) Each presents an opportunity for erroneous cause determination, and a significant number of erroneous determinations do occur. Even if fire investigators are correct 95% of the time, that allows for 3,000 incorrect determinations of arson each year. These calculations demonstrate the need for objective investigations based on the scientific method.
>
> Much has been written lately about the criminal justice system allowing guilty people to escape justice due to sloppy police work. Here is the case of the wrongful conviction of an innocent man, surely a worse result. The Lee case represents the ultimate triumph [in the courtroom] of junk science.

If arson investigation were the only area in which forensic science was in chaos, that might seem a containable problem—one that could be dealt with by stepping up the scientific training of arson investigators, many of whom have received none at all: they have learned the principles of their trade through training on the job under other, more experienced investigators, who unwittingly pass on dubious information. But a flurry of recent investigations has shown that much forensic science in other areas is of equally unsound basis, including even that old staple, fingerprints. It is not quite true that every fingerprint is unique; nuances of interpretation in fingerprint analysis can all too easily lead to false identification. Throughout, even in supposedly infallible DNA testing, forensic science is, like everything else, prone to corruption through simple human error, while personal traits such as arrogance among forensic scientists, or the desire to stay sweet with the DA's office, may also play their part. The fervor of prosecutors eager to obtain a conviction at any price, including that of the truth; the willingness of even the best-intentioned juries to see their biases confirmed; and the readiness of junk scientists and outright pseudo-scientists to bask in glory as handsomely paid "expert" witnesses in the courtroom—all of these corrupt the science yet further.

3

MILITARY MADNESS

Despite the vision and the far-seeing wisdom of our wartime heads of state, the physicists felt a peculiarly intimate responsibility for suggesting, for supporting and, in the end, in large measure, for achieving the realization of atomic weapons. Nor can we forget that these weapons, as they were in fact used, dramatized so mercilessly the inhumanity and evil of modern war. In some sort of crude sense which no vulgarity, no humor, no overstatement can quite extinguish, the physicists have known sin; and this is a knowledge which they cannot lose.

—J. Robert Oppenheimer, "Physics in the Contemporary World," lecture delivered at MIT, 1947

In the councils of government we must guard against the acquisition of unwarranted influence, whether sought or unsought, by the military-industrial complex. The potential for the disastrous rise of misplaced powers exists and will persist.

—President Dwight D. Eisenhower, Farewell Address to the Nation, January 17, 1961

A good case can be made that whenever science is controlled by nonscientists the result *is* a corruption of science. Yet, in our modern world, science almost always is controlled by nonscientists, sometimes by political edict but most significantly through selective funding—the allocation of money for research by governmental, commercial or military organizations. In 2001 in the US, even before the start of the "war on terror," an annual federal research budget of some $75 billion saw nearly $40 billion go to the Pentagon, primarily for research into weaponry, while a mere $4 billion went to the National Science Foundation for actual *science*. That disproportion is in itself an obscene corruption.

Yet consider the allocation of its vast budget by the Pentagon, with decisions being made by nonscientists as to which lines of research should be pursued, which discarded. To any criticism of the situation the reply is that, obviously, the military should decide which weapons show most promise. Yet is that so obvious? Turning the argument around, would the military think it reasonable if the scientist who'd created a new weapon for them then dictated the strategy for its deployment? Well, no: the military would regard that as wholly unreasonable—and they'd be right.

The meddling in science by the Nazi hierarchy, through their support of pseudoscience and through their imposition of a daft antisemitic ideology on the structures of scientific research, inadvertently went a long way toward ensuring the Reich never developed nuclear weapons; for that we must be thankful. Yet, in the decades since then, governments of all stripes, and their military institutions, have failed to recognize the lesson: untrained personnel are ill equipped to dictate the course of scientific research. Just as a politician or a religious demagogue lacks the tools to judge the desirability (or otherwise) of stem-cell research, so a soldier is incapable of discriminating between a sane line of weapons research and one that's loony.

In this chapter most of the madness we look at in the field of military technology is from the post-war US, simply because that nation has been by far and away the biggest spender in military R&D; indeed, any US politician who suggests a cut in spending in this area is likely soon to be an ex-politician. It is by popular vote that the military's enormous waste of public money on crazily corrupted science, or outright pseudoscience, continues. Quite how significant the waste is can be illustrated by the fact that, in the years after 2003, an Iraqi resistance whose typical weaponry was the home-made bomb could fight to a standstill the most technologically advanced army in the world.

EDWARD TELLER AND THE STRATEGIC DEFENSE INITIATIVE

One of the most dramatic examples of scientists fooling themselves concerns Edward Teller and the Strategic Defense Initiative (SDI, or "Star Wars"). There's no doubting Teller was a brilliant physicist; there's also no doubting he was a man of huge arrogance. From early in his career it was evident he was one of those scientists whose strength lay in his prolificity of ideas, often astoundingly unorthodox ones; it was left to others to do the detail work, the math, to evaluate those ideas. In the event, some 90 percent would prove valueless, but the remaining 10 percent still represented far more good ideas than the average theoretician could dream of.

After he fled the Nazis for the US (via, briefly, the UK) in 1934, Teller's ascent of the scientific ladder was rapid. In due course he became part of the Manhattan Project at Los Alamos under J. Robert Oppenheimer to create the atom bomb, although his irascibility saw him progressively sidelined from the project's main thrust. Even before the team's success in producing the A-bomb, which relies on the *fission* of radioactive isotopes of heavy metals, Teller was pushing for the development instead of the H-bomb, which depends for its working on the *fusion* of atoms of hydrogen (or, better, its common isotope deuterium, "heavy hydrogen"). The principal advantages of a fusion bomb are that its fuel is common and cheap and that the energy release is very much greater—considerably more bang for the buck, in other words. Although Teller later rejoiced in the honorific "Father of the Hydrogen Bomb," in fact the scientist whose conceptual breakthrough was primarily responsible for turning the device from a pipe-dream (or pipe-nightmare) into a reality was Teller's colleague Stanislaw M. Ulam; characteristically, in later life Teller was loath to mention Ulam's role.

In the early 1950s Teller, now at loggerheads with most of the scientific community at Los Alamos, persuaded the USAF that a second nuclear weapons lab should be founded, with himself as its scientific head. Although Oppenheimer and the Atomic Energy Commission vigorously opposed the move, the Lawrence Livermore National Laboratory, named for Ernest Lawrence, inventor of the cyclotron, was founded in 1952 at Livermore, California.

Here Teller initially produced a string of embarrassing flops. He had great difficulty in creating an H-bomb that would actually work; several tests fizzled before finally Teller's team succeeded in creating a detonation, long after the Los Alamos team had conducted their own successful H-bomb tests.

By now Teller regarded Oppenheimer as his bitter enemy, and he took advantage of the McCarthyite witchhunts to destroy the older scientist, testifying before the Atomic Energy Commission in April–May 1954 to devastating effect. The profoundly rightwing Teller did not say outright the moderate Oppenheimer had Communist leanings; instead he attacked Oppenheimer's opposition to the H-bomb, leaving others to draw their own conclusions: "I would like to see the vital interests of this country in hands which I understand better, and therefore trust more." Indeed, Teller's testimony to the FBI on this issue is widely thought to have been responsible for initiating the investigation of Oppenheimer in the first place.

In *Forbidden Knowledge* (1996), Roger Shattuck contrasts the introspection of the Oppenheimer remark cited at the head of this chapter with Teller's comment in a 1994 interview concerning the development of the H-bomb: "There is no case where ignorance should be preferred to knowledge—*especially* if the knowledge is terrible." While there is much to agree with in this *per se*, the difference between the moral worldviews of the two men could not be greater, and it was this difference which alienated them from each other and led Teller to seek, and achieve, Oppenheimer's downfall. Oppenheimer saw the shades of gray that make up genuine morality; Teller dealt with the stark blacks and whites of the *faux*-moralist. It was a view he stated forcefully in 1988 in Washington, DC, on meeting Andrei Sakharov, a major contributor to the Soviet H-bomb but later a leading Soviet dissident who became, through his campaigning for a nuclear test ban, the recipient of the 1975 Nobel Peace Prize. Sakharov denounced the practice of atmospheric H-bomb testing, on the grounds that untold thousands of people would be affected by the radiation, and he denounced SDI as a threat to world peace through upsetting the nuclear balance. Teller's response was that to suppress either would, aside from impairing the defense of the US, be to suppress the advance of human knowledge. Besides, the doses of radiation people might receive from the testing were minuscule, far too small to affect them adversely. In this latter contention he was at odds with the overwhelming conclusions of research regarding radiation exposure, but he let his ideology steer him. The cancer epidemic in downwind sites such as St. George, Utah, was just one tragic consequence of Teller's wrongheadedness regarding such exposure.

To return to the 1950s. Although the Livermore laboratory finally got its H-bombs to work, there were further embarrassments on the way. Teller convinced himself and President Dwight Eisenhower that such a thing as a "clean" nuclear bomb could be produced; he probably first got the notion as an argument to use against the imposition of a nuclear test-ban treaty, so that he might carry on detonating his beloved bombs. Teller persuaded Eisenhower and the US public that the development of a "clean" nuke was just around the corner; in fact, such a device has never been created.

With the concept of the "clean" bomb firmly implanted in everyone's minds, the possibility opened up of using nuclear detonations for peaceful purposes. Teller's first project was the creation of a new harbor near Point Hope in the remote north of Alaska. Six "clean" bombs, totalling some 2.4 megatons, would do the job, Teller claimed—and would herald the dawn of a glorious new H-bomb utopia for mankind. Substantial government money—a recurring feature of all Teller's projects—went into the scheme between the late 1950s and 1962, when the administration of President John F. Kennedy axed it. Typically, Teller had forgotten that the Inuit inhabitants of the Point Hope region might have something to say in the matter.*

Teller then turned his attentions to the notion of controlled nuclear fusion as a source of energy. Again, considerable amounts of US taxpayer dollars were bestowed on Livermore for a project that went nowhere. By the late 1980s, Teller himself was prepared to admit that his team's efforts were doomed, yet in 1989, when Pons and Fleischmann announced to a breathless world that they'd achieved cold fusion (see page 53–55), Teller was one of the first to put his weight behind their claims. He set his Livermore protege Lowell L. Wood, Jr., the task of reproducing the Pons–Fleischmann experiments. All Wood succeeded in doing was nearly blowing his laboratory to pieces—not for any fancy-schmancy nuclear-fusion reason, but because the hydrogen he was using caught fire.

Ever since departing Los Alamos back in the early 1950s, Teller had lacked the benefit of being surrounded by peers who could evaluate and challenge his ideas. He had made himself largely a pariah in the world of physics by instigating the split from Los Alamos and particularly by his malicious smearing of the widely respected Oppenheimer. There were difficulties recruiting top-caliber physicists to Livermore; some came, but Teller became surrounded more and more by handpicked sycophants. Hence the numerous failures amid the occasional successes of the Livermore team; hence, too, the untold billions of taxpayer dollars wasted on those failures.

By far the most expensive of his many grandiose ideas was SDI. Teller was probably not the first to conceive that the way out of the dead end represented by Mutual Assured Destruction between the two superpowers was to mount a space-based defense system that would destroy the enemy's missiles before they reached their targets, but he was certainly among the

---

* A few years later it was shown that in fact Inuit already suffer unduly from radioactive fallout, which is absorbed by the lichens that are eaten by the caribou that the Inuit in turn eat.

earliest. His initial vision was of a series of satellites armed with X-ray lasers whose beams would zap incoming missiles. The X-ray laser had been developed at Livermore in the late 1970s by George F. Chapline, Jr., with much of the theoretical work reluctantly done by the pacifist physicist Peter Hagelstein. Eventually, there were two rival designs.

The beam of an optical laser can do a certain amount of damage. Using shorter-wave (more energetic) electromagnetic radiation than light would create a laser capable of far greater destruction, and X-rays have almost the shortest wavelengths of all. (Even better would be gamma rays, a fact that gave rise to dreams of the gamma-ray laser, or graser.) The only trouble was that, in order to get an X-ray laser started, you had to use a nuclear explosion. Since the explosion would almost instantaneously annihilate the laser, this was obviously a one-shot-only weapon; it would also, to say the least, be somewhat cumbersome, especially if actually used.* Despite all the difficulties, in November 1980 in an underground Nevada test of Chapline's and Hagelstein's X-ray lasers, there were ambiguous indications that one or both had actually worked.

Teller and Wood, casting all doubts to the wind, proclaimed the test a triumph; a practicable X-ray laser was, they told their paymasters, just around the corner. Their proselytization did not fall on deaf ears, because the US election of 1980 brought Ronald Reagan to the White House. The popular image of Reagan as a genial old duffer conceals such less pleasant aspects of his presidency and politics as his support for some of the vilest dictatorships of the twentieth century (including that of Saddam Hussein in Iraq), his financing of Central American death squads, his fervent opposition to civil rights legislation, and so on. The image also obscures the fact that Reagan matched a passionate interest in science and technology with a profound scientific illiteracy. (He consulted astrologers for decades, and during his presidency was a client of astrologer Carroll Richter.) Put this alongside Reagan's paranoid detestation of communism and his gullibility, and there could hardly have been a US president better tailored as the mark for Teller's spiel.**

---

* These considerations would be less important if the detonation were performed in space, of course; further, a single explosion could be used to trigger a multiplicity of independently targeted lasers. However, you'd still have to lift a fairly massive payload into orbit.

** There were further factors involved. Reagan had since the 1960s taken literally the Book of Revelation prophecy of a final Armageddon, which he came to believe referred to a nuclear holocaust. He seems also to have believed around 1980 that the Soviet Union already had an effective anti-ballistic-missile shield in place, or nearly so.

Reagan put the full weight of his presidency—and unimaginable amounts of money—behind the development of SDI. Others, however, were considerably more skeptical: scientists who believed the system could never be made to work and strategists who, knowing nothing of the science, feared such a development would increase rather than reduce the likelihood of nuclear war. The reasoning of the latter was that any nation in possession of SDI could with impunity rain nuclear destruction upon its foes. The obvious course for said foes was to launch an assault upon the nation concerned while SDI was still in the process of development.

The Reagan administration disparaged such doubts. In March 1983, Reagan gave his famous "Star Wars" speech and the money began to flow.

In the event, the X-ray laser dropped from contention fairly early and other systems came to be preferred, among them chemical lasers and neutral particle beams. Nonetheless, without Teller's unjustified claims for the X-ray laser it is reasonably certain the program would never have gotten off the ground.

Over time, and particularly after the collapse of the USSR, the program's aspirations were trimmed back, the focus being less on defense against an all-out nuclear assault, more on protection against an accidental launch. Even here there were difficulties; in 1988, reporting to the House Democratic Caucus, Navy physicist Theodore Postol stated that at best the system might destroy a launch of up to five missiles. Furthermore, any space-based system would be remarkably easy to counter, since the first stage of a pre-programmed assault would be to knock out the satellites bearing the defensive weaponry.*

After the 1991 Gulf War, US attention shifted from SDI to the possibility of ground-based defensive systems. This was because the Patriot interceptor missile had supposedly shown astonishing success in destroying Scud missiles aimed by Iraq at Israel: over 50 percent of the Patriot's targets were nullified, with initial US estimates set far higher than that. As Richard Cheney, much later to be US Vice-President but in 1991 George H.W. Bush's Secretary of Defense, stated: "Patriot missiles have demonstrated the

---

* In 2000 Postol similarly took up the cudgels concerning an anti-ballistic-missile sensor manufactured by the company TRW, which had been declared successful after a 1997 test. In response to the claims of an earlier TRW whistleblower, a 1998 investigation reported that, though the sensor had indeed failed, TRW's software had functioned adequately. When Postol received a censored copy of this curious report in 2000, he rang the alarm bells, claiming possible "research misconduct." An investigation into this further claim was delayed by such tactics as classification of materials "for reasons of national security" until time ran out.

technical efficacy and strategic importance of missile defenses. This underscores the future importance of developing and deploying a system . . . to defend against limited missile attacks, whatever their source." Interestingly, however, Israeli observers, who had the advantage of being able to count the holes in the ground where the Iraqi Scuds had landed, were saying that the Patriot missiles had at best succeeded in a single interception, with zero being a more likely total. Later the House Government Operations Subcommittee on Legislation and National Security reported, with heavy understatement:

> The Patriot missile system was not the spectacular success in the Persian Gulf War that the American public was led to believe. There is little evidence to prove that the Patriot hit more than a few Scud missiles launched by Iraq during the Gulf War, and there are some doubts about even these engagements. The public and the Congress were misled by definitive statements of success issued by administration and Raytheon representatives during and after the war.

By midway through this century's first decade, the Missile Defense Agency—a modern analog of the abandoned Star Wars/SDI scheme—was spending $10 billion annually on such projects as the Space-Based Interceptor, which violates all existing treaties against the militarization of space, and that old favorite of SDI, the space-based laser weapon.* Alas, all of its tests were embarrassing duds; the Bush administration's response to this pattern of failure was to quietly remove the MDA's obligation to report on its progress to Congress . . . and increase its annual budget.

As many pointed out in connection with the MDA, for about $10 billion the entirety of the world's currently vulnerable depots of fissile material could be secured, thereby almost eliminating the risk of terrorists gaining access to it. There is an existing US program designed to do exactly that, at least for the former USSR: the Cooperative Threat Reduction (CTR) program, inaugurated in 1991 at the end of the Cold War. Since it is regarded as a foreign-aid rather than a defense program, however, it is vulnerable to foolhardy budget-trimming. In 2006 its budget was slashed to below $400 million per annum, its lowest level since the program's inauguration (and lower still, of course, because of inflation).

---

* Since US nuclear stockpiles dwarf those held by all of the rest of the world put together, and are many times more than required to render the entire earth uninhabitable, it's unclear who this defense is against.

The Obama administration saw the CTR program's budget annually cut yet further, although to be fair Obama was pursuing other lines of nuclear-weapons reduction.

It's not hard to see which of the two programs better serves national—and global—safety: MDA or CTR. It's equally easy to see that the ineffective, unproven program is vastly more expensive than the more effective one. We can only guess whether the disparity arises because boring old CTR lacks the whiz-bang sex-appeal of all that shiny new Buck Rogers weaponry.

HAFNIUM NO, BANANAS YES

If the protracted saga of SDI epitomizes the incredibly expensive denial of science, that of the proposed hafnium bomb beggars belief.

In 1998 a physics professor at the University of Texas at Dallas, Carl Collins, bombarded a few specks of the atomic isomer hafnium-178m2 with X-rays from a retired dental X-ray machine in an attempt to trigger an energy release.

Atomic isomers (AIs) are analogous to chemical (molecular) isomers: in the latter the atoms in the molecule adopt different spatial configurations than in the "standard" form, whereas in the former it is the particles within the atom that are differently configured. A property of AIs is that they can become "charged up" with energy—large amounts of it—which they then release gradually in the form of gamma rays. This made them of considerable interest to scientists following the grail of the gamma-ray laser, or graser—if an X-ray laser could be almost unimaginably powerful, just imagine a laser that uses the even higher-energy gamma-rays! The problem was how to trigger the AI such that it released its considerable stored energy quickly rather than through gradual decay.

The US government spent a lot of money on the graser before it finally cut funding; by the time Collins was ready to perform his experiment on Hf-178m2, he was forced to use accounting shenanigans merely in order to obtain a minute sample of the isomer—hence the retired dental X-ray machine, with the sample being placed atop an upturned styrofoam cup.

Despite the constraints, Collins reported success in triggering the Hf-178m2. Even though other experimenters comprehensively failed to replicate his results, the Defense Advanced Research Projects Agency (DARPA) seized on the experiment as grounds for a heavily budgeted pursuit of the hafnium bomb, a bomb that would have an explosive yield comparable to a smallish nuclear fission detonation yet with the advantage of scale: a two-kiloton bomb could be packed into a hand grenade!* That Hf-178m2's half-life of a mere 31 years would make this the "dirtiest" bomb in human history was an item glossed over by Collins's enthusiastic DARPA supporters. (To simplify hugely—because there are all sorts of modifying factors—the shorter the half-life, the more radioactivity emitted per unit time.)

What followed resembles an anxiety dream. Various investigations, both independent and government-commissioned, cast doubt on Collins's poorly documented and irreproducible (except by Collins himself and by an old friend of his, Patrick McDaniel) results and stated categorically that the physics was bunk. Several of the US's most expert and trusted scientists went to bat along the same lines. The response of DARPA was to proclaim to the world that research into hafnium triggering was going swimmingly, the only remaining hitch being the need to discover a cheaper way of producing the phenomenally expensive Hf-178m2.

In one instance, Martin Stickley of DARPA commissioned a report on Collins's experiments from the Oxford University physicists Nick and Jirina Stone, expecting it to be a glowing vindication of his claims; in fact it was damning. Rather than moderate the claims, Stickley swept the report under the carpet, where it remained until October 2004, when journalist Sharon Weinberger succeeded in prizing it from DARPA's grip under the Freedom of Information Act.**

At one point even Collins himself tacitly admitted the hafnium bomb was a non-starter: so far as he could establish, only about one in 600 of the X-ray photons with which he was bombarding his sample actually triggered anything, which meant more energy was required for the triggering than was being released by it. A bomb is not much use if you have to use a bigger bomb to detonate it.

Matters were further complicated by Collins's and DARPA's refusal to run a control experiment—for example, by bombarding a sample of ordinary hafnium with X-rays to see if the same results emerged, in which case the obvious suspicion would be that the results were a product not of triggering but of something else, such as faulty equipment.

---

* It seems it was some years before it occurred to anyone that a two-kiloton-yield hand grenade is useless except to suicide bombers—ineffective suicide bombers at that, because, as someone eventually pointed out, the hard radiation coming off the bomb would kill users before they got the pin out.

** See Weinberger's book *Imaginary Weapons* (2006) for a definitive account of the whole hafnium bomb fiasco.

The outflow of taxpayer money continued for several years, spurred ever onward by nonscientist military enthusiasts, by US paranoia over the (seemingly nonexistent) researches being pursued by other countries into the hafnium bomb, and by the sensationalist media, until finally reason prevailed: in late 2004, the relevant committees in House and Senate put their collective foot down, and funding was terminated. Yet a hard core of hafnium-bomb aficionados continue to pursue the dream, and it is suspected covert research is still being sponsored by the Pentagon.

RED MERCURY UNDER THE BED

No one is quite sure where the notion of "red mercury" came from, but it featured in any number of media scare stories from the late 1980s onward and was also—seemingly *is* also—taken seriously by various governments, even though the substance, if it exists at all, is of completely unknown nature. Popular notions are:

- it can be used to facilitate the production of enriched nuclear fuel for weapons
- it can perhaps be detonated itself as a sort of sub-nuclear but nevertheless very powerful bomb
- it is a ballotechnic (an explosive chemical) that can be used in place of the fission-bomb trigger for a fusion (hydrogen) bomb
- "mercury" is merely a code name for fissile material (plutonium, perhaps)
- it is a stimulated nuclear isomer along the lines of hafnium-178m2 and thus a potential explosive source of gamma rays
- it is a paint that, applied to stealth bombers and the like, helps them elude radar

The list could be extended considerably. Even the name of the putative material is confusing. Initially the substance was black, but it was called "Red mercury" because it supposedly came from the debris of the old USSR. Soon it lost the capitalization of the "R" and, obligingly, thereafter was red in color. It could be a solid, a liquid, or a powder.

One tempting theory is that the KGB "invented" red mercury as the basis for a sting operation aimed at finding out which rogue states and terrorist organizations were in the market for nuclear materials. A variation was that Russia was engaging in a scam designed to raise billions in foreign exports of a worthless substance, and that it was *western* intelligence agencies who opportunistically used the sales as a means of finding out who was seeking nuclear materials. A 1995 book, *The Mini-Nuke Conspiracy: How Mandela Inherited a Nuclear Nightmare* by Peter Hounam and Steve McQuillan, claimed South African scientists had created red mercury during the apartheid era, and using it had built countless tactical nukes which were now in the hands of rightist anti-government extremists.

Most certainly there were people in the market for red mercury, with money changing hands all through the 1990s and even later. In 1997, the *Bulletin of the Atomic Scientists* reported that the buying price ranged between $100,000 and $300,000 a kilo. Earlier, in 1992, Russian President Boris Yeltsin signed a decree licensing the Yekaterinburg company Promekologiya to produce and sell 84 metric tons of red mercury to the Van Nuys, California, company Automated Products International. The decree was canceled a year later and it seems likely no material changed hands; even so, Promekologiya's head reported his company received foreign orders for red mercury totaling over $40 billion. As late as 2004, in a sting mounted in the UK, three men were arrested for attempting to buy a kilo of the stuff for £300,000 (about $450,000). Although the IAEA made a public statement that red mercury was a hoax, the prosecution insisted, when the case came to trial in 2006, that this didn't affect the men's guilt: they were trying, however misguidedly, to obtain the material for terrorist purposes. (They were acquitted.)

With all of this activity regarding red mercury, still nobody knew what it was or what it could do! Well, some people thought they knew. In a paper published in 2003 in *Natsionalnaya Bezopasnost i Geopolitika Rossii*, A.I. Khesin and V.A. Vavilov claimed, according to a summary by the Center for Nonproliferation Studies, that "red mercury can be used to resolve the ills of the human race and planet earth by aiding in oil extraction, restoring exhausted mines to production, reviving unproductive agricultural land, recultivating nuclear test sites, cleansing land polluted with radionuclides, producing medicine, and creating environmentally clean fuel for new sources of energy." Wow!

OSMIUM-187

In a manner similar to red mercury, but a little later, the isotope osmium-187 became a much-sought item on the international terrorism market as an essential material in the manufacture of nuclear weapons, although its role in said manufacture was somewhat vague. Unlike red mercury, Os-187 certainly exists: it's one of the seven osmium isotopes that exist in nature, although in low concentrations by comparison with the standard form of the metal. It's nonradioactive, but extremely dense—osmium is the heaviest known element—and that density accounts for one of its purported weapons uses: it could be used as the tamper in a bomb, the tamper being the material that inhibits the explosion for as long as possible (not long) in order to increase the scale of the final bang. Unfortunately for such an argument, Os-187 is—at maybe $100,000 per gram—ludicrously more expensive than alternatives that are just as effective. Another putative use for the isotope might be as the bomb's neutron reflector, which increases the device's yield, but Os-187 is too dense to make a very effective neutron reflector; the lighter and far cheaper beryllium does a better job.

So why did rogue states and terrorist groups decide Os-187 was a must-have item? A clue might be found in the fact that the methods used to extract Os-187 from its parent metal are very like those used to enrich uranium. This similarity seems to have spread the misconception that, like enriched uranium, Os-187 must have nuclear applications. In 2002 Viktor Ilyukhin, a member of Russia's Security Committee, accused Kazakhstan of unlawfully producing and selling Os-187 for weapons purposes; it seems Kazakhstan itself has concerns about such a potential, because in that nation the substance is controlled.

The scam here is that Os-187 is a completely innocuous isotope. Although its presence in natural osmium is a mere 1.64 percent, that would still be enough to make you nervous of, say, the amount in the nib of your fountain pen or the filament of your incandescent lightbulb were Os-187 substantially radioactive. It isn't, and it has no sensible uses in nuclear weapons.

Psychotronic Warfare

Expenditure by the US government on research into "psychotronic weapons"—weapons that use psi powers—may seem on the face of it just plain barmy, but things aren't quite so simple. Consider this in context. One of the tasks of any government is to try to ensure national security. On the one hand, existence of psi powers is the longest of long shots; on the other, in terms of the overall defense and security budget, the government's investment in psychic research is just the tiniest of drops in the ocean, a minute fraction of a percent. Much of that enormous budget is spent on research into weapons and defenses that we know won't work, such as the hafnium bomb and SDI. If there's the remotest possibility that "there's something in" psi powers, and especially if there's good reason to believe the enemy is actively carrying out a research program in the field (as there was when the main perceived enemy was the USSR), then wouldn't any government be guilty of gross dereliction of duty if it didn't expend some effort to follow suit?

Or so the argument goes.

Even so, bafflement is the only possible response to some of the particulars. For example, one US military project in World War II focused on trying to telepathically influence seagulls to poop on the periscopes of U-boats, thereby obscuring the German submariners' view. Later, in the 1960s, the CIA investigated the possibility of mentally controlling cats so that, appropriately equipped with a microphone, the felines could eavesdrop on the enemy's conversations: two spies sitting in a park are unlikely to think twice if an affectionate kitty ambles up to them demanding to be stroked, are they? The CIA underestimated cats. When they first tried the system out the cat promptly deserted its post and had a fatal encounter with the traffic on a nearby road.*

The involvement of the CIA in psychic research apparently began in 1972 with a meeting between people from the Office of Scientific Intelligence (OSI) and Russell Targ, a maverick physicist who was co-founder with another such, Harold Puthoff, of the Stanford Research Institute (SRI).** The OSI—like the Defense Intelligence Agency (DIA)—was concerned about reports from the USSR of supposed psychics being investigated for their potential in intelligence gathering. Targ apparently showed them film of people moving objects around tabletops. OSI contacted other departments, including the Technical Services Division (TSD, which had already done some ESP research), to check out if they were prepared to contribute funding to further investigate Targ's claims.

The most fruitful area seemed to be remote viewing, the claimed clairvoyant ability of certain psychics that they can "visit" distant places; spies who could astrally explore the enemy's secret bases and report back would obviously be an invaluable asset. Accordingly Puthoff brought to the SRI the New York artist Ingo Swann, whose purported psychic abilities were

---

* A full account of the craziest US military researches into psychic warfare, including the formation and funding of a unit called the First Earth Battalion, is far outside the scope of this volume. Readers are referred to Jon Ronson's often hilarious book *The Men Who Stare at Goats* (2004).

** The Stanford Research Institute is not part of Stanford University.

already becoming widely known, and subjected him to a series of tests, in all of which, reported Puthoff, Swann performed spectacularly. Various agencies provided modest funding for further experiments involving Swann and others, notably Pat Price, a freelance building contractor who lived not far from the SRI.

In the initial experiments, Swann, Price and the rest were told to try to "visit" locales where SRI personnel had been sent as "beacons." It was apparently Swann who pointed out that this approach was, well, a bit useless: if the CIA could plant "beacons" in enemy installations, why bother with remote viewing? Instead, he proposed a technique dubbed "scannate" (scanning by coordinates), in which the remote viewer would be handed a set of map coordinates and told to describe what s/he "saw" there. In May 1973, an officer of the OSI gave Puthoff a set of coordinates that had in turn been given to him by a CIA officer. Putting this double barrier between the CIA officer and the remote viewer about what lay at the coordinates would seem to obviate cheating. In late May and early June, first Swann and then Price were shown the coordinates, and they offered quite similar descriptions of what they "saw" there—a military base of some kind—with Price's description being the more detailed. Subsequently Price was asked to "revisit" the site and seek yet further details; he produced an impressive list.

The CIA officer, on being told of all this, laughed: he'd given his OSI colleague the coordinates of his vacation cabin in the Blue Ridge Mountains. Puthoff's OSI contact, however, was not content to leave matters there. The similarity between Price's and Swann's descriptions, plus all the extra detail Price had supplied, niggled at him. In due course he found that the US Navy communications facility at Sugar Grove, West Virginia, which doubled as a covert National Security Agency (NSA) site, was near the cabin. The details of this facility did indeed appear to match what Price in particular had "seen." A report from the OSI to the CIA in October 1973 recounted this, plus the less accurate but nonetheless moderately impressive results of remote-viewing "excursions" to a couple of foreign sites.

The focus of the report's author was selective. The details Price had reported of the Sugar Grove site were largely correct, but many were out of date by about a decade, and those that were not were elements that had not changed during that decade. Had Price been fed his information by someone who'd known Sugar Grove but hadn't been there in a while? One possible source was Puthoff himself, who'd worked for the NSA during the early 1960s. This is not to accuse Puthoff of cheating. There's a technique known to conjurers and sham psychics as "cold reading" whereby a skilled interrogator can draw forth from people the most astonishing details through a series of apparently innocuous questions. It's at least possible that Price "cold read" Puthoff—or of course Price may have had another likewise somewhat dated source. Since none of the experiments were performed under proper scientific control, it's impossible to tell.

Further funding was provided by the Office of Research and Development (ORD) and the Office of Technical Service (OTS, the rechristened TSD) for a new program that sought means whereby remote viewing could be exploited for intelligence purposes. Almost at once scientists at the ORD began to kick up a fuss about the lack of scientific rigor with which the SRI experiments were done. The clamor spread to the scientific community in general; although in October 1974 Targ and Puthoff published a paper in *Nature* on their remote-viewing work, it was accompanied—most unusually for *Nature*—by a qualifying note from the editors expressing considerable doubts about the vagueness with which the experimental procedures were described.

By then there had been some personnel shifts in the leadership of the OTS and ORD, the newcomers being significantly more skeptical about the program than those they'd replaced. An experiment done with Price in July 1974, using as target a suspicious, newly discovered site in Kazakhstan, was essentially a fiasco: he provided a flood of details, but they bore little relation to the reality at the site. In his solitary success he correctly "saw" and drew a reasonably accurate representation of a gantry crane that was there, but then he did the same for three other gantry cranes that weren't. As one of the experiment's evaluators observed, with so many details being offered it was hardly surprising that Price would get *something* right. (Oddly, Puthoff and Targ continued for years to claim the experiment was a resounding success, and indeed that it was responsible for the SRI receiving continued government funding.)

A further trial with Price involved "visiting" the code rooms of a pair of Chinese embassies the CIA had succeeded in bugging. At first his results seemed sensational, describing the general aspects of the code rooms with great accuracy, but, when it came to the particulars, matters got rapidly vaguer. As later evaluators pointed out, the conditions of the experiment had been so sloppy that CIA officers who knew the details he was "seeing" were in the room with Price; once again the possibility of cold reading seems overwhelmingly likely.

Price's untimely death in 1975 of a heart attack rendered stillborn a new trial, this time to remotely view a Libyan installation. Since Swann had

departed some while before, that was more or less the end of the CIA's attempts to exploit remote viewing, and within a couple of years the agency
was effectively disowning it.

The DIA and Army Intelligence were not so timorous, however, and
their experiments in the field continued until at least the mid-1990s—and
may still be running. As an example, remote viewers were asked by the
Reagan administration to locate Muammar Gaddafi prior to the ill conceived 1986 US bombing raid on Libya. Even later, in 2002, the UK's Ministry of Defence spent £18,000 on a pilot project recruiting supposed remote
viewers to participate in the "war on terror"; the results were apparently
"too inconclusive" to take the project further.

4

The One True Book

Descended from the apes? Us? How awful! Let us hope that it is not true, but, if it is, let us pray that it will not become generally known!

—Emily Sargent (attributed), wife of Bishop Samuel Wilberforce, on Darwin's theory of evolution

I don't think we should be doing critical thinking. I think if you are conscious, if you're present with God at all times, then all things will be made clear to you. . . . You won't have to think it through or think about it, and then whatever you do, it always turn[s] out right. There's no second thoughts about it at all. And critical thinking sound[s] like people who are not conscious of God, they're not centered, and so they're trying to figure out things in their head. . . . [E]very thought you get is a lie. . . . And so if nothing in your mind is the truth, how can you think through things with . . . those thoughts?

That's why God said, Bring every thought into captivity. Every thought. Because every thought is a lie. . . . Because every thought is from the Deceiver.

—Rev. Jesse Lee Peterson, sermon, YouTube, May 21 2017

At the start of the twenty-first century, science is in many nations under threat from organized or quasi-organized religion. In many of those nations the situation is in reality less that science is under threat, more that it has yet to gain a secure foothold. Nations eager to take advantage of the products of science—i.e., technology—still reject large areas of science's essential underpinning, most publicly evolution but sometimes such even more important disciplines (at least on a day-to-day basis) as virology. One assumes those countries will catch up with reality in due course.

More alarming, though, is the corrosion of science by religion in those nations where it has been a central part of the culture for centuries. Nowhere is the attack more noisy than in the US, where the primary attacker is Christian fundamentalism, the belief that every word of both Old and New Testaments is literally true except the ones you disagree with. Other developed nations face the problem on a minor—but perhaps growing—scale. In the US the problem has become well nigh institutionalized, thanks in large part, perhaps, to the mistaken belief that the ideal of democracy can be applied to issues where it was never intended to be—and *cannot* be—relevant.

Consequently, the social status of science in many twenty-first-century US groups has reached a nadir incomprehensible in most other developed nations. In March 2006, Julia Reischel wrote, reporting in the *Broward–Palm Beach New Times* on research by Jesse Bering and David Bjorklund indicating that God is a psychological construct and a product of evolution:

> In the 19th century, scientific revelations about the age of the Earth and the development of animal species (and humans) led to the loss of faith of many intellectuals. But the twentieth century had a different legacy. While the technological sciences flourished, the end of the century saw science itself increasingly under attack by religious movements, business interests, and, in this country, at least, an antagonistic presidential administration. In a nation where most Americans don't accept evolution at all, science has been under an all-out onslaught.

What *is* exceptional is that Reischel made the statement not as a matter of controversy, but as one of simple fact: "[S]cience has been under an all-out onslaught."

The creationist broadcaster Ian Taylor, in an undated essay called "The Baconian Method of Science,"[38] exemplifies the tortured logic used by the anti-scientific in their dismissals of modern science. Anti-scientific broadcasters might be thought to face a particular dilemma for, if they beat the anti-scientific drum too hard, their audiences might start wondering just how the hell it is, if scientists are so stupid, TVs and radios work. The answer seems to be for the broadcaster to claim a fuller understanding of science than scientists have and to mount spurious demonstrations of how science itself can be used for its own demolition—as if all the world's secular scientists might collectively have been too obtuse to consider points that a lay thinker, guided by God, can come up with by the dozen.*

A key point of Francis Bacon's scientific method is that the scientist should approach evidence with as few preconceptions as possible; if we truly want to understand how things are, we should first of all strive to clear pre-existing hypotheses from our minds—otherwise all we're likely to find are "facts" that support our own biases. That appears to be good sense, and Taylor seemingly accepts it as such . . . but then runs smack into a wall: religious faith is surely one of the biggest of all disqualifying preconceptions. Time for a bit of squirming:

> The major difficulty with the inductive method is that it is an unachievable ideal since man cannot approach a problem with an unprejudiced mind.

This is a reasonable point, and good scientists are aware of it. They must factor into any conclusions they draw that their own conscious or unconscious biases may have played a part. But Taylor goes on:

> The insidious part of "clearing the mind of all preconceptions" is that the good will go with the bad and, if the Bible is the basis for one's worldview, that also will be forfeited. Even if it was possible to clear the mind, the immediate result would be that human reason would flood in like demons to a "house swept clean." The bottom line is that as human beings it is extremely difficult not to have a bias when approaching a problem, so that it becomes a question of which bias is the best bias to be biased with?

Taylor's argument seems to be that the only true way to advance one's knowledge is with a completely open mind, but that this may lead to Godless conclusions. Since Godless conclusions are by definition false ones, the mind should be simultaneously open and loaded with religious baggage—i.e., closed.

In another of his undated essays, "The Age of the Earth,"[39] Taylor tackles the matter of the earth's age. He concedes that Archbishop Ussher's figure of 4004 BCE may not be accurate, but is stalwart in defense of its being approximately correct. His motives are, obviously, to discount the long timescales necessary for modern lifeforms' evolution by natural selection.

---

* This is the same impulse that fuels many a crank theorist: orthodox scientists are too thick to have thought of the possibility that, say, the universe is made of vegetables—a hypothesis proposed by the German cult Vegetaria Universa in the 1960s.

In particular, he scoffs at science's rejection of the historicity of the Flood; in so doing, he must deny the sciences of stratigraphy and paleontology in particular, both of which he dismisses as unevidenced.

A few specifics:

(1) Taylor makes much of the erroneous estimate by the geologist Sir Charles Lyell of the rate of retreat (through erosion) of the Niagara Falls. Lyell, eager to show the earth was far older than Biblical estimates, put the rate of retreat at about 30cm (1ft) per year, despite contemporary estimates that the rate was more like 60cm (2ft) annually. Lyell's figure gave him the result that the falls, to have carved out an 11km (7-mile) gorge, must be about 35,000 years old, putting a 4004 BCE date for the creation of the earth out of the question. Taylor reports that a more recent measurement of the rate of retreat shows the falls retreating by about 1.8m (6ft) annually, giving them an age of about 6,000 years—neatly within Ussher's timescale. What this proves is hard to establish, unless Taylor believes the falls are necessarily the same age as the earth.*

(2) Taylor states quite correctly that comets lose mass each time their orbits bring them close to the sun, and calculates that a body like Halley's Comet loses enough mass at each encounter that it can be no more than a few thousand years old. Periodic comets like Halley's indeed have a limited lifespan; but this has no bearing on any arguments about the age of the universe, since periodic comets are merely those that are knocked by chance gravitational encounters within the Oort Cloud into orbits that bring them relatively close to the sun. To deal with this point, Taylor just flatly denies the existence of the Oort Cloud: "There is not a shred of evidence for it."

(3) The gravitational fields of planetary bodies attract a steady infall of space dust, as one might expect. For this reason, using then-current estimates of the amount of dust scooped up annually by the earth, it was anticipated by many that the early lunar landers would discover the surface of the moon to be covered in dust perhaps hundreds of meters deep. In the event, the dust on the moon was found to be mere centimeters deep, and various models had to be revised. In Taylor's view, however, the shallowness of the moon's dust layer is further proof that the moon—and hence the universe—can be only a few thousand years old.

---

* Current estimates reckon the average rate of retreat has been about 1m (3.3ft) annually, further disqualifying Taylor's argument . . . as if further disqualification were needed.

(4) The Big Bang theory is, according to Taylor, straightforward bunkum: "One obvious difficulty with the theory is the evident order in the universe, galactic walls, precise distances of earth, sun and moon etc. that cannot have arisen from an explosion! Then there is the great problem of there being insufficient mass in the universe and the need to appeal to 'dark matter' to account for the supposed accretion of the sub-atomic particles."

I am not at all sure what Taylor means by "galactic walls," and it's equally difficult to understand his point about the "precise distances of earth, sun and moon etc." Is he saying that the various distances within the solar system are precisely fixed? If so, he's wrong: all the orbits are subject to decay, and there are infinitesimal, but measurable, orbital variations for each body caused by the ever-changing gravitational influences of the other bodies.

In referring to dark matter Taylor does point to what was at his time of writing a genuine cosmological puzzle. It is less of a puzzle now, not so many years later. The fact that something is as yet not fully understood is no indication that orthodox science is in tatters. Science is *all about* taking things that are not yet fully understood and working to understand them.

(5) Kelvin, in one of his arguments against Darwinian evolution, said the earth could not be nearly as old as the theory demanded.* Basing his calculations on the known rate of heat escape from the earth's core, he stated the planet to be no more than 25 million years old. What Kelvin did not know about was radioactivity; it was beyond his (or, at the time, anyone else's) conception that radioactive decay within the core could be a major source of heat energy. Taylor believes such claims are nonsensical. Radioactive decay produces helium, he says, and this "would have diffused through the solid rock to fill the earth's atmosphere so that today our atmosphere should be mostly helium with traces of oxygen and nitrogen. Helium would not be lost to outer space." The last part of this porridge is perhaps the most puzzling, since Taylor offers no support for his ringing declaration. *Why* does he think helium wouldn't be lost into space? It's much *lighter* than the nitrogen and oxygen that comprise 99 percent of the atmosphere.

(6) The Dead Sea can be shown to be no more than a few thousand years old. Precisely how this affects calculations of the age of the earth is not explained.

---

* He also said that the sun was not old enough—see page 67.

(7) Arbitrarily assuming an average rate of 2.4 children per couple, it would take about 5,000 years to build up the current population of the world starting with a breeding pool the size of the Noah family. What Taylor ignores, *inter alia*, in his calculations is that, for vast swaths of human history, most people born did not reach breeding age.

(8) There are uncertainties in radiometric dating techniques because no one really knows if the decay rates of radioactive atoms are constant over time. It is more likely, says Taylor, that physicists have ascribed very long half-lives to certain isotopes purely in order that extrapolation will give the earth an antiquity measurable in billions of years. Taylor points out that, for example, nearby supernovae can affect the measurements, but ducks the question as to how nearby those supernovae might be . . . a pity because, if under his cosmology the most distant galaxy can be a mere 6,000 light years away (otherwise its light would take longer to reach us than the 6,000-year age of the universe), space must be so exceptionally crowded that many of those exploding stars would be grazing the top of our planet's atmosphere.

(9) Occasionally a lifeform believed to have been extinct for millions of years turns out to be still extant. He's perfectly correct in this: think of the coelacanth. Taylor claims that even the occasional dinosaur turns up, proving false the generally accepted dating of the extinction of the dinosaurs to some 66 million years ago. Really? Show me your dinosaur.

Overall, Taylor's essay—used here as a type example for countless other modern young-earth texts—is such a mixture of misconception and misinformation that it might seem hardly worth the effort of public dissection, any more than a child's seriously flawed school science essay might be. But this would be to forget that Taylor was not a child in urgent need of private coaching but, as a broadcaster, a figure of some influence.

AN INEVITABLE CONFLICT?

The Judaic tradition, which encompasses also Christianity and Islam, seems always to have had an ambivalent attitude toward the gaining of knowledge, it being ever assumed there are things beyond the bounds of human comprehension. At the same time, it seems to accept that some knowledge crucially useful for humankind must be gained *despite* the apparent censure of God. Cain murdered Abel and was denied the favor of the Lord . . . yet it was Cain who, according to the Tanakh/Bible, thereafter founded the first city and so gave humankind the valuable gift of civilization, a gift without which the Tanakh/Bible itself could never have been written.

The confusion deepens when it comes to the response of God to the building of the Tower of Babylon—or Babel. Nowhere can one find a divine prohibition issued against the building of mighty towers, yet God retaliated by inflicting upon humankind the curse of countless mutually incomprehensible languages. The "sin" of the Babylonians was one of presumption: building an edifice which in its mightiness rivaled the works of the Lord. That God is intolerant of human advancement is echoed by many fundamentalists today, although they seem uniformly disinclined to give up their cars, TV sets, microwave ovens and assault rifles. The art is to cherry pick the bits of science and technology of which God approves or disapproves. The morning-after contraceptive pill is accursed of God, but civilian-targeting weapons like cluster bombs, landmines, nukes and white phosphorus are okay.

A further point here is that, while the Tower of Babel might have been seen in its own age as so impressive as to rival God's own works, today it would be regarded, at least in scale, as trivial. What today might seem like advancements of science and technology into the realm properly reserved for God will likewise, eventually, be seen as humdrum—as basic stuff. It's just that sometimes "eventually" can mean a depressingly long time. It's a century and a half since Darwin established the basic mechanism of evolution, and yet many citizens of supposedly developed nations still resist this essential truth on strictly irrational grounds.

In so doing they are, of course, perpetuating a longstanding Christian tradition. St. Paul and St. Augustine regarded curiosity as the instigator of the original sin (disobedience), the one responsible for Adam and Eve being expelled from the Garden, and so they warned against curiosity. The Devil was everywhere, and human curiosity would sooner or later lead the possessor of an inquiring mind into his embrace. Francis Bacon's greatest philosophical breakthrough was his expanding hugely the accepted territory that human beings could legitimately investigate without trespassing into the domain of the Lord. While Bacon still reserved strictly theological matters as the rightful province of God alone, he proclaimed that, not only was it humanity's right to probe the workings of the natural universe, it was divinely approved to do so. God had not put us here to remain ignorant (but worshipful) brutes: the Lord wanted us to explore His wonders.

There is another fundamental difficulty that religion—all religion—has with science. It's hard to think of any aspect of science that does not have *time* as an important component; the time may be very short, as in a quantum event, or it may be exceedingly long, far longer than we relatively short-lived creatures can easily comprehend. When we consider the creation of the universe, the development of the oceans, the evolution of life, or even the formation of the Grand Canyon, then even the longest "human" unit of time measurement, the generation, becomes uselessly small. Within science's relationship with time lies the concept of an exceptionally long past and, equally, an exceptionally long future. Judaism and Christianity have at their heart a problem with this in that they are essentially deniers of time: the future will last only until the coming (or second coming) of the Messiah, and thereafter all the rules will change such that the idea of time's passage no longer has any real meaning. The Muslim's dilemma is similar, in that everything since the coming of the Prophet—including the future—is a bit irrelevant.* Other religions seek transcendental timelessness, which is again an obviator of one of science's most important foundations. And many religions incorporate a similar short-term approach when it comes to the past: if a deity or deities created the universe and our planet solely in order to be a home for us, then this surely cannot have happened very long ago; it cannot have been so long ago as not to be realistically measured in human generations.

Perhaps the most damaging consequence of religious short-termism about the future is the attitude of many of the devout that global warming, whatever its causes, simply does not matter: there is not enough time left for its full impact to make itself felt. Even were that not the case, we can trust in God to look after us (or at least the members of whichever sect we belong to), to avert catastrophe. Similar arguments have been advanced by Christian fundamentalists in various other attacks on environmentalists: God told us to harvest the fruits of the earth, so we should go ahead and destroy the rainforests secure in the knowledge that, since we're following His plan, no harm can ensue. Perhaps these people derive a comfort from their short-term beliefs that's denied to the rest of us: they never have to undergo the thought experiment of being confronted by their great-grandchildren, who will have to live with the results of their criminal irresponsibility, because in their worldview the future will be over and done with by then.

---

* Of course, the vast majority of Jews, Christians and Muslims are intelligent enough to have worked their way round this theological dilemma.

# # #

The clash between creationism and science is often portrayed as one between, instead, religion and science. But is this really the case? Has it been a universal perception? No. Take for example the school of thought called Natural Theology. In high vogue during the seventeenth and eighteenth centuries, Natural Theology was the study of the attributes of God as revealed through the study of nature. It initially arose far earlier, in the writings of medieval scholars such as Thomas Aquinas, who essentially spliced Aristotelian and Platonic ideas onto prevailing Christianity to advocate a marriage of reason with faith. To these medieval theologians, the very existence of creation necessitated there being a creator, so the study of nature was really the study of the consequences of the initial designs of that creator. Similarly, faith-based "knowledge" of the attributes of the creator would assist in the deciphering of what was discovered in nature. From this latter interaction arose such concepts as the "chain of being,"* which again dated originally back to Plato and Aristotle; this notion was important to the medievals, and flowered in the seventeenth and eighteenth centuries as an explanation of the world's profusion of lifeforms. Books such as *Wisdom of God in the Creation* (1691) by John Ray and especially *Natural Theology* (1802) by William Paley were influential in promoting the concept of Natural Theology; alarmingly, they're still cited approvingly by some creationists and Intelligent Design proponents today.

But the myth is deeply entrenched that science and religion have been mutually antagonistic throughout history. The myth underpinned Andrew Dickson White's pro-science diatribe *History of the Warfare of Science with Theology in Christendom* (1896). In many ways, White was correct: science and religion *should* be at loggerheads, because inevitably improvements in human knowledge reveal that more and more of the basic doctrines of religion are untenable. In fact, however, institutionalized Christianity and scientific progress have for the most part co-existed and even cooperated quite happily. The numerous clashes between the two incompatible forces that White described in his book were almost without exception illusory,

---

* Life exists at every level of complexity from simplest "animalcules" to the most complex creature of all, the human; beyond humankind lie angels and God; below the "animalcules" lies inanimate matter, such as fossils. Each level could be viewed as an essential link of a continuous chain.

many of the tales being products of the two main Christian sects, Catholicism and Protestantism, spinning the facts to make each other look bad.

For example, by the time Columbus "sailed the ocean blue" in 1492 it was widely accepted in Europe, including by the Catholic Church, that the earth is spherical. Aside from anything else, there are extant European globes that predate Columbus's voyage. The myth of universal medieval belief in the flat earth seems to have begun with the imaginative biography *The Life and Voyages of Christopher Columbus* (1828) by Washington Irving. Thereafter it was picked up by countless other authors, including White in his *History*.

And it is less than the whole truth to say that Galileo Galilei was persecuted by the Catholic Church for insisting that the earth was not the fixed center of the universe. The Church was not especially antagonistic to this notion; at the same time, it was not immediately prepared to publicly embrace the new Copernican cosmology. In short, it was hedging its bets until the picture became a bit clearer. If the supposedly infallible pope, having received the information directly from God, proclaimed the earth went around the sun and then later it proved not to be the case, the Church would be embarrassed. But Galileo, who was a pugnacious type, kept harassing the Church to make a decision. In the end, reluctantly, the Church brought a legal case against him in an effort to shut him up.

Similarly, one of the classic tales within the history of science is that of Giordano Bruno, burnt at the stake in 1600 for his support of the new Copernican cosmology. Again the tale is a tad misleading. Bruno was condemned for his heretical mysticism. His cosmology, if such it can sensibly be called, reads like the wildest flights of fantasy, and only *in passim* mentions that he thought, for reasons immured in his mysticism, that the earth went around the sun rather than vice versa; indeed, he seems to have despised Copernicus as a mere mathematician. While of course it's unforgivable that the Church burned someone to death for disagreement with orthodoxy, the claim of Bruno as a martyr in the name of science is overstated.

Our perceptions of both the Galileo and Bruno incidents have been, then, molded more by later propagandists than by the actual historical record. Only a few decades after Galileo's confrontation came an example of the Church in more typical mode. The seventeenth-century Danish geologist Niels Stensen (Nicolaus Steno) announced that studying the rocks and fossils of the earth indicated a far longer history for our planet than Archbishop James Ussher's estimate of a creation date of 4004 BCE could encompass. Far from there being riots in the cloisters, Steno was promoted up the ecclesiastical ranks, eventually becoming a bishop (and finally, in 1988, being beatified by Pope John Paul II). The vast majority of clerics and the faithful had paid little attention to Ussher's calculation, because for some centuries it had been assumed Genesis was allegorical rather than a literal history.* Even when Biblical fundamentalism began to rear its head in the late nineteenth and early twentieth centuries, there was at first no conflict: one of the early fundamentalist pioneers, William Bell Riley, stated outright that no "intelligent fundamentalist . . . claims that the earth was made six thousand years ago, and the Bible never taught any such thing."

After the initial shock of Darwinism—as clerics like Bishop Samuel Wilberforce made fools of themselves with their antiscientific protests—the various Christian churches settled down to a peaceful coexistence with science's new theory. And why not? The churches had already accepted that a Genesis "day" was not the 24-hour period we're accustomed to—why should God obey the rules of Man?—and the course evolutionary theory painted of the history of life on earth was much the same as that outlined in Genesis. By the end of the nineteenth century, the few remaining young-earth creationists might shout a lot but they were regarded as very much on the fringe of mainstream Christian thought.

Andrew White's *History*, despite its multiplicity of false conflicts, laid down a gauntlet on behalf of science, but at first there were not many theologians prepared to pick that gauntlet up . . . and *there still aren't*. While, especially in the US and increasingly in Africa, there are plenty of populist preachers and self-styled Christian demagogues who argue for a young earth and the damnation of Darwinian evolution, the vast majority of theologians *per se* steer clear of such controversy, seeing in evolution no threat to their faith, and in some instances perhaps even a strengthening of it.

The French scientist Pierre Simon de Laplace presented the Emperor Napoleon with a copy of the latest volume of his monumental five-volume *Mécanique Céleste*, a treatise on celestial mechanics. After studying it, Napoleon asked Laplace why there was no mention of God in his treatise. "Sir," Laplace is famously reported to have responded, "I have no need of that hypothesis."

The truth is that science does not in fact preclude or discount the exis-

---

* Ironically, Steno faced stronger criticism from scientists than he did from the clergy, since his ideas flew in the face of the established theory of fossils.

tence of the irrational or supernatural: it simply ignores it because it has no need of it. As science progresses, more and more that was once believed to require a supernatural explanation becomes explicable in perfectly rational terms. Science is, if you like, progressively explaining the supernatural and turning it into the natural. There are plenty of things science does not yet understand, yet, unlike irrationalism, science does not immediately *identify a specific cause* for the unknown, a cause that is itself unknown; that would be no explanation at all.

But some theologians refused to compromise. These scriptural fundamentalists tended to belong not so much to the nineteenth century as to the newly dawning twentieth. Perhaps that had something to do with the aftermath of the turning of centuries, times that are especially popular for predictions of the end of the world. Afterwards, there's inevitably bitter anticlimax among those who'd confidently anticipated the destruction of all, and a tendency to redouble their efforts to counter the appalling sin of rationalism. And now, of course, we're not so far past the turning of another century.

> If one criticism could be leveled against the book, it is that the author does not emphasize with sufficient force that the Scriptures are so completely our rule of faith, also in the matter of creation, that the doctrine of creation does not (and ultimately cannot) depend upon scientific evidence or the ability to answer scientifically all the notions of scientists, it rests on faith alone. The battle between us who believe in the truth of God's Word in Genesis 1 and 2 and those who have adopted some form of evolutionism, particularly theistic evolutionism, is a spiritual battle between faith and unbelief, and must never be construed as a battle over the scientific evidence supporting the one position or the other.

Astonishingly, that passage comes not from the opening years of the twentieth century, but from the dawn of the twenty-first. It is in fact from a book review by Herman C. Hanko of *Green Eye of the Storm* (1998) by the creationist John Rendle-Short; the review was published in the *Standard Bearer* in February 2001.

The war by religionists against evolution is, in truth, largely a modern phenomenon, and can be regarded as more a political than a religious war—as per the Wedge (see page 145), the strategic document created by the Discovery Institute that reveals the intent of using the Intelligent Design hypothesis as a means of not just subverting the public's comprehension of science but of transforming society as a whole. Nevertheless, while the creationists' venom and dirty tricks came to a head in the last quarter of the twentieth century and after, the modern creationist movement in the US has long antecedents.

THE CRUSADE TO MISEDUCATE THE YOUNG

Where US creationists triumphed in the early part of the twentieth century was in having the teaching of Darwinian evolution banned in public schools—at least in some states. (In others, many teachers and school boards were intimidated into omitting evolution from the curriculum, a practice that persisted right up until the 1960s, and has recently returned to haunt us.) There have been several battles on this issue, by far the best known being the celebrated Scopes Monkey Trial of 1925. In fact, this was very much a staged contest.

In March 1925, the State of Tennessee had passed the Butler Bill, outlawing the teaching of evolution in Tennessee schools. The American Civil Liberties Union (ACLU), recently formed, saw a chance to live up to its charter. They took out advertisements in Tennessee seeking anyone who'd offer themselves up as a defendant; the town council of Dayton, Tennessee, recognizing the opportunity for publicity, persuaded science teacher John Scopes to be the sacrificial goat.

The ACLU hired Clarence Darrow as Scopes's defense lawyer; he was most famous for having successfully pleaded against the death penalty in the Leopold–Loeb murder case in 1924. The prosecutor was William Jennings Bryan, a colorful populist politician who had three times run as the Democratic candidate for the US Presidency—in 1896, 1900 and 1908—being three times roundly defeated. He had tried to push through a bill in Kentucky much like Tennessee's Butler Bill, but had failed.

There's no doubt Darrow was the brighter bulb of the two. There's also no doubt that the judge, John Tate Raulston, a devout Baptist, was biased against the defense, for one of his first rulings was that Darrow be not permitted to bring to the stand, as he'd planned, a bevy of distinguished biologists to explain the principles of evolution and how harmless they would be if taught to Tennessee's youth. (Bryan had parallel difficulties: he'd asked a bunch of scientists whom he knew to have creationist sympathies to appear as expert witnesses on his behalf, and they'd all declined.)

Darrow's retaliation to the court's ban was to call Bryan himself to the stand and interrogate him on the principles of creationism in a—successful—attempt to show them up as nonsense. Bryan made a pathetic exhibition as he stumbled and evaded his way through the examination. When asked to list geologists with whose earth-history hypotheses he agreed, Bryan could name just two: the maverick amateur geologist George Mc-Cready Price, who unfortunately could not be there because he was lecturing in the UK, and the recently deceased George Frederick Wright, who had written extensively on creationist geology, but whose qualifications were in divinity, not geology.

Bryan's performance would, in short, have convinced few outside Judge Raulston's court. Although Scopes was convicted (as part of the whole staged-trial scenario, he did not have to serve his token sentence, and Bryan paid his legal costs), the writing was on the wall for the legal enforcement of creationist teaching—at least for a few decades.

# # #

After the launch of the first artificial satellite in 1957, the Soviet *Sputnik 1*, the federal government became belatedly more insistent on, among other things, the proper teaching of science, including evolution, in schools: matters had reached such a pass that most school biology textbooks, for fear of canceled orders worth millions from various populous states where evolution was disliked, either omitted discussion of the subject or blurred it.

As part of its effort in the late 1950s to bring the teaching of biology out of the Middle Ages, a National Science Foundation-funded group called the Biological Sciences Curriculum Study (BSCS) produced a set of textbooks that aimed to bring US school biology up to the level of that of the rest of the developed world. This naturally alarmed creationists all over the US. Under the direction of Walter E. Lammerts, the Creation Research Society (CRS), one of several influential US creationist groups, produced a competing school textbook, *Biology: A Search for Order in Complexity* (1970), edited by John N. Moore and Harold Schultz Slusher. After some delay—the text was rejected by all the major US publishing houses—the book was issued by the Christian publisher Zondervan. Although there was considerable difficulty in having it accepted in schools (it was overtly religious, therefore its use in a science class would have been in violation of the Constitution), the book sold very healthily, and was undoubtedly influential.

The book's preface was written by Henry Morris, the CRS's president. By the time the third edition appeared, in 1974, Morris seems to have come to believe that the way forward for creationism was to wage a propaganda war, the opening gambit in which should be to face head-on the charge that creationism was not scientific but a purely theological fantasy. Changing the fundamental tenets of the belief was out of the question, but a simple semantic trick might do the job. Morris therefore titled a new textbook he edited, aimed at high school teachers, *Scientific Creationism* (1974); the degree to which the change in focus was purely cosmetic can be recognized in the fact that this book was issued in two versions, one from which all theological references had been carefully stripped out and another, tailored for the more devout, in which they had been preserved. We can view this book as marking the birth of Creation Science, which was not so much a variant of creationism as a mere retitling of the same old stuff.

Creationism, or Creation Science, as a public concern in the US might almost have died had it not been for Ronald Reagan. While he was still Governor of California, he oversaw a push to encourage the teaching of creationist ideas in the public schools, and when he began campaigning for the presidency leading up to the 1980 election he maintained the theme. It seems certain this was a reflection of his own views rather than mere political opportunism. Once elected he appointed a presidential science advisor, George Keyworth, who was a minor physicist unqualified in the biological sciences, but with creationist leanings, and an education secretary, William Bennett, who was at least amenable to creationist notions. Whatever the original motivations, the Republicans discovered that pandering to the scientific ignorance of the Christian Right minority was electoral gold. So began that party's war on science, a war that has continued to this day; there is much more about it in Chapter 9.

Eugenie C. Scott of the National Center for Science Education reported in 1996 on the reactions of a number of teachers. A science teacher in Knoxville, Tennessee, gave this infinitely depressing response: "I'd probably skip the theory of evolution as part of the origin of mankind or the earth . . . We live in the Bible Belt, and it's offensive to some students to hear the theory that man came from monkeys"—this despite the fact that a science education which omits "the theory of evolution as part of the origin of mankind" is not a science education at all.

In 1996 a parent in Lincoln County, West Virginia, complained that the curriculum for a high school genetics course included the theory of evolution: "Even if it is scientific theory, I don't think it should be taught." Astonishingly, the response of the school board was to scrap the course entirely, and to forbid the teacher concerned from any further teaching of evolution. The explanation for this profoundly anti-educational decision by a body supposedly designed to promote education was offered by one of the teachers involved: "They know a vote for evolution is a vote out of office." It's somewhat unsurprising that in December 2004 Lincoln County's school superintendent Tom Rinearson was ruefully reporting:

> I'm disappointed in where we are academically overall. In almost every category, our students perform at a lower level than the average across the state. To me that's unacceptable. We need to look below the surface to find out why. Small changes could make a big difference.

They could indeed.

Intelligent Design

A more recent variant of creationism is Intelligent Design (ID), called by its proponents a theory although in fact, technically, it's barely a hypothesis, lacking any empirical underpinning. To simplify, ID accepts evolution has occurred, but brings into the equation the notion of "irreducible complexity," which holds that the components of certain organs, bodily structures, etc., would be useless except in combination with other components, and therefore could not have evolved in isolation—at least not by natural selection. For instance, there is no use for the lens of the eye, and certainly no survival advantage in having one, unless you have the rest of the eye to go with it. Therefore, such complex structures must owe their existence to the periodic intervention of an Intelligent Designer, whom ID proponents are careful for political reasons not to identify as God.

The fallacy of the basic idea lies in its deliberate avoidance of the fact that natural selection (a) produces massive redundancies, with all sorts of useless mutations appearing and then disappearing, and (b) encourages opportunism, whereby structures that originally served one purpose, or even no purpose at all, can be co-opted into serving another. In the instance of the eye, the fossil record—and indeed lifeforms extant today—show all sorts of proto-eyes, which can be regarded as partway stages along the route to the modern mammalian eye. The eye that stares dolefully up at you from your grilled trout is an example of a more primitive form of eye than your own.

ID represents a marginally more sophisticated form of pseudoscience than Creation Science and, because its proponents are sufficiently scientifically literate that they can utter superficially convincing terminology at appropriate moments, it has persuaded many that it is indeed a valid scientific hypothesis.

The prime mover behind the ID campaign is the avowedly rightwing Discovery Institute, founded in 1996. In connection with the claim by ID proponents that their hypothesis has everything to do with science and nothing to do with religion, it's worth noting that the original logo/banner used by what was initially called the Discovery Institute's Center for the Renewal of Science & Culture showed the classic Michelangelo image of a bearded God reaching out to touch fingertips with Adam. As the National Center for Science Education waspishly pointed out, "The image was entirely appropriate, since the Discovery Institute's president, Bruce Chapman, explained that the Center seeks 'to replace materialistic explanations with the theistic understanding that nature and human beings are created by God.'"

What is most sinister about the whole ID movement is that its intent goes far beyond mere science; it's a political/theocratic crusade. A strategy document known as the Wedge, compiled by creationists at the Center for the Renewal of Science & Culture in the late 1990s, spells this out. Its "Five Years Strategic Plan Summary" reads in part:

> The social consequences of materialism have been devastating. As symptoms, those consequences are certainly worth treating. However, we are convinced that in order to defeat materialism, we must cut it off at its source. That source is scientific materialism. . . . If we view the predominant materialistic science as a giant tree, our strategy is intended to function as a "wedge" that, while relatively small, can split the trunk when applied at its weakest points. The very beginning of this strategy, the "thin edge of the wedge," was Phillip Johnson's critique of Darwinism begun in 1991 in *Darwinism on Trial*, and continued in *Reason in the Balance* and *Defeating Darwinism by Opening Minds*. Michael Behe's highly successful *Darwin's Black Box* followed Johnson's work. We are building on this momentum, broadening the wedge with a positive scientific alternative to materialistic scientific theories, which has come to be called the theory of intelligent design (ID). Design theory promises to reverse the stifling dominance of the materialist worldview, and to replace it with a science consonant with Christian and theistic convictions.
>
> The Wedge strategy can be divided into three distinct but interdependent phases, which are roughly but not strictly chronological. We believe that, with adequate support, we can accomplish many of the objectives of Phases I and II in the next five years (1999–2003), and begin Phase III . . .

> Phase I: Research, Writing and Publication
> Phase II: Publicity and Opinion-making
> Phase III: Cultural Confrontation and Renewal

That scientists and rationalists should object to this plan for the demolition of truth is not surprising; the fact is that many Christian theologians are equally horrified by it, not just from a scientific but from a theological viewpoint. In the *Sydney Morning Herald* for November 15, 2005, Neil Ormerod, Professor of Theology at Australian Catholic University, summed this up in an article titled "How Design Supporters Insult God's Intelligence." He describes ID as "an unnecessary hypothesis which should be consigned to the dustbin of scientific and theological history":

> Much depends on what its proponents mean by the term "intelligent design." If they mean that the universe as a whole displays a profound intelligibility through which one might argue philosophically that the existence of God is manifest, their position is very traditional.
>
> However, if by intelligent design they mean that God is an explanation for the normal course of events which would otherwise lack scientific explanation, then this is opposed to a traditional Christian understanding of divine transcendence. In seeking to save a place for God within the creation process, the promoters of intelligent design reduce God to the level of what the early theologian Thomas Aquinas would call a "secondary cause."
>
> This is just a more sophisticated version of so-called "creation science," which is poor theology and poor science.

Further bad news for those who argue that ID is a God-blessed hypothesis is that the Vatican doesn't agree with them either—and its officials have frequently said so very publicly. Typical was the article by the Vatican's Chief Astronomer, George Coyne, in June 2005 in the Catholic magazine *The Tablet*:

> If they respect the results of modern science, and indeed the best of modern biblical research, religious believers must move away from the notion of a dictator God or a designer God, a Newtonian God who made the universe as a watch that ticks along regularly.

At a conference in Florence later that same year he stressed the point more specifically and more robustly, commenting that ID "isn't science, even though it pretends to be."

Cardinal Christoph Schönborn has been perhaps even more dogged in his criticisms of Creation Science and ID. He also has some criticisms of what he calls "evolutionism," which he defines as a quasi-religious belief system that invokes evolution as a sort of universal cure-all and regards any criticisms of evolutionary theory as "an offense to Darwin's dignity": "The theory of evolution is a scientific theory. What I call evolutionism is an ideological view that says evolution can explain everything in the whole development of the cosmos, from the Big Bang to Beethoven's Ninth Symphony." It's slightly unclear who these "evolutionist" true believers actually *are*, although some scientists might well give that impression as they respond heatedly to the millionth scientifically gormless attack on the theory of evolution, but one takes Schönborn's point.

Around the turn of the twenty-first century, the Templeton Foundation, whose aim is to seek reconciliation between science and religion, financially sponsored some conferences and courses to debate ID, then asked the proponents of ID to submit grant proposals for research projects the foundation might sponsor. But, reported the foundation's senior vice-president, Charles L. Harper, Jr., "They never came in." If even the proponents of ID themselves can't come up with any possible areas of research that might help test their hypothesis, this surely gives the lie to their claims that it is scientific. Or is it simply that they're *uninterested* in experiment and research, accepting the principle of ID as an article of faith without need for such encumbrances as proof? Rather like, say, a religious belief . . .

One criticism often made by ID proponents (and orthodox creationists) is that, if evolution by natural selection is a reality, why don't we see any evidence of evolutionary changes going on around us? In fact, there's plenty of such evidence—Darwin's original Galápagos observations showed the effects of evolution over a relatively short period, which was one of the reasons he proposed his theory in the first place—but every time it's offered to the creationists they raise the bar: that piece of evidence is somehow *not enough*.

As an example of evolution visibly in action, the "standard" wing color of various moth and butterfly populations can be observed to change over a very few years in response to environmental factors such as soot-laden urban air, the wings darkening in response to the darkening of the places an insect might perch. Such changes can—with difficulty—be explained away within the context of creationism, but more difficult to discount are researches published from late 2005 onward in which the effects of continuing human natural selection can be observed directly at the genetic level. A team led by Jonathan Pritchard at the University of Chicago reported in *PLOS–Biology* in March 2006 on their experiments surveying DNA from four different populations: Yoruba (Africa), Japanese and Han Chinese (Asia), and a Utah community (North American, originally of European stock). In the various populations scanned, most genes are universal, but a few are still at the stage where they're possessed by some individuals in a population but not by others; the ratio between these two groups gives a measure of how long it has been since the gene arrived on the scene and began to be selected for. Various of the genes chosen for examination concern digestion, and a correlation can be found between their spread through these populations and the shift from hunted/gathered foods to domesticated ones—in other words, to the invention and spread of agriculture among the relevant populations. Similarly, the genes that code for the pale skin color of Europeans seem to be only about 6,600 years old, which means Europeans retained the dark skin of their African ancestors for some 40,000 years after migrating to the new continent. Again, we're directly observing the fruits of evolution.

A final, more readily observable, everyday example of evolution concerns antibiotics. Bacteria, which reproduce very quickly, thereby evolve at a prodigious rate. By the time we're halfway through a course of antibiotics, natural selection has already acted to ensure that the remaining bacteria, though perhaps relatively few in number, are considerably hardier than the original stock. That is, incontestably, Darwinian evolution in action. It is for similar reasons that, in the longer term, use of antibiotics can lead to the emergence of "superbugs," which have evolved to become resistant entirely to a particular antibiotic or, most dangerously, to a whole range of them.

We'll talk about superbugs in more detail in Chapter 6.

# # #

The events leading up to what was hailed as "the new Scopes Trial"— more correctly, *Kitzmiller et al. versus Dover Area School District*—began on October 18, 2004, when the members of the Dover [Pennsylvania] School Board voted by 6 to 3 to add the following statement to the area's biology curriculum:

Students will be made aware of the gaps/problems in Darwin's theory and of other theories of evolution including, but not limited to, intelligent design. Note: Origins of life is not taught.

A month later the Board prescribed that biology teachers must read, at the start of any lesson covering evolution, the following statement:

The Pennsylvania Academic Standards require students to learn about Darwin's theory of evolution and eventually to take a standardized test of which evolution is a part.

Because Darwin's theory is a theory, it continues to be tested as new evidence is discovered. The theory is not a fact. Gaps in the theory exist for which there is no evidence. A theory is defined as a well-tested explanation that unifies a broad range of observations.

Intelligent Design is an explanation of the origin of life that differs from Darwin's view. The reference book, *Of Pandas and People*, is available in the library along with other resources for students who might be interested in gaining an understanding of what Intelligent Design actually involves.

With respect to any theory, students are encouraged to keep an open mind. The school leaves the discussion of the origins of life to individual students and their families. As a standards-driven district, class instruction focuses upon preparing students to achieve proficiency on standards-based assessments.

The *eminence grise* behind the move was a rightwing organization called the Thomas More Law Center, which bills itself as the "Christian answer to the ACLU"—with whom, in various areas, from gay marriage to public displays of the Ten Commandments, it has since its foundation in 1999 been picking fights. In order to precipitate a showdown on evolution, the center had been combing the US since at least early 2000 for a school board that would be dumb enough to dictate the teaching of ID in science classrooms, and Dover's board swallowed the bait.

On December 14, 2004, the ACLU filed suit on behalf of 11 parents, and soon it was joined by Americans United for Separation of Church and State and by the National Center for Science Education. The latter put out a plea for a legal firm to take on the case *pro bono*, and within hours Eric Rothschild had volunteered the company in which he is a partner, Pepper Hamilton: "I've been waiting for this for fifteen years."

In earlier trials involving attempts to teach Creation Science alongside science, the ACLU's technique had been to demonstrate that Creation Science was not science at all. Obviously ID was vulnerable in the same way. Also, the ACLU team was convinced that the book *Of Pandas and People* (1989) which the Dover School Board had recommended was merely a creationist text adapted at some early stage to bring it into line with ID; playing a hunch, they subpoenaed the Foundation for Thought and Ethics (FTE), publishers of the book, for any early versions of the text that might exist. This long shot paid off. Most unusually for the publishing world, FTE had kept several earlier versions of the manuscript, whose successive titles alone were revealing: *Creation Biology* (1983), *Biology and Creation* (1986), *Biology and Origins* (1987), and finally *Of Pandas and People* (1987); the final version was the one published in 1989 (revised 1993). Not only were early versions straightforwardly creationist, much of their content had been transplanted verbatim into the final version with the term "Intelligent Design" being directly substituted for "Creation." These texts were in themselves powerful evidence that, far from being a brand-new scientific hypothesis, ID was merely creationism under a different name.*

Meanwhile, the legal team for the school board was having difficulties, primarily because of squabbling between the Discovery Institute and the Thomas More Law Center. In consequence, three of ID's heavyweights, William Dembski, John Angus Campbell and Stephen Meyer, declined to take any part in the proceedings.** The main expert witness left to them was Michael Behe, but in the event his arguments wilted when placed under the spotlight. For example:

Q: Please describe the mechanism that Intelligent Design proposes for how complex biological structures arose.

A: Well, the word "mechanism" can be used in many ways. . . . When I was referring to Intelligent Design, I meant that we can perceive that, in the process by which a complex biological structure arose, we can infer that intelligence was involved.

Q: What is the mechanism that Intelligent Design proposes?

A: And I wonder, could . . .? Am I permitted to know what I replied to your question the first time?

Q: I don't think I got a reply, so I'm asking you. You've made this claim here: "Intelligent Design theory focuses exclusively on the proposed mechanism of how complex biological structures arose." And I want to know, what is the mechanism that Intelligent Design proposes for how complex biological structures arose?

A: Again, it does not propose a mechanism in the sense of a step-by-step description of how those structures arose. But it can infer that in the mechanism, in the process by which these structures arose, an intelligent cause was involved.

---

* The book was written anonymously by Dean Kenyon and Percival Davis. Author Kenyon insists that the book's thrust and motivation are purely toward better science; his coauthor, Davis, is a little franker: "Of course my motives were religious. There's no question about it."

** This was an uncanny echo of the difficulty Williams Jennings Bryan had had, during the Scopes Trial, in persuading creationist geologists to appear on his behalf.

Behe also got into deep water over his claim that his *Darwin's Black Box* (1996) had been even more thoroughly peer reviewed than the average science-journal paper because its subject matter was so controversial. One of the reviewers he mentioned was Dr. Michael Atchison of the University of Pennsylvania. The ACLU's team produced in court an article written by Atchison in which he said his sole involvement had been a phone call with the book's editor on general matters of content.

Another claim of Behe's was that there had been no serious work done on the evolution of complex biochemical structures like the bacterial flagellum (the example most often used by ID proponents), of the immune system, or of the complicated sequence involved in the clotting of blood. When a stack of peer-reviewed books and papers on exactly these subjects was put in front of him, Behe had to admit that, yes, well, they existed, but he hadn't read them.

His cause was not helped by the fact that his colleagues at the Department of Biological Sciences, Lehigh University, while having no argument with his moral or academic right to embrace the ID hypothesis, at the same time were less than keen about the publicity it was attracting to their department, and especially about the way in which his testimony in the Dover trial, bound to be reported internationally in the media and to feature in books (like this one!) for years and probably decades, was likely to forge an indissoluble link between his name and that of the university. Accordingly his departmental colleagues took the unusual step of posting a disclaimer on the department's website:

> The faculty in the Department of Biological Sciences is committed to the highest standards of scientific integrity and academic function. This commitment carries with it an unwavering support for academic freedom and the free exchange of ideas. It also demands the utmost respect for the scientific method, integrity in the conduct of research, and recognition that the validity of any scientific model comes only as a result of rational hypothesis testing, sound experimentation, and findings that can be replicated by others.
>
> The department faculty, then, are unequivocal in their support of evolutionary theory, which has its roots in the seminal work of Charles Darwin, and has been supported by findings accumulated over 140 years. The sole dissenter from this position, Prof. Michael Behe, is a well-known proponent of "intelligent design." While we respect Prof. Behe's right to express his views, they are his alone and are in no way endorsed by the department. It is our collective position that intelligent design has no basis in science, has not been tested experimentally, and should not be regarded as scientific.

The difficulties of their star witness were not the defense's only worry. One of the linchpins of their argument was that this wasn't a dispute between science and religion but between two different schools of science, and to this end former Dover School Board president Alan Bonsell and the chairman of the Board's curriculum committee, William Buckingham, staunchly denied having any religious motive. Unfortunately, there were first-person accounts of their talking about how they wanted to see evolution "balanced" with creationism in the curriculum, and even video of Buckingham saying precisely this on a local TV station. The local newspapers had carried reports of a meeting back in June 2004 when Buckingham had said of the evolution/creationism "balance": "Two thousand years ago, someone died on a cross. Can't someone take a stand for him?"

The final decision, when it came on December 20, 2005, represented a triumph for rationality. US District Judge John E. Jones III issued a 139-page judgment that effectively demolished ID as a purported science. He was blunt about the dishonesty of the pretense whereby a religiously based scheme was being dressed up as something scientific:

> We find that the secular purposes claimed by the Board amount to a pretext for the Board's real purpose, which was to promote religion in the public school classroom. . . . It is ironic that several of these individuals, who so staunchly and proudly touted their religious convictions in public, would time and again lie to cover their tracks and disguise the real purpose behind the ID Policy.

He added some harsh words for the tactics of the ID movement in general, words which naturally fell on deaf ears, and noted:

> Those who disagree with our holding will likely mark it as the product of an activist judge. If so, they will have erred as this is manifestly not an activist court. Rather, this case came to us as the result of the activism of an ill-informed faction on a School Board, aided by a national public interest law firm eager to find a constitutional test case on intelligent design, who in combination drove the Board to adopt an imprudent and ultimately unconstitutional policy. The breathtaking inanity of the Board's decision is evident when considered against the factual backdrop, which has now been fully revealed through this trial. The students, parents, and teachers of the Dover Area School District deserved better than to be dragged into this legal maelstrom, with its resulting utter waste of monetary and personal resources.

Truth in science is not something that can or should be decided in a court of law; reality does not obey court orders any more than it does democratic votes. At the same time, it cannot be denied that the Dover trial did much more than settle a particular legal case: in open court and in Jones's judgment, the doctrine of Intelligent Design was effectively shredded, and very publicly so.

CREATIONIST BILLS

Although since the Dover case the Discovery Institute has somewhat shrunk in prominence and, it seems, funding and staffing levels, its influence hasn't entirely evaporated—in fact, in terms of the corruption of US school science education, it has in some states increased. This has come about through the advocacy of forms of the institute's so-called "Academic Freedom" draft bill in various state legislatures.

The "Academic Freedom" draft can be traced back to 2001, when the institute's guiding spirit, Phillip Johnson, crafted an amendment for then Representative Rick Santorum (R–PA) to add to the education bill that would become the No Child Left Behind Act, signed into law in January 2002. Fortunately the amendment didn't make it into the final bill, but we can get its gist from this extract:

> Where topics are taught that may generate controversy (such as biological evolution), the curriculum should help students to understand the full range of scientific views that exist, why such topics may generate controversy, and how scientific discoveries can profoundly affect society.

The sneaky tactic deployed here has become known as "Teach the Controversy," and it's a prime example of the Discovery Institute's charade of scientific objectivity. The tactic involves pretending there's an academic controversy over something and then demanding that luckless school science teachers be given the "freedom" to discuss both sides of the issue with their students, who will then be at liberty to make up their own minds on the issue.

"Is the moon made of green cheese, as many people claim? The science is still unsettled . . ."

In other words, inexperienced students are to be told there's a scientific controversy over topics such as evolution and climate change, and then instructed to decide for themselves who's right. If you cast your mind back to your own adolescence and pre-adolescence, you'll see one of the problems with this approach immediately: if you were like all the rest of us, you were still trying to work out how to pop your zits, not pick sides in a scientific debate. The other problem with the approach is, of course, that its premise is a lie: there *is* no scientific controversy over the reality of evolution and climate change. Because science is vibrant rather than ossified, there's plenty of discussion among biologists and climate scientists, respectively, over the kind of details that are too esoteric to make it into school science classes, but the only outright dissent within the respective disciplines comes from a few scattered individuals politely called mavericks.

The peril of teaching both sides of an issue as if they were of equal merit was exposed in a completely different context by President Trump in August 2017 in the wake of the violence at Charlottesville, Virginia, that included a terrorist attack on anti-fascist protesters which injured 19 people, some seriously, and killed one, Heather D. Heyer. Both sides, Trump told an appalled nation, were at fault. If Heyer hadn't been there protesting against Nazis . . .

Most of the attempts to get "Academic Freedom" bills through state legislatures have been unsuccessful, although often, depressingly, they've died for solely bureaucratic reasons rather than because representatives didn't vote for them. In June 2008, however, then-Governor Bobby Jindal signed into state law the Louisiana Science Education Act, a version of the Discovery Institute's bill; this was the first creationist act to become law since Tennessee's 1925 Butler Act.

In May 2017 the Alabama Senate passed a non-binding resolution that "urged" state and local education authorities to encourage Teaching the Controversy in the state's classrooms. This could in fact be more damaging to schoolchildren than the Louisiana Act. While it doesn't have full legal force, it may very well have the same effect as a law, while at the same time, for the very reason that it isn't a law, be difficult to litigate against.

A few months before that, in February 2017, the Indiana State Senate had by a huge majority passed its own resolution, again non-binding, where the creationist senators piously claimed they weren't really trying to get creationism taught in the schools, oh no:

The Indiana General Assembly understands that an important purpose of science education is to inform students about scientific evidence and to help students develop the critical thinking skills they need in order to become intelligent, productive, and scientifically informed citizens . . .

So it's really all about the development of critical thinking, you see. It might be useful if some of those Indiana state senators would practice what they preach.

Yet more pernicious was Florida's House Bill 989, signed into law by Governor Rick Scott in June 2017. This allows any Florida resident to complain to his or her local school board about the instructional materials being used in the classroom, or the books in the school library. The door this (deliberately) opens for endless disruption of Florida education by stupid people and fundamentalist malcontents is obvious. So far as I can ascertain, the Act makes no provision for people to complain about the *omission* of material from the curriculum or the school library. In other words, if you live in an area where creationists rule the roost and you want your kids to get an education, your only real option is probably to move elsewhere.

No Jobs for the Boys?

A frequent plaint of creation scientists is that they're discriminated against by the powers-that-be of orthodox science: their papers are rejected by the scientific journals, or they are even dismissed from their jobs because of their beliefs. These are serious charges.

In fact, there are plenty of examples of creation scientists publishing papers in peer-reviewed scientific journals, so their beliefs alone are clearly not grounds for discrimination. Where creationists have failed to gain publication in the journals is *when they have been advocating creationism,* and there the "discrimination" is on the grounds of bad science—the same "discrimination" whereby countless other papers are rejected. As a single example of a creationist widely published in the journals there's Robert V. Gentry, who did important work on the radioactive "haloes" found in primordial igneous rocks. These microscopically sized rings are produced by radioactive decay during the rock's formation. Gentry convinced himself they offered proof that the delay between God's creation of the chemical elements and His creation of the rocks could be at most a matter of minutes. Gentry's research was published in several eminent science journals, although the editors of those journals naturally balked at his theological inferences, which had no place in a scientific paper.

In 1982, Arkansas's "balanced-treatment" law—whereby schools were forced to offer both evolution and Creation Science side-by-side—came up before federal judge William R. Overton, and during the trial the creationists were challenged on their claim that the scientific journals discriminated against them. Asked to produce examples of papers that had been rejected solely because of their writers' creationist beliefs, they failed to produce even one.

Cases of unfair dismissal are similarly hard to find. One classic example often cited is that of Forrest Mims, who in 1988 was reportedly fired by the magazine *Scientific American* on the grounds of his creationism. This sounds like discrimination pure and simple . . . except that the story isn't true. Mims was not an employee of *Scientific American*, but a freelancer who'd sold articles to the magazine's "The Amateur Scientist" column. There was talk of him taking over the column full-time, but that failed to materialize. As to whether or not his creationist views played a part in his rejection, it's hard now to tell. Mims—a very widely published author in the field of electronics—certainly believed this to be so. Whatever the case, the fact remains that he was not employed in the first place, so could not have been fired.

Then there's the case of Jerry Bergman, who in 1979 was denied tenure at Bowling Green State University, Ohio, supposedly because of creationist beliefs. Well, perhaps. Bergman also wrote a published letter to the newsletter of the racist National Association for the Advancement of White People that the reason for the denial was affirmative action/reverse discrimination. (To hammer home the message, he cited two further examples where he believed he had been the victim of affirmative action, one relating to color and the other to gender.) Leaving aside the oddity of the letter's place of publication—Bergman was vehement that writing to a racist publication concerning a racially sensitive accusation does not necessarily imply racism in the writer—it's clear that he himself, despite his own later claims, at the very least regarded his denial of tenure at the time as being for other than religious/creationist reasons. Bergman went on to write a book, *The Criterion: Religious Discrimination in America* (1984), about people whom he claimed had suffered unfair discrimination or persecution because of their creationism.

One such was Ervil D. Clark, the son of the pioneering creationist Harold W. Clark. He failed twice to gain a PhD at Stanford University, both times through a split committee decision over his oral defense of his dissertation. While the creationist literature regards this as evidence of discrimination, Clark himself in later life was far less sure, apparently agreeing at least in part with the committee's strictures concerning his narrowness of knowledge in his chosen field, ecology. Later he gained a PhD without mishap from Oregon State University.

But by far the best known creationist to have claimed academic discrimination against him is Clifford Burdick, famed for his false interpretation of the dinosaur and "human" fossil footprints at Paluxy River.* He served as a geologist with the Creation Research Society and later the Institute for Creation Research. Missing from the "official" story is that Burdick's creationist colleagues likewise expressed significant misgivings about his work.

In 1966, Burdick announced he had found modern conifer pollens in strata at the Grand Canyon that had been dated to the Precambrian; clearly such an anachronism could cast doubt on generally accepted geological dating. Even Walter Lammerts, the Creation Research Society's director, had qualms about the report. In 1969, consequently, the CRS deputed two independent scientists to go to the Grand Canyon with Burdick to check his results; they concluded that the anomalous findings were a result of Burdick's incompetence at taking soil samples. This was presented by Burdick to the CRS as his discovery having been confirmed!

Earlier, in 1960, Burdick had claimed to the CRS that the University of Arizona had refused him a PhD solely on the grounds of his creationism. The truth of the matter emerged during and after the lawsuit he brought and lost against the university for religious discrimination. Although he had studied at the university, he wasn't pursuing a degree course; even so, he did sufficient work of a suitable caliber to be permitted to sit a comprehensive examination that would have, if he'd passed it, brought him that coveted PhD. All this time he had been keeping his creationist views hidden from his professors, but a few days before the oral exam one of them mentioned having come across a creationist article Burdick had written. According to Burdick, this threw him into such a state of indigestion and insomnia that, on orals day, he was too exhausted and mentally confused to perform adequately. He failed the exam. He also failed in the defense of his Masters thesis at the university, this time because, he said, he had reacted adversely to the medicines he was taking for hepatitis acquired on a field trip.

Lammerts initially intervened on Burdick's behalf with the university; however, when it was explained to him what had actually happened, he backed off hastily. Burdick did go on to make a second attempt to gain a PhD from the University of Arizona, but this time (according to Burdick)

---

* For an account of this fiasco, see my *Discarded Science* (2006) or, for a far more exhaustive discussion, Ronald L. Numbers's *The Creationists* (1992).

his course was disrupted by the discovery by one of the professors that Burdick was the source of the Paluxy River footprints story. Burdick lied his way out of that dilemma, but felt the writing was on the wall for him at the university. In 1966 he gained a PhD in geology from an outfit called the University of Physical Science, based in Phoenix, Arizona. This proved to be not a university at all, just a sort of geological club. Lammerts suggested Burdick henceforth use the rather bizarre "Hon. PhD" rather than "PhD" after his name—even though various other members of the CRS quite blithely listed their "honorary degrees" without any such caveat. The CRS continued to make use of Burdick's services, but cautiously.

Henry Morris, Lammerts himself, and Duane Gish, to name but three, have been among the important creationist figures who have gained degrees despite their open fundamentalism. More recently, in 1989, the openly creationist Kurt Wise gained his PhD at Harvard despite the fact that he studied under no less a personage than Stephen Jay Gould, the *bête noire* of creationists everywhere.

The complaints about job discrimination are part and parcel of a more general claim by creationists and fundamentalists that US society persecutes them. It's arguably cruel to pick on a single example from so many, but a particularly loopy piece of statistical reasoning was expressed by the religious media pundit Cal Thomas at the end of 2006. He asked, "Why is it 'in vogue' to disbelieve in a Creator of the universe, who loves us and wants to have a relationship with us, and not 'in vogue' to believe?" On this ground he bewailed the fact that so many Americans espouse atheism—but declared with wide-eyed wonder that it's really quite possible to engage civilly with atheists so long as you know a few tricks of the trade!

How many people would have to believe in God for Thomas to start regarding belief as being "in vogue"? At the time of his writing, polls showed that approximately 90 percent of people in the US believed in God, with agnostics and atheists together comprising a mere 10 percent. One cannot credit that he was ignorant of these poll results, which were publicly available and fairly consistent with previous polls over decades.* The only possibilities are that either (a) he was incapable of understanding the statistics they represented, or (b) he was committing a "rhetorical truth."

---

* The nonreligious percentage of the population has risen in recent years, and in 2016 sat at 23 percent, according to both Pew and Gallup.

OTHER WORLDS, OTHER CREATIONS?

SETI stands for "Search for Extraterrestrial Intelligence," but in practice today the term is usually taken to refer to the scientific speculations clustered around the possibilities of searching for and communicating with intelligent civilizations elsewhere in the universe. SETI became a focus of serious interest in the latter half of the last century, since when the various creationist schools of thought have been at pains to disparage it. Newer creationist organizations, like Answers in Genesis and the Discovery Institute, have continued the tradition.

That tradition received a major jolt when the first exoplanets—planets of another star—were discovered in the early 1990s, but at least the creationists could console themselves that these (or at least the ones orbiting sun-like stars) were huge "super-Jupiters," not earthlike worlds, and usually orbited so close to their stars that life as we know it was exceptionally unlikely. That the first-discovered exoplanets had these characteristics was simply because big, close-orbiting planets are a lot easier to find than smaller ones in more distant orbits. As our technology and our skills have improved, we've discovered lots of smaller worlds, too.

The creation in January 2003 of the European HARPS (High Accuracy Radial-velocity Planet Searcher), a spectrographic device attached to the largest of the three telescopes at La Silla Observatory, in Chile, increased the rate of discovery of new exoplanets considerably, and the launch of the Kepler space telescope in March 2009 led to a yet more dramatic increase. To date we've detected over 3,500 exoplanets, and quite a number of these match the specifications to be potential homes for life as we know it: they're of roughly earthlike size, probably rocky rather than gaseous, orbit a single star that appears to be relatively stable, and do so in what's known as the Goldilocks Zone, where it's not too hot and not too cold for liquid surface water to exist.

Saying that a planet is habitable doesn't mean it's actually inhabited. Even in the most favorable of circumstances, there's no guarantee that life will arise. Some scientists, like Steven Vogt of the University of California at Santa Cruz—who was part of the team that in September 2010 announced the discovery of the first apparently habitable exoplanet, Gliese 581g, which orbits a star about twenty light years away—believe that life will almost inevitably emerge wherever the conditions are right. Others look at the mathematical odds more skeptically and conclude that the emergence of life may be extremely rare, but note that with the sheer scale of the universe and the fact that, as we now know, most stars have planets, we may still be talking about a lot of life-bearing worlds. Both schools of thought, and of course all of those that lie somewhere in between, are scientifically respectable, as is the speculation that we are indeed alone.

It's easy enough to understand why creationists, whose ideas on the subject are inspired solely by their theology, would be so concerned about the possibility of life having emerged elsewhere. If their religious teachings insist that their god (or gods) created the people here on earth uniquely, then this leaves no room for ETs. If He created us in His own image, according to the teachings, then the problem becomes even deeper.

For supposedly scientific schools of creationist thought, though, like ID, there's more of a dilemma. Scratch the surface of the ideas promoted by the Discovery Institute and you find fairly orthodox Christianity, but the IDers themselves are never allowed to admit this. Any attack they mount on SETI must be grounded, therefore, in arguments that at least their supporters will believe are scientific. An added problem for the IDer is that there's nothing in ID itself that actually forbids the emergence of life on other worlds. If the Designer isn't the Christian god, his actions circumscribed by what it says in the Bible, then he (or she or it) can go about creating life on other planets whenever the fancy strikes.

Even so, the IDers regard any successful outcome to SETI as a threat to their belief system. As long ago as 1992, William Dembski, a pillar of the Discovery Institute's early proselytizing efforts, attempted to dismiss scientific ideas about exoplanets as mere figments:

[Richard] Dawkins, to explain life apart from a designer, not only gives himself all the time Darwin ever wanted, but also helps himself to all the conceivable planets there might be in the observable universe (note that these are planets he must posit, since no planets outside our solar system have been observed, nor is there currently any compelling theory of planetary formation which guarantees that the observable universe is populated with planets).

The latter part of this pronouncement was, perhaps unsurprisingly, a flat-out lie. At the time when Dembski was telling his audience this, the near-universally accepted theory of planetary formation was (essentially) the nebular hypothesis. Built into the very foundations of that hypothesis, for which supporting evidence by 1992 was very good, is that the formation of planetary families around stars is common—and this, by the simple law of averages, implies that habitable planets are likewise frequent.*

When this was pointed out, the response of the Discovery Institute was to clumsily change the subject. It's a technique they employ fairly often in discussions of exoplanets and SETI. For example, Steven Vogt's comments noted above about the likelihood of life emerging wherever conditions are favorable drew an attack from the Discovery Institute's Casey Luskin that falsely attributed Vogt's views—which are a bit out on a limb—to evolutionary biologists as a whole.[40] We should anyway stress that the theory of evolution says nothing about the origin of life, including the likelihood or otherwise of its doing so.

In 2011, Discovery Institute fellow David Klinghoffer claimed[41] that the lack of results from SETI endeavors to date is somehow an indication that life is rare in the universe:

> If it were true that the conditions for life are common around the galaxy, yet no intelligent life has checked in with us so far, that would seem to suggest that life itself—as opposed to the conditions that might make it hypothetically sustainable somewhere—is very special indeed.

As Klinghoffer was presumably aware, we've had the technology to go looking for signs of life elsewhere for only a tiny fraction of the billions-of-years-long history of life on earth. Moreover, the instruments we've been using for SETI are limited; as example, the Arecibo dish (300m/1,000ft in diameter) could pick up radio and TV "noise" comparable to Earth's out to a distance of only about one light year—one-quarter the distance to the nearest star.

So the failure to date of SETI efforts to pick up a signal tells us almost exactly nothing about how common life is in the universe. It tells us only a tiny amount more about how common technological civilizations are in our neck of the woods, because we haven't been looking for them long enough to come to any sensible conclusions and because we're making the rash assumption that any alien technological civilization might be remotely interested in getting in touch with its neighbors.

---

* Count in suitable moons, like the one in the movie *Avatar* (2009), and current estimates of the number of habitable worlds in our galaxy alone lie anywhere between hundreds of millions and tens of billions. Now multiply this number by the number of galaxies in the universe . . .

Bearing in mind all these factors, Klinghoffer's conclusion was a piece of outright sophistry—or, to use the technical term, bullshit.*

The Discovery Institute website *Evolution News & Science Today* also published Tom Bethell's October 8, 2013, essay "The Anxious Search for Extraterrestrials"[42] in which Bethell seems to think that the failure to find life on Mars is a bit of a deal-breaker for SETI and for evolution:

> If we can't find life on Mars, we are unlikely to find it anywhere else in our solar system. Maybe life takes a designer?

He also seems to believe that a science-fiction writer is a more significant authority on the possibility of life elsewhere in the universe than scientists such as Lawrence Krauss, Seth Shostak, and Carl Sagan. Bethell continues:

> The late novelist Michael Crichton turned this into comedy in a lecture at Caltech in 2003—"Aliens Cause Global Warming." There is "not a single shred of evidence for any other life forms and in forty years none has been discovered. SETI is a religion," he said. Make that fifty years now.
>
> Crichton added a brief tour of nuclear winter, second-hand smoke and global warming, wherein science always defers to politics. We are seeing a "loosening of the definition of what constitutes legitimate scientific procedure," he concluded.

Since Crichton was comprehensively wrong about nuclear winter, second-hand smoking,** and climate change, the conclusion to be drawn is surely that he was probably comprehensively wrong, too, about extraterrestrial life!

There are countless other examples of the institute's attacks on SETI and belittling of the significance of exoplanet discoveries. Browsing around on the *Evolution News & Science Today* site offers you one of those sobering experiences that's likely to leave you reaching with trembling hand for an antidote.

The hostility toward exoplanets and SETI likely to be espoused by a more fundamentalist creationist outfit such as Ken Ham's young-earth Answers in Genesis (AIG) might seem more predictable.** In fact, Ham and AIG

---

* I'm not joking about "bullshit" being a technical term—see Harry G. Frankfurt's 2005 book *On Bullshit*. Klinghoffer quite evidently did not care that his argument was nonsensical.

** It's interesting that Bethell should, like Crichton, include this on his list. See pp. 215–218 for more on the links between smoking denialism and scientific denialism in general.

have tried to play it both ways: life doesn't exist anywhere else except here on earth and, even if it does, evolution is still bunk. Here's AIG's Elizabeth Mitchell discussing the activities in early 2013 of the Mars rover Curiosity:

> No complex organic molecules, such as amino acids from which proteins are built, have been found, though even those would not be definitive evidence that life ever existed on Mars, since such compounds are also found in the absence of life.
>
> The scientists hasten to point out that no evidence of life, past or present, has actually been found. Furthermore, the equipment on Curiosity is not designed to detect living organisms.
>
> So why all the excitement? As we've discussed recently, it is an article of faith among evolutionists that given liquid water, a few chemicals, and sufficient time, life can evolve through natural processes. Yet even on earth where we can make much more direct and abundant observations, no evidence that life can evolve from non-living substrates through random natural processes has ever been found. And the likelihood that the Martian environment may have once been different than it is today does not demonstrate that the solar system formed billions of years ago.
>
> If evidence of microbial life—past or present—on Mars is eventually found, evolutionists will of course claim such life as evidence for evolution. But in reality, such a finding would simply be evidence that life is (or once was) there, not proof of that life's origins. The Bible does not say whether God created any life on other planets, but the Bible does tell us God created all life on earth during the first six days of Creation week, the same week in which He created the rest of the universe, about 6,000 years ago. Discovery of evidence that a "habitable" environment containing liquid water once existed on Mars or even life itself would neither disprove nor undermine biblical truth.[43]

---

* Ham and AIG are perhaps most famous for Ark Encounter, which opened in July 2016 in Grant County, Kentucky, and takes the form of a creationist museum housed inside a supposedly full-sized model of Noah's Ark. Both the state and county made hefty tax concessions to the project on the grounds that it would bring major tourist business to the area. A year after its opening, however, attendance figures appear to be, while far from negligible, well below those projected. Meanwhile there have been a lot of questions over various tax-related issues, including a brief attempt in July 2017 by AIG to avoid safety tax on ticket sales by selling the land on which Ark Encounter stands to Ark Encounter LLC's non-profit parent company, Crosswater Canyon, for a princely $10. On discovering that, while this would save an annual safety-tax bill of some $500,000 it would incur a cost, through lack of tax kickbacks elsewhere, of some $18 million, AIG swiftly reversed its decision.

There is so much wrong and/or muddleheaded here that it's hard to know where to start, so I won't. We should note, though, that it *isn't* "an article of faith among evolutionists that given liquid water, a few chemicals, and sufficient time, life can evolve through natural processes" for the very good reason that, to repeat, evolution says nothing about life's origins. Yes, most evolutionary biologists would agree that, given the right circumstances, natural processes might very well lead to the emergence of life, but the theory is concerned with what happens to life *after it has formed*.

Regular fans of Mitchell's musings on exobiology might find some of her discussion of the Mars exploration oddly familiar . . . as indeed they are. A couple of paragraphs are recycled almost verbatim from her discussion two years before of the implications of the 2011 discovery of the planet Kepler-22b, the first planet known to orbit in its star's Goldilocks Zone. In that instance she added:

> [E]ven if life were to be indisputably found on another world, its existence would not prove molecules-to-man evolution ever occurred. Such life would simply be another demonstration of God's creative power to create life where He chooses. Meanwhile, recognizing the whole of creation was corrupted due to man's sin, we continue to doubt the existence of sentient life elsewhere . . . The discovery is fascinating, but if we wish to know we are not alone, we'll get more reliable results by getting to know the God of the Bible.[44]

While AIG maintains at least the pretense of an open mind, its main rival in the competition for young-earth-creationist allegiance, the Institute for Creation Research (ICR), is significantly more doctrinaire on the matter:

> As far as science "knows," the planet earth is unique in the entire universe. . . . Nothing that we have observed leads us to believe that there is any other planet like earth.[45]

5

IDEOLOGY OVERRULES SCIENCE

In order to be a conservative person in America you have to be anti-abortion, pro-guns, pro-death penalty, small government, no regulation—and climate is in there too. If you look at that set of beliefs, that's an identity. And you can't change an identity with facts.

—James Garvey, Secretary of the Royal Institute of Philosophy, cited by Will Storr in *The Unpersuadables* (2014)

The theory is widely accepted within the scientific community despite a lack of any conclusive evidence . . . It should be noted that these scientists are largely motivated by a need for grant money in their fields. Therefore, their work can not be considered unbiased. Also, these scientists are mostly liberal athiests [*sic*], untroubled by the hubris that man can destroy the Earth which God gave him.

—"Global Warming," *Conservapedia*, 2007

Perhaps the most dramatic example of the ideological corruption of science is book-burning, the deliberate destruction of knowledge for political or religious reasons.

In 645 CE, Caliph Uthman ordered the destruction of the Library at Alexandria on the grounds that either the books agreed with the Quran, in which case they were redundant, or they disagreed with it, in which case they were worthless or evil. In an act that has rightly been vilified throughout the centuries since, the Library's books were used as fuel to heat Alexandria's public baths. There were reportedly so many books the burning took six months.

What is less known is that Uthman was far from the first to burn the Library's books, and the crime was far from exclusive to Islam. The first three times the Library felt the flames were through acts of criminal negligence: Ptolemy VIII (88 BCE), Julius Caesar (47 BCE) and the Emperor Aurelian (273 CE) simply didn't care that their burning of large parts of Alexandria would destroy the Library along with the rest. After each desecration the Library was reconstituted to a greater or lesser degree. By 391 CE, a little over a century after Aurelian's crime, some 40,000 scrolls housed in Alexandria in a temple to Serapis were destroyed in another act of religious madness, this time by a Christian, Archbishop Theophilus, on the grounds that anything contained in a pagan temple must be by definition diabolical.

It was to a great extent because of Uthman's destruction of the Library at Alexandria that Europe suffered a Dark Age for the best part of the next millennium. The destruction still affects us today. Imagine what our world would be like had there not been that long hiatus.

Indeed, mass book burning—which in more recent times has come to include mass burnings of CDs and the like, as in the wake of the Dixie Chicks' public denunciation of President George W. Bush—have been throughout history almost a commonplace among regimes or mobs whose ideology tells them they're in the exclusive possession of rectitude . . . or, perhaps, who become nervous at the threat presented by reality, as enshrined in books, to ideological certainty. The Nazis' declared reason for their mass book-burnings in the 1930s was that, since the Reich was going to last for at least 1,000 years, it did not need intellectual pollution by the "falsehoods" of such people as Proust, Wells, Freud, and Einstein. In truth, of course, the burning was defensive: the ideas of these and many, many other writers whose works met the same fate could far too easily prick the frail bubble of the absurdity that was Nazi ideology. Mao Zedong and Josef Stalin attempted similar exercises, but with less success. Book-burning's digital equivalent, though less dramatic, is now also practiced. At the time of writing, the Trump administration in the US, frightened by the conflict between its ideology and environmental realities, has recently started a rapid program to make the archives/libraries of the EPA unavailable; the George W. Bush administration and, in Canada, the Stephen Harper administration acted similarly. As always, suppressive acts in defense of an untenable ideology are presented as if done, rather, because the strength of that ideology dictates it.

But it was because of confidence in an ideology that the other great book destruction of ancient times occurred—the Asian equivalent to Europe's loss of the Alexandria Library. In 213 BCE the Emperor Shih Huang-ti, founder of the Ch'in Dynasty, ordered all the books in China to be destroyed because his dynasty, which was set to last 10,000 generations and unify all the world—the claims for the Third Reich seem modest by comparison—would have no need of the irritating disruption that contemplation of the past might cause. Exceptions were made for books on agriculture (which were practically useful), medicine (likewise) and fortune telling (because the Emperor was really keen on fortune telling). All the Chinese literature we possess from before 213 BCE is, accordingly, essentially ersatz: copies done by much later writers, relying on memory.* In the event, the Ch'in Dynasty endured for rather less than 10,000 generations—just 15 years, in fact.

The New World was not spared. Precisely four books survive from the Mayan Empire. All the rest were put to the torch in the sixteenth century by Christian zealots, led by the Franciscan Diego de Landa. The aim was to destroy the Mayan culture entirely and permanently, and in essence that aim was achieved. Scholars today know, because the virtuous monks kept self-congratulatory records of the destruction they wrought, that the Mayans wrote down their culture in thousands of books; but our knowledge of the culture itself is sketchy, based on deduction from desperately scant evidence—writings preserved on pottery and walls. The Serbs made a similar concerted attempt in Sarajevo in 1992 to annihilate Bosnian culture by destroying the National Library and University Library.

Today, it's almost impossible to guess toward which book the attentions of the book burners and banners might turn next. The innocuous Harry Potter children's novels have frequently drawn the hostility of Christian fundamentalists because of their "glorification of witchcraft"; plenty of other completely harmless children's fantasies have been similarly pilloried, albeit with less publicity. Still, one can predict that a book with wizards in it is likely to generate heat under some bigot's collar, but who would have predicted that *The Diary of Anne Frank* (1952) might have a similar effect? Yet there have been several local attempts in the modern US to ban this book. And the next book to be banned could be your kid's biology textbook, on the grounds that it deals with evolution.

---

* We can see where the scenario of Ray Bradbury's novel *Fahrenheit 451* (1953) came from. It's interesting that this novel grew to be one of the best known in the world—to the point where the falsehood that book paper burns at 451°F is routinely repeated. So powerful has the specter of book-burning become to us that the book's premise alone has assured it a place in our cultural iconography.

What do I mean, "could be"?

The destruction of knowledge is of course the very antithesis of science, whose aim is to increase it. This is true even if the knowledge threatened with destruction is demonstrably false. Without a comprehension of false knowledge we cannot properly appreciate what we have so far uncovered of the truth. The fallacies of past misunderstanding help us to become better detectives in tracking down possible fallacies in our own worldview.

RACIST PSEUDOSCIENCE

Elijah Muhammad, leader of the US movement Nation of Islam from 1934, taught—claiming the teaching came from his predecessor, Wallace Fard Muhammad—that black people were the original inhabitants of the world, with whites a race of degenerate brutes created later by a mad scientist called Yakub, working for 600 years about 6,000 years ago on the island of Patmos. These "grafted" people—i.e., whites—would rule the earth until the blacks regained their rightful position, which process had begun in 1914.* The tale of Yakub is of course scientific nonsense, as is the racism of the Nation of Islam—although at least we can see where Elijah Muhammad's anti-white racism came from: before he was out of his teens, he reportedly witnessed three lynchings of blacks in his Georgia homeland by degenerate white brutes.

"Every part of the body of a non-Muslim individual is impure, even the hair on his hand and his body hair, his nails, and all the secretions of his body." Thus the Ayatollah Khomeini. Again: "This is why Islam has put so many people to death: to safeguard the interests of the Muslim community. Islam has obliterated many tribes because they were sources of corruption and harmful to the welfare of Muslims."

The notion of racial purity, linked often with religious considerations, has ever brought out the worst in humanity. The pages of history are littered with the corpses of those slaughtered simply because they were of the "wrong" race. The pattern is simple, and oft-repeated: first you declare an identifiable minority as inferior; then, in times of trouble—especially of economic recession—you announce the problems are the responsibility of this minority; then you start the pogroms. The theory's insult to humanity is its refusal to recognize individuals as individuals: they are "blacks,"

---

* Since Elijah Muhammad's death, Nation of Islam followers have generally come to regard Yakub as allegorical, like Adam and Eve.

or "Jews," or "gays," or "gypsies," or "Catholics," or "women"—members of categories, not people in their own right.

Some of the pronouncements of the racial purists are truly risible, or would be if they weren't so repellent—like those of the Holocaust deniers. Around the end of the nineteenth century, Charles Carroll—who was far from alone in his views—produced three books, *Negro Not the Son of Ham, or Man Not a Species Divisible into Races* (1898), *Negro a Beast, or In the Image of God* (1900) and *Tempter of Eve, or The Criminality of Man's Social, Political, and Religious Equality with the Negro, and the Amalgamation to Which These Crimes Inevitably Lead* (1902), seeking to show that God had created black people as merely a higher animal whose sole function was to act as a servant to whites. When Genesis refers to the tempting serpent it is *really* referring metaphorically, Carroll told us, to Eve's black maidservant.

Ellen Gould White, writing in 1864 in fundamentalist vein, sought to answer the question: Why are there diverse human races? Only one "type" of human being could have been saved from the Flood—because, of course, Noah's family could not have been made up of more than one racial stock. Since the Flood, however, people have committed the unforgivable sin of mating with animals; all races except the aptly named Mrs. White's are the results of such miscegenation.

Prewar German theorists produced some of the more bizarre "scientific" racist ideas. The attack was two-pronged—not only the attempt to prove the inferiority of, say, the Jews, but also the more positive struggle to prove the superiority, bordering on perfection, of themselves. Hans Günther, a leading Nazi anthropologist, gave us a few illuminating clues as to the natural superiority of the Nordic type. For example, Nordics wash themselves and brush their hair more often, are better athletes and, if female, keep their legs rigidly locked together when traveling by bus. In the mid-1930s, Julius Streicher proposed that Jewish blood-corpuscles are totally different from those of the rest of us. Why have biologists failed to report this? Because so many biologists are Jews, of course!

The racial-purity theory is, in its details, both untenable and dangerous; but what of the broader brush? In fact, Mrs. White unwittingly pointed up its fundamental fallacy. We all came from common stock; ever since, breeding between diverse human populations has been commonplace. There is no such person as a genetically "pure" individual: we're all mongrels.

# # #

The early-nineteenth-century French physician Julien-Joseph Virey, who first put forward the notion of biological clocks and pioneered the study of what's now called chronobiology, had some curious ideas when it came to other subjects. Blithely unaware that the aesthetics of the human form might vary from one culture to the next, he maintained that all the ugly peoples of the world were "more or less barbarians" and beauty was "the inseparable companion of the most civilized nations," while confidently proclaiming that black women possessed a greater degree of lasciviousness than their white counterparts, a difference he attributed to black women's greater voluptuousness. As for intellectual differences between the races, he produced this piece of dazzling illogic:

> Among us [Europeans] the forehead is pushed forward, the mouth is pulled back as if we were destined to think rather than eat; the Negro has a shortened forehead and a mouth that is pushed forward as if he were designed to eat instead of to think.

Similarly Georges Cuvier, one of the fathers of geology and paleontology, mapped his own racist preconceptions onto his studies of comparative anatomy. In his *Elementary Survey of the Natural History of Animals* (1798) he wrote that the "White race, with oval face, straight hair and nose, to which the civilized peoples of Europe belong and which appear to us the most beautiful of all, is also superior to others by its genius, courage and activity," adding that there was a "cruel law which seems to have condemned to an eternal inferiority the races of depressed and compressed skulls." In the phrase "which appear to us the most beautiful of all" he might seem, rare among his contemporaries, to have at least a foggy awareness of the role of cultural relativism in determining our ideas of human beauty, but he crushes this assessment with his further remark that "experience seems to confirm the theory that there is a relationship between the perfection of the spirit and the beauty of the face"—a remark that implies he knew nothing whatsoever about his fellow human beings.

The degree to which racism was endemic among US whites in the nineteenth century is hard for us to comprehend today, blinded as we are by the myth that the North fought primarily to free the Southern blacks from slavery. In fact, at the time it was widely held that Abraham Lincoln brought Emancipation into the picture solely in hopes of inspiring the Southern blacks at last to rise up against the slave owners; and Lincoln's stated ideal was that, after Emancipation, the freed slaves would be "repatriated" *en masse* to Africa. In debate against Stephen Douglas in 1858 Lincoln made his position clear:

> I will say, then, that I am not nor ever have been in favor of bringing about in any way the social and political equality of the black and white races—that I am not, nor ever have been, in favor of making voters or jurors of negroes, nor of qualifying them to hold office, nor to intermarry with white people; and I will say in addition to this that there is a physical difference between the white and black races which will ever forbid the two races living together on terms of social and political equality. And inasmuch as they cannot so live, while they do remain together, there must be the position of superior and inferior, and I, as much as any other man, am in favor of having the superior position assigned to the white race.

Even after signing the Proclamation of Emancipation on January 1, 1863, Lincoln felt compelled to say that "I can conceive of no greater calamity than the assimilation of the Negro into our social and political life as our equal." Remember that Lincoln was, for the society in which he lived, one of the most enlightened of men.

# # #

Of course, there's no reason why creationism should foster racism, but nonetheless the two often seem to go hand in hand. A central cause is the belief that, after the Flood, the earth was repopulated by the descendants of Noah's three sons Japheth, Ham, and Shem. It has proven far too tempting to many creationists to connect these three lineages with the races of mankind. A general creationist concurrence appears to have been that the lineage Ham sired is represented by the dark-skinned races, although the definition of "dark-skinned" varies a little, sometimes including the Asiatics and sometimes not. Thus we have the famous creationist Henry Morris writing: ". . . all of the earth's 'colored' races—yellow, red, brown, and black—essentially the Afro-Asian group of peoples, including the American Indians—are possibly Hamitic in origin." This would seem harmless enough, if anthropologically nonsensical, but we can note in passing that the specious correlation between blacks and Hamites was used in the nineteenth century as one of the many justifications for slavery. Further, as soon as the creationist mind fastens upon such a hypothesis, it's driven to start extrapolating from it, and therein lies a danger. Here's Morris again:

Often the Hamites, especially the Negroes, have become actual personal servants or even slaves to the others. Possessed of a racial character concerned mainly with mundane matters, they have eventually been displaced by the intellectual and philosophical acumen of the Japhethites and the religious zeal of the Semites.

In other words, the Hamites are the intellectual inferior of the other two races. To repeat, the fault here lies not in creationism *per se*, which has plenty of faults of its own, but in the minds of its proponents—just as there are those who attempt to use (usually flawed) knowledge of evolution to promote racist ideas. But it seems especially obvious in the instance of creationism because, as numerous surveys have shown, there is in modern times a correlation between low educational level and creationism and another between low educational level and racism. Put the two correlations together and the reasoning becomes obvious. In past eras, of course, the highly educated and the poorly educated believed in the Creation alike.

Even without invoking the lineages of Noah's three sons, it was perfectly possible to bend fundamentalist views into supporting racist ideas. Considerably before Morris and his ilk there was the prevalent notion (monogenesis) that all human beings were descended from Adam and Eve: humans were once very close to the perfection God had intended, but after the Fall, as they spread to different parts of the globe and diversified into races, they degenerated. Obviously (to the racist eye) the blacks were the ones who'd degenerated the most, and a frequent reason given for this was climate: the hotter the climate, the greater the degeneration. Even the liberals of their day could subscribe to this entirely unevidenced pseudoscience, as when Samuel Stanhope Smith, president of what would later become Princeton, wistfully hoped that blacks in the US, since now exposed to a colder clime, would in due course turn white and be able to take their place alongside their more fortunate compatriots. Georges-Louis Leclerc, Comte de Buffon, a towering figure in the history of French science and a staunch advocate of the abolition of slavery, nevertheless regarded whites as the superior race:

The most temperate climate lies between the 40th and 50th degree of latitude, and it produces the most handsome and beautiful men. It is from this climate that the ideas of the genuine color of mankind, and of the various degrees of beauty, ought to be derived.*

* As cited in Stephen Jay Gould's *The Mismeasure of Man* (1981).

But not everyone agreed that climate could be responsible. The English anatomist William Lawrence, in *Physiology, Zoology and the Natural History of Man* (1828), a book which forcefully put forward the then-startling claim that there were so many similarities between humankind and the rest of the animal kingdom that it was folly to regard human beings as other than animals, noticed that African Americans, despite having existed in a different climate from Africa's for centuries, had not changed color. He advanced the idea of domestication as a cause for racial variation, although later in life, seeing how this hypothesis seemed unworkable, opted instead for environmental factors, climate included. The thinking of the English anthropologist, anatomist and staunch abolitionist James Prichard, author of *Researches into the Physical History of Man* (1813), followed a similar trajectory from domestication to environmental factors. Interestingly Prichard, because of his abolitionist views, went out of his way to stress that apes and humans were of different lineages, for fear that pro-slavery advocates might claim blacks as the missing link and thereby justify their abhorrent slave-owning ways.

An alternative explanation for the perceived differences between the races was polygeny, the notion that the races had been created separately and were in effect different species; the fact that they could and did interbreed made no difference to this argument—after all, couldn't donkeys and horses interbreed to produce mules? If the colored races weren't true races of mankind but really just humanoid animals, then there was surely nothing morally reprehensible about the whites, the "genuine" strain of humanity, enslaving them—and indeed some theologians argued it was the whites' moral duty to do so, for had not God issued instructions that the beasts of the field were there to serve humans? The latter part of this equation was excoriated memorably by the French anatomist (and non-polygenist) Antoine Etienne Serres in his *Principes d'Emryogénie, de Zoogénie et de Teratogénie* (1860) as "a theory put into practice in the United States of America, to the shame of civilization." In fact, it had, not all that long before, been the shame of European civilization as well.*

Polygeny and attempts to justify slavery did not necessarily go hand in hand, though. Equally robust in rejecting slavery was the UK physician

---

* Serres is notorious for arguing that Africans were more primitive than Europeans because the distance between their navel and their penis remained relatively small throughout life, whereas that of Europeans started small but increased as the individual male grew older: the rise of the navel relative to body height was a mark of civilization.

Charles White, author of the "manifesto" of polygeny: *Account of the Regular Gradation in Man* (1799). White included this eulogy to the fortunate race:

> Where shall we find, unless in the European, that nobly arched head, containing such a quantity of brain, and supported by a hollow conical pillow, entering its center? Where the perpendicular face, the prominent nose, and round projecting chin? Where that variety of features, and fullness of expression; those long, flowing, graceful ringlets; that majestic beard, those rosy cheeks and coral lips? Where that erect posture of the body and noble gait? In what other quarter of the globe shall we find the blush that overspreads the soft features of the beautiful women of Europe, that emblem of modesty, of delicate feelings, and of sense? What nice expression of the amiable and softer passions in the countenance; and that general elegance of features and complexion? Where, except on the bosom of the European woman, two such plump and snowy white hemispheres, tipt with vermilion?

Polygeny's biggest villain was undoubtedly the UK philosopher David Hume, who used the hypothesis as an excuse to promote the most extreme racism. The colored races should realize and accept their inferiority, and look upon the whites as their saviors, gratefully accepting colonialism, slavery and in some instances even genocide as blessings bestowed upon them. In his essay "Of National Characters" (printed in *The Philosophical Works of David Hume* [1854]) he wrote:

> I am apt to suspect the negroes and in general all the other species of men (for there are four or five different kinds) to be naturally inferior to the whites. There never was a civilized nation of any other complexion than white,* nor even any individual eminent either in action or speculation. No ingenious manufactures amongst them, no arts, no sciences. On the other hand, the most rude and barbarous of the whites, such as the ancient GERMANS, the present TARTARS, have still something eminent about them, in their valor, form of government, or some other particular. Such a uniform and constant difference could not happen, in so many countries and ages, if nature had not made an original distinction betwixt these breeds of men. Not to mention our colonies, there are Negroe [*sic*] slaves dispersed all over EUROPE, of which none ever discovered any symptoms of ingenuity; tho' low people, without education, will start up amongst us, and distinguish themselves in every profession. In JAMAICA indeed they talk of one negroe as a man of

---

* Um . . . China? Egypt? Japan? Persia? India? . . .

parts and learning; but 'tis likely he is admired for very slender accomplish-
ments, like a parrot, who speaks a few words plainly.

If we compare this passage with what Hume himself had to say in the
Introduction to his *Treatise of Human Nature* (1739–40) we don't know
whether to laugh or weep: "And as the science of man is the only solid
foundation for the other sciences, so the only solid foundation we can give
to this science itself must be laid on experience and observation . . ."
When it came to the crunch, as he sought to justify his irrational hatred of
the "inferior species" and his support for the slave trade that granted him
and his social class a luxurious life, Hume, like so many pseudoscientists,
was unwilling to let mere facts get in the way.

Hume was hugely influential through the eighteenth century and beyond, and it was likely because of his influence that the 1798 edition of the
Encyclopedia Britannica bore this shameful entry, in which, note, blacks
are given a different taxonomic name from that of true *Homo sapiens*:

> NEGRO, *Homo pelli nigra*, a name given to a variety of the human species,
> who are entirely black, and are found in the Torrid zone, especially in that
> part of Africa which lies within the tropics. In the complexion of negroes
> we meet with various shades; but they likewise differ far from other men in
> all the features of their face. Round cheeks, high cheek-bones, a forehead
> somewhat elevated, a short, broad, flat nose, thick lips, small ears, ugliness,
> and irregularity of shape, characterize their external appearance. The negro
> women have the loins greatly depressed, and very large buttocks, which give
> the back the shape of a saddle. Vices the most notorious seem to be the por-
> tion of this unhappy race: idleness, treachery, revenge, cruelty, impudence,
> stealing, lying, profanity, debauchery, nastiness and intemperance . . . They
> are strangers to every sentiment of compassion, and are an awful example of
> the corruption of man when left to himself.

Eric Morton, who cites this passage and others reproduced here in his
essay "Race and Racism in the Works of David Hume" (2002), adds that
"the editors continued publishing entries in this vein until well into the
twentieth century."

With such attitudes being, if not universal, at least widespread in the
nineteenth century and even after, novels like H. Rider Haggard's *King
Solomon's Mines* (1885) and John Buchan's *Prester John* (1910), in both of
which at least some blacks are regarded as equal or superior to the finest
the Caucasian race has to offer, must have landed like a bombshell in the
average Victorian or Edwardian living room. And we should not forget
Kipling's "You're a better man than I am, Gunga Din." How much greater

must have been the impact in the US, where the influence of this and similar racist pseudoscience was far slower to wane than in Europe? There one found anthropologists of the highest distinction spouting what must have seemed baloney even to the racists among their European colleagues. No wonder Harriet Beecher Stowe's *Uncle Tom's Cabin* (1852) was so startling.

One of those US anthropologists was the Swiss-born Louis Agassiz. A disciple of Cuvier, Agassiz made his name while still in Europe as a paleontologist and glaciologist; he emigrated to the US in 1846. He was appointed Professor of Natural History at the Lawrence Scientific School, Harvard, in 1847, a position he held for the rest of his life. Agassiz appears to have shown no particular signs of militant racism before he came to the US, almost certainly because he had never met any black people up to then. He recorded his immediate gut revulsion on meeting blacks for the first time, and this primitive reaction proceeded to engender in him a profound detestation of the colored races. The idea of miscegeny seems to have become an outright phobia: "The production of half-breeds is as much a sin against nature as incest in a civilized community is a sin against purity of character." (To his credit, it should be added that at the same time he vociferously opposed slavery.) Of course, as a scientist, he quickly sought some means, however desperate, of justifying his purely irrational emotion.

Polygeny was the obvious answer. It was, as an added bonus, in good accord with his earlier work, because Agassiz was a taxonomist on steroids. Where other taxonomists would note the strong similarities between widely distributed creatures and realize they were all merely different variants of the same species, Agassiz would blithely name each variant, however minor the distinction, as a separate species. In order to make sense of his prodigious species-generation, he evolved the hypothesis that creatures were brought into being at various "centers of creation," near which they tended to stick; he very much downplayed the possibilities of large-scale migration from a species' point of origin. Thus the idea that the various races of mankind were distinct, separately created species seemed a natural one—although there was one problem: Agassiz, like many other scientists of his time (this was before 1859 and the explosion of Darwinism), was a devout creationist, and it was hard to equate polygeny with the notion that we were all descended from Adam and Eve. However, he rationalized himself out of this dilemma by deciding that Genesis spoke only of the region of the world known to that book's writers; beyond their ken could have been the creation by God of multiple Adams and Eves elsewhere around the globe—"centers of creation" again.

This hypothesis would not in itself necessarily have fostered racism; however, Agassiz felt driven to add to it, basing his extrapolations on his own profound ignorance of the human races:

> It seems to us to be mock-philanthropy and mock-philosophy to assume that all races have the same abilities, enjoy the same powers, and show the same natural dispositions, and that in consequence of this equality they are entitled to the same position in human society.

There was no point in educating blacks for other than hard labor; in his astonishingly unscientific ranking of the moral and intellectual qualities displayed by the various races, blacks were at the bottom of the pile in every possible category, and that was the place they should occupy in society, too. Reading some of his writings on the subject of the races—there's a good selection in Stephen Jay Gould's *The Mismeasure of Man* (1981), along with a fuller account than here of Agassiz's thinking—it's difficult to conceive how a supposed man of science was able to persuade himself that what he was propounding was science, or anything like it: most of his "evidence" has the status of urban legend.

An important contemporary of Agassiz was the US physician Samuel George Morton. He early adopted the principle of polygeny and its "natural" extension, that the races could be placed in order of rank—the usual whites-at-the-top-and-blacks-at-the-bottom ranking, of course. In order to prove his hypothesis, Morton amassed a huge collection of human skulls—hundreds—whose cranial capacities he measured as an indication of brain size, filling the skull cavities with sifted white mustard seed and then pouring the seed into a measuring cylinder. Surprise, surprise, he found this metric supported his notions, and he published extensive tables to back up the contention. His research was for decades regarded as conclusive. Much later, however, Stephen Jay Gould went through Morton's research with a fine-toothed comb and discovered the dramatic extent to which Morton had, most likely subconsciously, tailored his results in order to fit his hypothesis. Re-analyzing Morton's own raw data, Gould found they indicated only minor differences between the races—so minor they're likely a mere statistical effect. Several hundred skulls may seem a large sample to have measured—and of course in a way it is—but as a representation of humanity's billions it is infinitesimally small.

Morton's polygenistic flag was kept flying by his student Josiah Nott and the Egyptologist George Gliddon in their stunningly racist and profoundly pseudoscientific *Types of Mankind, or Ethnological Researches* (1854); much of Gliddon's Egyptological work has been, and was even then, regarded as similarly suspect.

It might be thought that, with the advent of our understanding of evolution through the publication of Darwin's *On the Origin of Species* (1859), the idea of mutually inferior and superior races would ebb dramatically. However, evolution too can be used as a justification for racism if the evolutionist is so inclined. Even Darwin himself was prone to misinterpret the consequences of his own theory, *inter alia* forgetting (in *The Descent of Man* [1871]) that one of the factors in future human evolution is likely to be human free will:*

> At some future period, not very distant as measured by centuries, the civilized races of man will almost certainly exterminate, and replace, the savage races throughout the world. . . . The break between man and his nearest allies will then be wider, for it will intervene between man in a more civilized state, as we may hope, even than the Caucasian, and some ape as low as a baboon, instead of as now between the negro or Australian and the gorilla.

Darwin clearly had a very Eurocentric view of the meaning of the term "savage races." And T.H. Huxley, "Darwin's Bulldog," was little better, writing in 1871:

> No rational man, cognizant of the facts, believes that the average negro is the equal, still less the superior, of the white man. And if this be true, it is simply incredible that, when all his disabilities are removed, and our prognathous relative has a fair field and no favor, as well as no oppressor, he will be able to compete successfully with his bigger-brained and smaller-jawed rival, in a contest which is to be carried out by thoughts and not by bites.

It's interesting to notice that Huxley was yet another on whom weighed heavily the notion that the supposed over-prognathosity of the African skull was an indication that Africans had evolved for eating and biting rather than, as per their European cousins, intelligence. Again, it's hard to comprehend how Huxley, of all people, did not realize that, even if the speculation were true concerning the jaw size being an adaptation toward eating and biting, this was totally irrelevant to the adaptation for intelligence.

---

* If there is such a thing, and whatever precisely the term means. Let's adroitly sidestep that particular mare's nest.

Just because Darwin had changed our views on human ancestry didn't mean there weren't hold-outs. Among these was the Edinburgh anatomist Robert Knox, known primarily to us today as the best customer of the serial murderers Burke and Hare. His association with the murderers, although he was publicly exonerated of any guilt or complicity, led to his widespread ostracism from Edinburgh society, and in 1842 he moved to London. It was there that he began his anatomical studies of the peoples of Southern Africa, developing and becoming near-obsessed by ethnological hypotheses that lacked all evidential support. To judge by the absence of scientific objectivity in this passage he must have suffered some kind of mental collapse after his experiences in Edinburgh:

> Look at the Negro. Is he shaped like any white person? Is the anatomy of his frame, his muscles or organs like ours? Does he walk like us, think like us, act like us? Not in the least! What an innate hatred the Saxon has for him!

What's truly frightening is how easy it is to find people in the twenty-first century who say very much the same thing, as if it formed some sort of rational argument.

Knox's student James Hunt, a profound racist, founded the Anthropological Society of London in 1863 to advance his polygenistic views. An important early member was the ethnologist and translator Richard Burton, of *Kama Sutra* fame.

# # #

For a long time, as we saw, it was believed the three main races—European, African, and Asiatic—must reflect the three sons of Noah who had survived the Flood. Columbus's opening up of the New World, and the subsequent discovery that its inhabitants were of a fourth, hitherto-unknown race, threw the standard model into some disarray.

One hypothesis advanced—by, for example, Isaac de la Peyrère in *Men Before Adam* (1656)—was that the Native Americans, along with all other non-white people, were the descendants of a separate, earlier creation of man from the one in the Garden of Eden; these people, having been created at the same time as the beasts, were really likewise to be regarded as beasts. Somehow they had escaped drowning in the Flood, so their descendants were still alive, co-existing with "true" human beings. De la Peyrère also proposed that Genesis was not really a history of humankind as a whole, just a history of the Jews. But these notions were a little too daring for their day, suggesting as they did both that the account in Genesis was incomplete and that all good, God-fearing Christians were in fact Jews. He was thrown in prison for his dangerous ideas.

The general conclusion was that the Native Americans must have colonized the Americas from *somewhere else*, and the debate began as to where that somewhere else might be.

Ironically, the true explanation for the origin of the Native Americans—that they had migrated from northeastern Asia to the northwest of the Americas, and spread southward from there—was one of the earlier ideas advanced, purely on the basis of the physiological resemblances between Native Americans and Asiatics, such as the broadness of the face and most especially the epicanthic fold. The Italian navigator Giovanni da Verrazano (or Verrazzano) came to this view after his famous 1524 voyage, in which he became the first European to see Manhattan Island; he did not land there, but he landed in various other places and noticed the Asian-ness of the natives' features. The Portuguese seafarer Antonio Galvaño, author of *Discoveries of the World* (1555) and Governor of Ternate, arrived at the same conclusion, and for the same reason, about thirty years later.

Yet the best exposition of this hypothesis was undoubtedly that by the Jesuit friar Joseph de Acosta in his *The Natural and Moral History of the Indies* (1590). Acosta noted the same facial similarities as others before him, but he went far further. He observed that many of the wild animals found in the Americas strongly resembled their counterparts in the Old World. He reasoned that, while immigrant humans might have brought with them their domesticated animals, they would not have brought predators like bears and wolves. The only conclusion must be that the animals had been able to reach the New World under their own steam—i.e., on foot. People had presumably done likewise. Acosta therefore speculated there must be a place where the New World met the Old, that the two landmasses could not be entirely separate. This had to be in some today little-frequented area, and he proposed that in the far northeast of Asia there must be a crossing-place allowing ingress to the northwest of the American continent. Not until decades later, in 1648, was the narrow Bering Strait discovered, by the Russian explorer Semyon Dezhnev, and in fact it wasn't until the 1741 expedition of the Danish navigator Vitus Bering that the world became aware of how close the two landmasses approached. Acosta's deductions were thus over a century and a half before their time.

But these rational voices were drowned in the chorus of wilder speculations. The Lost Tribes of Israel were a frequent choice as ancestors for the

Native Americans, especially after the Dutch scholar Manasseh ben Israel published his popular *The Hope of Israel* (1650). The idea originated far earlier, however, possibly with a suggestion by the Spanish priest Diego Duran in 1580. It was regurgitated to dramatic effect in the writings of the founder of Mormonism, Joseph Smith. The claim in the Book of Mormon that Native Americans were the Lost Tribes was to cause significant embarrassment to the Church of Jesus Christ of Latter-Day Saints over 150 years later when testing of the genomes of Native Americans during the 1990s proved beyond all possibility of doubt that they were of Asiatic descent; how could God have got it so wrong when transmitting the information to Smith? At the start of the twenty-first century the more liberal wing of the Church was calling upon its leaders to apologize to the millions of Native Americans it had converted under false pretenses, as it were; far from agreeing, Mormon leaders described as heretical the very notion that the DNA studies undermined the validity of the Book of Mormon, and presented the public face that the genetics and the scripture were not in fact mutually incompatible. Mormons trying to strike a middle course put forward such hypotheses as that the Lost Tribes first went to Asia, where their genes were "swamped" through interbreeding with the locals; it was sometime thereafter that the hybrid population came to the Americas.

Even so, such ameliorative efforts did not stop the Mormon leaders from beginning excommunication proceedings against a Mormon anthropology professor, Thomas W. Murphy of Edmonds Community College, Washington, who came out into the open in 2002–2003 about the difficulty of believing the historical veracity of the Book of Mormon in light not just of the DNA results but also of its claims that dark skin indicated sinfulness and that a new human life has its sole origin in the father's sperm. The case attracted widespread media attention—some dubbed Murphy the "Galileo of Mormonism"—and in light of the attendant publicity the Church backed down.

Atlantis was also popular as the supposed source of the Native Americans. This was the belief of, among many others, the Spanish cleric Francisco Lopez de Gómara, who presented an argument in 1552 based on the fact that there is a word, *atl*, in the Nahuatl language that means "water." What clearer reference could there be to an Atlantean origin? The Dominican Gregorio Garcia also favored the Atlantean hypothesis, according to his *Origins of the Indians* (1607); on the other hand, in the same book he also proposed that they descended from voyaging Greeks, Chinese, Vikings and/or others, not to mention their being the Lost Tribes of Israel—he described the Native Americans as craven Jews, fit only for manual labor—so who can tell how much credence he gave to the Atlantean conjecture?

Astonishingly, the Atlantean hypothesis still had its adherents in the latter part of the twentieth century. In *Psychic Archaeology* (1977), Jeffrey Goodman, basing his claims on paranormal methods, predicted that a particular site in Arizona would soon reveal indisputable evidence that it was first populated by refugees from the Atlantis catastrophe. His prediction goes unfulfilled.

The Native Americans themselves didn't help much in sorting out all this muddled thinking, for their own myths-of-origin declared they were created in the very territory they inhabit. Again, this hypothesis is still presented on occasion. Jeffrey Goodman, once more, proposes in *American Genesis* (1981) that the Native Americans were the creators, in prehistoric times, of the first human civilization, which then spread from the New World to the Old. Vine Deloria, Jr., in his *Red Earth, White Lies* (1995), roots his arguments for a separate North American human creation more in indigenous myths and legends than in the kind of psychic means favored by Goodman. He also argues that the onetime existence of the Bering land bridge is merely a matter of supposition, not proven fact. It would appear Deloria's motivations are primarily political.

# # #

In 1824, Antoine Fabre d'Olivet published his *Histoire Philosophique du Genre Humaine*, in which he announced a brand new theory of human evolution. Rather than consider that there is a single species of man, whose races may have minor physiological differences but are otherwise of similar nature and antiquity, he theorized that the various races—distinguished by their colors—have succeeded each other. Thus the Native Americans are the relics of the primordial red human race, largely extinguished when Atlantis was lost; then came the blacks (although there seem rather a lot of survivors of this race); and most recently the whites. No mention in d'Olivet's scheme of the yellows and browns.

While the hypothesis is ridiculous—and was so even at the pre-Darwinian time he advanced it—this did not deter the Theosophists from adopting large parts of d'Olivet's theory. According to Helena Blavatsky, founder of the Theosophical Society in 1875, there will be in all seven "root races" of Man, of which we are the fifth. Each of these has seven subraces, and it is from the subraces of one that the next root race is born, with the original root race perishing as its continent sinks. The first and second root races were, respectively, totally ethereal and partly ethereal; it is unclear, in this light, why they should have drowned as their continent sank. The third root race inhabited Lemuria; their continent was inundated when they discovered sex. The fourth root race occupied Atlantis; its survivors became the Mongolians. They got into the habit of miscegenation, so their offspring were monsters. The descendants of these monsters are still visible among us as the "lesser races" of mankind. So far our own root race, the Aryans, has produced only five subraces out of its allotted seven, but a sixth distinct subrace is apparently now making its appearance in California, so our time is running out.

The evidence in favor of the root-race theory is all around us, of course. As Blavatsky pointed out, the Easter Island statues definitely depict members of one of the other root races—probably the Atlanteans.

Another of Blavatsky's more bizarre hypotheses concerning human evolution, expressed in *The Secret Doctrine* (1888), reversed conventional wisdom by saying that in fact apes descended from man. The Mongolians, the previous root race, had been replaced by the Aryans. Apes owed their origin to inbreeding among the Mongolians. She extended this reasoning to claim that all mammalian life owed its origin to the human variety.

While such ideas seem ludicrous to us now, we should realize they were merely part of a spectrum of nineteenth- and early-twentieth-century racist pseudoscience, and by no means at the most extreme fringe of that spectrum. A titan of society like Thomas Alva Edison was one of the early members of the Theosophical Society; even though he remained so but briefly, he clearly didn't find their ideas entirely ridiculous or he'd never have joined in the first place. With so much nonsense being so widely accepted in even the topmost social strata, it's hardly surprising a movement like eugenics sprang up.

EUGENICS

Although often perceived as a pseudoscience or even perhaps as scientifically oriented, eugenics was and is not so much either of these as a belief system. The idea that the species can be improved either through stopping "undesirables" from breeding (negative eugenics) or the encouragement of breeding between "ideal" partners (positive eugenics) dates back long before any ideas of evolution came onto the scene—Plato mentioned notions along these lines—but really came to prominence after the publication of *On the Origin of Species*. A prime early advocate was a cousin of Darwin's, Francis Galton, who also coined the term "eugenics." His statistical researches into human heredity convinced him intelligence and other qualities—such as courage and honor—were inherited characteristics, and this encouraged him to become a proselytizer on behalf of "racial improvement" through both positive and negative selection. Galton's ideal of the human species was the Anglo-Saxon model; on the European continent, like-minded thinkers opted for the Nordics, supposed descendants of the once-great Aryan race.

In the US, the idea of eugenics found fertile soil. People of every nationality, creed and color were arriving in the country, while the end of the Civil War caused a further mixing of populations as the poor from the South sought a living in the North. In places the melting-pot effect was successful, but often, as in so many instances where communities feel vulnerable, people sought scapegoats whom they could blame and, best of all, regard as inferior. Racism was rampant in a complex of forms, as was religious discrimination; and eugenics, with its "scientific" veneer, was perfectly tailored to be a socially acceptable way of expressing such bigotry. In particular, since the eugenicists believed blacks were a separate and inferior strain of humanity, miscegenation was to be discouraged as deleterious to the white lineage. (Strangely, no one seemed to wonder why all the best and most intelligent dogs seem to be mongrels.) The ideal scapegoat group was relatively powerless, so the immigrant Irish Catholics were a good choice; in turn, the Irish blamed and hated the blacks, the single easily identifiable group in society who had even less power than the Irish; and so on.

The most prominent eugenicist in the US was the lawyer and Aryan aficionado Madison Grant, author of the books *The Passing of the Great Race* (1916) and *The Conquest of a Continent* (1933), among others. Passage of the Johnson/Reed Act of 1924 (not repealed until 1952) was a notable success for Grant and his cronies; it selectively restricted immigration of people of "undesirable" stock. Grant also advocated racial segregation (primarily as a way of avoiding miscegenation) and the sterilization of members of "inferior" races and "feeble-minded" whites. In 1907, Indiana passed a state sterilization law, and 31 other states followed suit over the next two decades. The definition of "undesirable" could be very broad indeed: under the 1913 Iowa state law, "criminals, rapists, idiots, feeble-minded, imbeciles, lunatics, drunkards, drug fiends, epileptics, syphilitics, moral and sexual perverts, and diseased and degenerate persons" were all eligible for forced sterilization. Even so, few of the states in fact carried out many sterilizations, since the popular mood was not wholly in favor and since there were questions as to whether the operation constituted "cruel and unusual punishment."

In 1927, however, the *Buck vs. Bell* case came before the Supreme Court. The State of Virginia was determined to sterilize both teenage mother Carrie Buck—deemed feeble-minded solely because she had conceived out of wedlock—and her child on the grounds they were a drain on the state economy and their offspring would be a further drain. In the ensuing court case, the state enlisted the "scientific" support of the Eugenics Record Office (ERO), a laboratory founded in order to research ways of "improving" the US population by Grant ally Charles Davenport—author of such books as *Heredity in Relation to Eugenics* (1911)—and funded by the Carnegie Institution. When the case reached the Supreme Court it was, shamefully, found in favor of the State of Virginia. The degree to which the judges bought into the eugenicists' pseudoscience can be assessed by the majority statement of Judge Oliver Wendell Holmes, son of the great essayist:

> We have seen more than once that the public welfare may call upon the best citizens for their lives. It would be strange if it could not call upon those who already sap the strength of the state for these lesser sacrifices, often not felt to be such by those concerned, in order to prevent our being swamped with incompetence. It is better for all the world, if instead of waiting to execute degenerate offspring for crime, or to let them starve for their imbecility, society can prevent those who are manifestly unfit from continuing their kind.

The judgment was the green light for forced sterilizations, which during the 1930s averaged 2,200 per year in the US; by 1945 over 45,000 people had been compulsorily sterilized, of whom about half were inmates of state mental institutions. Almost half of all such operations were carried out by California. The consequences of the Supreme Court decision were not just national, however, but—and tragically so—international. Over the next few years, sterilization laws were passed in Denmark, Norway, Sweden, Finland and Iceland.

And then there was Germany. There was a strong US connection to the ghastly happenings there, too. In *Mein Kampf* (1925) Adolf Hitler promoted eugenics-based sterilization heavily, and in that same year the US Rockefeller Foundation gave $2.5 million to the Munich Psychiatric Institute as well as further money to Berlin's Kaiser Wilhelm Institute for Anthropology, Human Genetics and Eugenics, all in order to promote eugenics-oriented research. On Hitler's accession to power in 1933, one of the first acts his government passed was a sterilization law; the onus was placed upon physicians to report to a Hereditary Health Court any time they came across someone who was "deficient." The German law was to a large extent based on the existing law in California. In the following year, the American Public Health Association publicly praised the German law as a prime example of good science-based health policy that would benefit society, while the *New England Journal of Medicine* and the *New York Times*—a strong supporter of the US sterilization laws—were effusive in their approval. By 1940, nearly 400,000 Germans had been sterilized according to the country's law. Rather than being appalled, US eugenicists were concerned their nation was lagging behind Germany's sterling example.

From about 1940, Germany simply murdered the insane and other "defectives." This idea was, again, a product of the enthusiasm for eugenics in the US, in particular of Madison Grant's 1916 book, *The Passing of the Great Race*, of which Hitler was a great fan. Grant's position was that, if killing the unfit was the only way to stop them breeding, then it was preferable to allowing them to stay alive. Grant was far from alone in this view—and far from the most extreme. Alexis Carrel, winner of the 1912 Nobel Physiology or Medicine Prize and employed by the Rockefeller Institute for Medical Research, wrote in *Man the Unknown* (1935):

> Gigantic sums are now required to maintain prisons and insane asylums and protect the public against gangsters and lunatics. Why do we preserve these useless and harmful beings? The abnormal prevent the development of the normal. This fact must be squarely faced. Why should society not dispose of the criminals and insane in a more economical manner?

His answer was that they should be "humanely and economically disposed of in small euthanasic institutions supplied with proper gases." That was advice the Nazis took to heart; between 1940 and 1941, when Hitler discovered a new, antisemitic use for his chambers, the Nazis slaughtered some 70,000 of the mentally ill, mainly in Poland, and mainly for the sake of saving money.

After the end of World War II, when the full horrific scale of the atrocities at the German death camps became known to the US public, the ideas of the eugenicists, including enforced sterilization—which at times had enjoyed a 66 percent approval rating among that same public—took an abrupt nosedive, and fortunately they have remained marginalized ever since.

# # #

One of the books that helped fuel the eugenics movement in the US was *The Jukes: A Study in Crime, Pauperism, Disease and Heredity* (1877) by Richard L. Dugdale. Dugdale was a volunteer inspector for the New York Prison Association, and in 1874, while visiting the prison of Ulster County, in New York State's Hudson Valley, he noticed that no fewer than six of the prisoners there were blood relatives. Intrigued, he probed further, and discovered that, of 29 male blood relatives, 15 had been convicted of crimes— an extremely high rate for any extended family. He then traced the family concerned, the Jukes, back as far as he could in the Hudson Valley, identifying 709 Jukes under a diversity of surnames, and finally reaching an ancestor called Max, who was born sometime around 1720–40. The branch of the family that had caused all the trouble had begun with a daughter-in-law of Max's, a woman Dugdale called "Margaret, Mother of Criminals." (All the names he used, including "Juke," were pseudonyms.) Tracing her descendants, he was able to show the family displayed an extraordinarily high incidence of criminality, mental defects and the like, and calculated the financial burden they'd placed on society as an astonishing $1.3 million (equivalent to about $220 million today). He speculated as to whether this catalogue of miscreancy and misfortune was due to heredity—"bad blood"—or environment, coming to no firm conclusion but tending to believe the responsibility was, as it were, the inheritance of a bad environment by each generation from the one preceding it: in other words, the Juke kids always had a lousy upbringing.

The burgeoning eugenics movement ignored Dugdale's tentative conclusions about environmental influences and declared the Jukes' failings to be exclusively hereditary, seizing on the Juke family as an example of the kind of people who might justifiably be weeded out of society to society's benefit, either by sterilization or by euthanasia. In 1911, Dugdale's original notes were discovered and sent to the ERO, which gave a researcher named Arthur H. Estabrook the job of updating the study. Over the next few years, Estabrook claimed to have tracked down a further 2,111 Jukes; there were 1,258 alive at the time, and many of them were—oh horrors—reproducing to produce yet *more* Jukes, at vast potential cost to the taxpayer. In his book *The Jukes in 1915* (1915), Estabrook estimated this further cost at over $2 million (about $48 million in today's terms). Estabrook's researches did show, however, that the Jukes were becoming less problematic—a point the ERO, in its official pronouncement, blithely ignored. At the 1921 Second International Congress of Eugenics, held in New York at the American Museum of Natural History, a full display was devoted to the Jukes as prime targets for eugenic removal. Right up until the general demise of the US eugenics movement at the end of World War II, the Jukes were held up as an example of the kind of problem "sensible" eugenics could cure.

In 2001, however, a poorhouse graveyard was unearthed in Ulster County, and some of the graves there were discovered to be of members of the Juke clan. Further, some of Estabrook's papers became available to researchers, including his charts of the pseudonyms he and Dugdale had used for the various individuals in the extended family. It emerged that, while indeed there had been plenty of bad hats in the lineage, there had also been some pillars of society, a fact neither Dugdale nor Estabrook had thought worth noting. Further, it appeared the real problem besetting the Jukes was in most instances nothing more than straightforward poverty, which had the effect not only of, in the usual way, enticing or forcing some family members into criminality but also of making others vulnerable as scapegoats. The eugenicists' idea that the family suffered an inheritable biological flaw was simply untenable, as one might gather from the title of a more recent book on the subject, *The Unfit: A History of a Bad Idea* (2001) by Elof Axel Carlson. This hasn't, of course, stopped some modern-day self-appointed moral arbiters pointing to the Jukes as a classic example of the way in which "immorality" can be inherited.

The Moral Compass

Sylvester Graham was a crusading dietician and vegetarian in the US; the sugary, over-processed object known today as the Graham Cracker is a bastard descendant of the nutritious wheat-rich biscuit he devised and recommended. A Presbyterian minister, he published books not only on diet and vegetarianism but also on religion and morality. In his 1834 work *A Lecture to Young Men on Chastity, Intended Also for the Serious Consideration of Parents and Guardians* he let rip a full-scale tirade of anti-sexual pseudoscience:

> Those LASCIVIOUS DAY-DREAMS, and amorous reveries, in which young people too generally,—and especially the idle, and the voluptuous, and the sedentary, and the nervous,—are exceedingly apt to indulge, are often the sources of general debility, effeminacy, disordered functions, and permanent disease, and even premature death, without the actual exercise of the genital organs! Indeed! This unchastity of thought—this adultery of the mind, is the beginning of immeasurable evil to the human family. . . .

Beyond all question, an immeasurable amount of evil results to the human family, from sexual excess within the precincts of wedlock. Languor, lassitude, muscular relaxation, general debility and heaviness, depression of spirits, loss of appetite, indigestion, faintness and sinking at the pit of the stomach, increased susceptibilities of the skin and lungs to all the atmospheric changes, feebleness of circulation, chilliness, head-ache, melancholy, hypochondria, hysterics, feebleness of all the senses, impaired vision, loss of sight, weakness of the lungs, nervous cough, pulmonary consumption, disorders of the liver and kidneys, urinary difficulties, disorders of the genital organs, weakness of the brain, loss of memory, epilepsy, insanity, apoplexy,— and extreme feebleness and early death of offspring,—are among the too common evils which are caused by sexual excesses between husband and wife.

In other words, sex—not just real sex but, even more terrifyingly, merely *thinking* about it—rotted the mind and body to almost equal, but certainly enormous, degrees. To many readers of Graham's day this must have seemed nothing more than sound common sense, and a suitably dire warning to wave in the faces of the lascivious, ever impressionable young. What's truly alarming is that among some communities these or similar ideas are still taken seriously today, at least as a matter of public morality, even if private standards of behavior are significantly different. When such attitudes become molders of national and international policy, and are presented as if the conclusions of science, then we need to be severely worried.

Numerous scientific studies have shown that the best way of reducing the number of teenage pregnancies is good sex education coupled with an emphasis on widespread access to contraception, such access to be as little hindered as possible. Similarly, education, this time in techniques of "safe sex," lies at the heart of any scientifically based social campaign to reduce the incidence of AIDS, coupled with the easy availability of condoms; condoms cannot guarantee protection from HIV infection (and other STDs), but do very considerably reduce the risks.

Fundamentalist Christians in the US and elsewhere are generally appalled by the way in which the proven approaches to both problems assume the people involved will actually, you know, *have sex*, and have been active in pushing sexual abstinence as a means of reducing both unwanted pregnancies and the spread of AIDS. The situation in connection with AIDS is further complicated by the fact that the official position of the Roman Catholic Church is to reject all forms of contraception, condoms included; obviously, even if the motive in using a condom is to protect from infection, it's still a contraceptive as well.

Clearly complete sexual abstinence would indeed reduce the risks of both pregnancy and the transmission of STDs, AIDS included, by a full 100 percent. The tricky part is the actual abstaining. Even those supposedly most enthusiastic about pushing abstinence programs have been revealed time and time again to have difficulty with that half of the equation.

That the problems exist is not at issue. According to the CDCs 2015 figures, each day in the US alone 630 babies are born to teenage mothers and over 25,000 people between the ages of 15 and 24 contract an STD, while about 22 percent of all new HIV cases are of people in the age group 13–24. What is at issue is the best means of tackling those problems, and here science is clear: abstinence-only programs are ineffective by comparison with the education/contraception mix, and are probably the main reason the US has the highest levels of teen pregnancies and STDs in any of the developed countries. When the resolution of the iron-willed abstinent youngsters finally cracks, intercourse is likely to be on the spur of the moment and far less likely to involve condoms—assuming the participants can obtain condoms at all. So they risk it, "just this once," and then again, and . . . Among adults, where the primary concern is AIDS, the situation is similar, but worsened by the attitude of especially the males in some communities that condom use is non-macho; this latter is a problem that can only be cured through education. Yet today as much as one-third of all the money given by the US for AIDS prevention is spent, and mandated by Congress to be spent, on abstinence-only programs—i.e., is very largely wasted. Further, in order to pander to the prejudices of the electorally important self-styled "moral majority," US funding is systematically withheld from anti-AIDS organizations that advocate condom use.

To put this in some context, many of the victims of AIDS are infants, infected by their mothers at birth or while breast feeding. Some 60 percent of the infants thus infected die before reaching the age of three. This mortality rate can be cut by half through inexpensive treatment with anti-retroviral drugs. In many instances, in poor countries, money is not available for such treatment because it has been channeled away to be used instead on ineffective but "morally" prescribed abstinence-only programs. The victims of this sort of "morality" are thus innocent babies.

Of course, the victims of such ideological prohibitions could be the mothers instead. In the early 2000s, the pharmaceutical corporation Merck developed a vaccine called Gardasil, with seemingly 100 percent effectiveness against the most prevalent virus (human papilloma virus—HPV) responsible for cervical cancer; it was accordingly approved in 2006 by the

FDA. No wonder. Cervical cancer hits about 14,000 US women each year, killing nearly 4,000 of them; worldwide, some 270,000 women died of cervical cancer in 2002. Anything to reduce that nightmarish toll would surely be welcome.

Well, not according to religious groups like the Family Research Council, which boasts that it "promotes the Judeo-Christian worldview as the basis for a just, free, and stable society." Presumably the justice and freedom don't extend to the mainly poor women who suffer the highest rates of cervical cancer since they can't afford regular checks for HPV. The FRC's president, Tony Perkins, presented the moral justification for the organization's savage resistance to Gardasil: "Our concern is that this vaccine will be marketed to a segment of the population that should be getting a message about abstinence. It sends the wrong message."

In recent years, convincing evidence has emerged that HPV is a significant cause of oral cancer—the CDC reports 16,400 HPV-induced cases annually—with 80 percent of those cases occurring in *men*. One wonders if the views of Perkins and the Family Research Council about HPV vaccination will change as a result.

# # #

Although the vast majority of people regard syphilis as a scourge, this was in the past by no means a universal opinion. Some, like John Bunyan, thought it was God's just punishment for mortal sin. Even after western society had drifted away from the belief that God micro-managed every aspect of the human condition, ideological morality—tritely summarized by the notion that, if it was fun, it must be sinful—still played a part. Scientists were not immune to these misguided ideas. As late as 1860, we find the UK physician Samuel Solly, President of the Royal College of Surgeons, claiming that syphilis was a blessing: without the disease as a check, "fornicators would ride rampant through the land." The fallacy here lay in the fact that fornicators rode rampant through the land *anyway*; the task of scientists is to deal with the world as it actually is, not the world as they'd like it to be.

Similar prudery was to be found in many of the other medical textbooks of the day. In John Hilton's *Rest and Pain* (1863) male masturbation—Hilton seems to have been unaware of female masturbation—was accused of causing serious illness and thus something to be prevented at all costs. Hilton's proposed cure was draconian: applying iodine to the penis until it blistered, so that handling it became an agony too far for even the most dedicated masturbator.

In general, preventives against masturbation were even worse than the various awful afflictions supposedly associated with the practice. John H. Kellogg, in his *Embracing the Natural History of Organic Life* (1892), passed on the fruits of his practical experience:

> A remedy which is almost always successful in small boys is circumcision, especially when there is any degree of phimosis [tightness of the foreskin]. The operation should be performed by a surgeon without administering an anaesthetic, as the brief pain attending the operation will have a salutary effect upon the mind, especially if it is connected with the idea of punishment, as it may well be in some cases.

For nonphimosal or already circumcised boys, the prescribed operation "consists in the application of one or more silver sutures in such a way as to prevent erection." Luckily for all concerned, this operation had fallen out of fashion by the era of the airport-security metal detector. But if Kellogg's young male patients should have been petrified, pity the plight of the females:

> In females, the author has found the application of pure carbolic acid to the clitoris an excellent means of allaying the abnormal excitement, and preventing the recurrence of the practice in those whose will-power has become so weakened that the patient is unable to exercise self-control.

It seems not to have occurred to Kellogg that any sensible woman would simply stop reporting her "problem" to him.

It comes as something of a shock to find nineteenth-century prudishness in the medical profession as late as 1959. That was when the British Medical Association withdrew and pulped all quarter-million copies of its annual publication *Getting Married* because an article in it by prominent sexologist Dr. Eustace Chesser suggested that premarital chastity was nowadays only optional—even though Chesser's conclusion was to recommend it.

# # #

Sexual pleasure is not the only sin that has attracted the sharper end of the moral compass. With better reason, so has booze. The US Prohibition experiment of 1920–33 was a dramatic example of the folly of legislating according to what abstract morality says people *ought* to do, rather than what people actually do.

Perhaps unsurprisingly in that, for decades before it was regarded as a medical condition, alcoholism was dismissed as a sin of the weak-willed, there has been ideology-driven controversy over the condition's treatment. The longstanding approach, as adopted by Alcoholics Anonymous and others, is that alcoholism is a physiological or metabolic disease, possibly congenital, whereby a single drink can trigger the sufferer into uncontrollable drinking. The only treatment, in this view, is total abstinence. There is no such thing as a cured alcoholic, merely an alcoholic who declines to drink. A converse view is that alcoholism is not so much a disease *per se* as a behavioral ailment, and that a better system of treatment involves individual counseling alongside training in the art of controlled drinking.

Leaving aside the debate as to what alcoholism is, the disadvantage of the total-abstinence approach is, obviously, that its prediction can all too easily become self-fulfilling: the abstinent alcoholic, if tempted into that single fateful drink, "knows" there's no escape from plunging back into the habit, and therefore plunges. This is not to say the work of the AA has been without its successes—they are legion—but, equally, there has been a high level of failure. Any approach that would reduce that failure level would obviously be welcomed by all.

Or not. Various scientific studies have indicated that the counseling/controlled drinking approach has a higher success rate. One of the fullest of these, published in 1982, was by Linda and Mark Sobell of the Addiction Research Foundation, Toronto: it (and a post-controversy follow-up by the Sobells) showed that after a decade the traditionally treated group had a mortality rate of about 30 percent while that of the counseled group was significantly lower (although still appallingly high), at 20 percent. This paper caused outrage in some circles. Almost immediately a team led by Mary L. Pendery published a rival paper that in effect accused the Sobells of fraud—of fudging the figures.

The consequence was not just one but two investigations of the Sobells' claims—by the Addiction Research Foundation and by the US House of Representatives' Committee on Science & Technology. Both found there was absolutely nothing wrong with the Sobells' work—and most certainly not the slightest suspicion of fraud. Pendery's team had simply set out to attack the conclusion, whatever the validity of the science. In his discussion of the affair in *False Prophets* (1988), Alexander Kohn cites the physician D.L. Davies, writing in 1981:

Yet so strong are entrenched ideological views on this issue that the argument waxes even more fiercely, recalling the 19[th-]century battles between wets and drys, using indeed the very language and thoughts of early 19[th-]century temperance workers with the same preoccupation with the morals and religious aspects of the "first drink" and the role of divine help.

In other words, the total-abstinence approach must be the uniquely right one because God says so, whatever the scientific evidence. And, if one is empowered by Divine Authority, any means of attack, no matter how venal, is virtuous. This, of course, makes no sense even in its own terms.

The enforcement of celibacy among the priesthood of the Roman Catholic Church by Pope Gregory VII eventually gave rise to a still surviving form of "hidden discrimination" within the sciences.

In the first millennium or so of the Church, celibacy wasn't much of an issue. It was only with the rise of the monastic movement—initially regarded as an extremist fringe—that the idea of celibacy's desirability came to the fore, and by the end of that millennium it became official Church policy, although only patchily obeyed. Gregory, seeing in priestly celibacy a way of ensuring Church property stayed Church property—celibate priests could have no offspring likely to raise troublesome arguments about inheritance—cracked down harshly. While Gregory's motives were not misogynistic, his move encouraged the development of misogynistic attitudes among the male clergy. When the first universities arrived on the scene around 1200 they did so as offshoots of the cathedral school system, and were for the benefit of the male clergy; the rule of celibacy, and its associated misogyny, thus became institutionalized in them. Women were not admitted,* and thus they missed out—and for some centuries continued to miss out—on the entirety of Europe's renaissance in mathematics and philosophy (the equivalent of science). Such subjects became, in the eyes of society, "not womanly." As late as the twentieth century, some western universities barred women.

We're still suffering the cultural hangover from such attitudes. In the US in the late twentieth century only 9 percent of physicists were female.

---

* Although apparently the rise of the universities closely parallels that of the brothel, it being discovered that students provided a constant and eager clientele.

(A mere 3 percent of full physics professors were women.) Matters were considerably better in mathematics (36 percent), chemistry (27 percent) and especially the life sciences (41 percent), but still not good. Hopefully this social bias will soon disappear: the more good physicists we have, the better.

CORRUPTION OF SCIENCE BY THE IDEOLOGY OF SCIENCE

Science is itself capable of generating its own scientific ideologies. As example, there's the idea put about by Thomas Szasz in books like *The Myth of Mental Illness* (1961) and its successors that there is really no such thing as mental illness, rather, it is an invention of psychiatrists eager to earn a quick buck by creating a profession where none is needed. In Szasz's view, the people we regard as mentally ill are merely unusual human beings: there is no ailment there to treat. Insanity is nothing more than a social construct, decided by the majority. It's rather startling to find that these very postmodernist-sounding notions—which seem based on an ideological rather than a scientific agenda—are being advanced by someone who was for decades Professor of Psychiatry at Syracuse University. His position might have seemed more reasonable in previous centuries, when indeed glorious eccentrics, unless rich enough, might be classified as mentally ill (as still, of course, sometimes happens)—or, even earlier, burnt at the stake as witches—but it appears somewhat fantasticated today, when glorious, and even inglorious, eccentrics are more likely to be booked on Oprah.*

In a quite different area, there are various ideologies involved in the search for extraterrestrial life. Leaving aside the ideologies of ufologists and creationists, we're far too prone to regard our own *modus vivendi* as the one likely evolved by organisms elsewhere. In recent decades discoveries on earth have had a rather sobering effect on this preconception: lifeforms have adopted all sorts of ways of coping with environments that would have earlier been thought impossible—and which are certainly far removed from what we'd regard as the terrestrial norm.

Recognizing this earth-chauvinism, the National Research Agency now has a panel devoted to "weird life" that advises NASA on what else its Mars

---

* That said, it's true that in our modern world vast numbers of people are, thanks to aggressive marketing techniques, being prescribed mind-influencing drugs for which they have no need.

missions might look out for rather than earth-type, water-based cells. As was pointed out in 2007 by geologist Dirk Schulze-Makuch of Washington State University, the Viking missions of the 1970s, in testing for life, may actually have drowned Martian microbes through the assumption that living cells must be filled with salty water and thus would respond positively to the addition of a richer water supply. In the environment of Mars, Schulze-Makuch reasoned, where temperatures plummet far below freezing, a much more likely—and perfectly workable—constituent of living cells would be a mixture of water and hydrogen peroxide. Such a mixture does not freeze until below about –55°C (–67°F); moreover, on doing so, unlike water it does not expand, and thus would not necessarily rupture cell walls.

# # #

Today the great talk is of globalization, the process, rapidly increasing in pace, whereby science and technology is permitting—indeed, encouraging—such a large-scale integration of culture between all the peoples of the world that a single, truly global culture is no longer just a visionary pipe dream. At the moment the benefits of globalization are, notwithstanding enthusiasts such as Thomas Friedman, somewhat intangible. To a great extent the effect of globalization has been one of leaching employment from the wealthier countries to the poorer ones, which would be a good thing were it not that the economic priorities of the corporations of the wealthier countries simultaneously seek ever cheaper sources of workers, thereby encouraging the spread of what is tantamount to slave labor. The shades of those German corporations which discovered before and during World War II (and even, shamefully, for a short while after it) the economic bonus of using slave workers are not so far behind us.

There's a further problem with the emergence of a global economy and culture. Civilizations collapse. History is filled with examples of major civilizations being wiped from the face of the planet. Remember Shelley's Ozymandias. Always, however, as one civilization has collapsed there has been another, or at least the potential for another, waiting in the wings. In the case of a unitary global civilization, though, there can be no such backup: all our eggs will be in a single, very large basket. What will happen if—or, more realistically, *when*—the global civilization collapses?

Science has no answer, for the very good reason that science has yet, aside from a few lonely and distant voices, to address the problem. In large part this is because of the structure of modern science, which structure, while immensely valuable, has an unrecognized corruptive influence on the very science it is supposed to protect and enhance. Science progresses in general through the publication of scientific papers, which are reviewed by other scientists, who accept, reject or amend their conclusions. Modern science is thus primarily an accretional process; complete paradigm shifts, or "scientific revolutions," are an extreme rarity, and anyway generally have plenty of precursors. Even such a radical paradigm shift as the Darwin–Wallace theory of evolution by natural selection did not come out of nowhere: its precursors can be traced back to ancient times. This accretional, collegial process is the strength of modern science. In general, loony notions do not last long. While science's record in accepting genuine new truths is a bit more patchy, sooner or later those make it into the canon; for example, the theory of continental drift proposed by Alfred Wegener was rejected and even ridiculed for decades until other knowledge emerging through the collegial system led to its eventual acceptance.

The seriously debilitating weakness of the process is that the measure of modern scientific success has too often come to be the publication of those papers. Academic and commercial employers alike, not to mention those who award government grants, want to see a plentiful bibliography attached to the name of any scientist. As we saw earlier, this has led to many cases of fraud and to the rise of the predatory journal, but it is also systemically damaging. The onus is on scientists to publish early and often. The short-term experiment becomes more valuable than the more major one that could take years or even decades to produce a result. Yet the longer-term experiments are very often the ones that are more important for the advancement of human knowledge. When Gregor Mendel, for example, conducted the experiments that laid the groundwork for the whole science of heredity, he had to observe numerous generations over a period of years. What modern scientist, worried about tenure and with the dean breathing down her neck, could consider an experiment that would take that long before a paper could be published?

Of course, some scientists still do. But they do so *in spite of* a system that rewards the frequent publication of trivial knowledge. Perhaps they're placed such that they can ignore the system's imperatives—perhaps they are the dean!—or perhaps the experiment is one that, albeit long-term and important, can be done in, effectively, spare time while other things are going on. Fortunately a few commercial entities have noticed the problem, and are beginning to take appropriate steps to solve it. They recognize that, while the bean-counters are clamoring that the important thing is to sustain profits for the next quarter, it's actually pretty useful to ensure, too, that you'll still have customers in twenty years' time.

This is of course a role that democratic governments used to perform, and many still do. (Totalitarian regimes sometimes think in the longer term too, although often for motives of self-glorification—the Ozymandias syndrome.) Increasingly, however, governments are being lured into the trap of short-term thinking: if the voters are going to the polls in a year's time, you want some Big Result to show them *now*, not just the news that you've started a project that may benefit their grandchildren. An example was the George W. Bush administration's hogtying of stem-cell research, seeking to capitalize in the short term on the religious right's antipathy toward such research. One argument was that, while there are plenty of claims about the potential of stem-cell research, the benefits have not been proven and cannot be so until at least a decade or two down the line. The fact that *this is what good science is all about* was ignored in the rush to appeal, however spuriously, to a powerful voting bloc.

We're probably right to have some confidence that the scientific establishment, once the problem of short-termism has been properly confronted, will find a way to solve it. At the moment, though, it remains an example of science in effect corrupting itself.

# # #

One bizarre form of ideological corruption of science has been nationalism—or, rather, the stereotyping of a country's science along nationalistic grounds. The arch-exponents of this sort of thing were of course the Nazis, whose doomed attempts to create a purely "German science" we'll consider in Chapter 7. But that was merely one end-point of a story that had been going on in Europe throughout the eighteenth and nineteenth centuries, the principal antagonists being Germany and France. Some of the chauvinism was positive (our country's science is the best because . . .) and some of it negative (your country's science is lousy because . . .). As an example of such stupidity, in *The Undergrowth of Science* (2000) Walter Gratzer cites (but alas does not name) the French Minister of Education, speaking in 1852:

> Does not our tongue appear especially suited to the culture of the sciences? Its clarity, its sincerity, its lively and at the same time logical turn, which shifts ever so rapidly between the realm of thought and that of feeling—is it

not destined to be not merely [scientists'] most natural instrument but also their most valuable guide?

Matters heated up during World War I.* Clearly scientists took sides during that conflict—and some of them served and died—but this was a somewhat different issue than the importation of nationalist enmities into science, which reached such a level that *Nature* was moved to remind its readers that science is not a matter of politics and transitory human preferences.

It would be pleasing to think that such childish follies are behind us, but there are still plenty of cultures, even in the west, that regard science as a whole as intrinsically evil simply because it is of western origin, preferring instead to look to other traditions . . . specifically those that lead the credulous off into woo-woo land. It can certainly be argued that the physiognomy of what we know might have been different had science followed a different cultural course—biology might have advanced at the expense of physics, perhaps—and this might have been a good or a bad thing, but that's a different *what if?* game. Reality is unaffected by the order in which we discover its secrets. To claim that the nature of reality would somehow have been *other* if we'd gone about the task of exploring it differently is patent nonsense.

One of those who with particular venom denounced western science as corrupt was Mao Zedong. At the time that Marx and Lenin were writing, it was reasonable to believe matter was infinitely divisible, as they said, and that the universe was infinite in both space and time. Since Marx and Lenin must be right in everything they'd written, Mao declared newfangled notions like particle physics, the expanding universe, and the Big Bang to be merely the corruptive fictions of bourgeois western scientists. Despite Mao's strictures, the Chinese managed to develop the atomic bomb and nuclear power, so perhaps the Chairman had two sets of belief, one for public and the other for private consumption.

---

* In an amusing prefiguring of US attempts in the 2000s to popularize the neologism "freedom fries" for "french fries," there was a move during World War I, again in the US, to rename German measles "liberty measles"!

Media Muffins

A further corruption of science that's sometimes ideological, sometimes not, occurs through the fact that almost all of the information most adults receive about developments in science and technology has been filtered through the media—newspapers, magazines, websites, radio, and television. While some science journalists are marvelously gifted at their jobs, explaining complexities and significances to the lay audience with dazzling skill, far too many are not. All too often one gets the impression that the job of discussing science issues has been fobbed off onto whoever was slowest to leave the room.

Further, some media pundits feel it is their prerogative to make pronouncements on science, and often said pundits display a definite streak of anti-scientism. Richard Dawkins, in *Unweaving the Rainbow* (1998), suggests, probably correctly, that much of this anti-scientism is born from the primitive habit of disliking things we don't understand, and cites several examples of such folly. Here, for instance, is Bernard Levin in *The Times* in 1994 mocking the notion of quarks: "Can you eat quarks? Can you spread them on your bed when the cold weather comes?"

This particular fit of lunacy drew a prompt response, in the form of a Letter to the Editor, from metallurgist Sir Alan Cottrell: "Mr. Bernard Levin asks 'Can you eat quarks?' I estimate that he eats 500,000,000,000, 000,000,000,000,001 quarks a day . . ."

On the other side of the Atlantic, anti-scientism is rife among the political pundits. Who could forget the 2011 interview in which Bill O'Reilly lectured David Silverman on the ineffability of the tides:

> I'll tell you why [religion's] not a scam, in my opinion: tide goes in, tide goes out. Never a miscommunication. You [scientists] can't explain that.

Elsewhere O'Reilly, like his spiritual sibling Rush Limbaugh, has expressed a reverence for scientific studies that have proved, to put it mildly, a little hard to track down. Here's a December 2006 exchange from *The Radio Factor with Bill O'Reilly* in which O'Reilly discusses with co-host Edith Hill a significant statistical correlation:

> *O'Reilly*: [H]ere's something [they] didn't poll but I know: that most women who like artificial trees—

Hill: Yeah?

O'Reilly:—have artificial breasts.

Hill: What?

O'Reilly: Did you know that? Yeah, there's a correlation. Yeah. There was a study done—

Hill: You know . . .

O'Reilly: It was, it was done at UCLA in LA. All right—

Hill: I don't believe you . . .

O'Reilly: We gotta take a break.

Also from December 2006 there's the *Washington Post* pundit Charles Krauthammer using spurious science to enter the debate concerning the murder of disaffected ex-KGB agent Alexandr Litvinenko with a dose of polonium-210. Was the murder committed under the auspices of Vladimir Putin, or could responsibility lie elsewhere? Krauthammer has no doubts:

> Well, you can believe in indeterminacy. Or you can believe the testimony delivered on the only reliable lie detector ever invented—the deathbed—by the victim himself. Litvinenko directly accused Putin of killing him. Litvinenko knew more about his circumstances than anyone else. And on their deathbed, people don't lie.

Not only is it obviously the case that people *do* sometimes lie with their last words—"I have always been faithful to you, my darling"—and that it's a great venue for the settling, honest or otherwise, of old scores, but there's the equal possibility that perfectly truthful people might simply be wrong. (That said, one suspects Litvinenko was right.)

Far more pernicious is a relatively recent example (chosen from among a treasure house of others) of an ideological agenda blinding the pundit to established science. In January 2007, commenting on the freak weather that was devastating the US Midwest, Fox News's Neil Cavuto commented:

> Twenty-four degrees [F] in Fresno, twenty-nine degrees in Phoenix, down to nine in Amarillo, Texas, and on and on. It is some of the coldest air in this part of the country in twenty years. Proof that all of this hype over global warming could be just that—hype?

It's inconceivable that any journalist could be so ignorant of the science concerning climate change that s/he would be unaware that localized, short-term weather extremes of any kind—cold or hot, wet or dry—are perfectly compatible with global warming.* Whether he was deliberately corrupting the science to mislead viewers in the hope of promoting his ideology or whether his own comprehension of the relevant science had been corrupted by that same ideology, the result was the same: the scientific understanding of his viewers was being grossly corrupted. Had he been as inaccurate about, say, the latest football scores, he might well have been fired; as it was, his baloney seemed to go unnoticed (or was even encouraged) by his employers—a sad comment on their own attitude toward scientific truth.

# # #

The inevitability of the popular media's contribution to the corruption of science may seem obvious: if a piece of science were capable of being explained to the scientifically uneducated within a one- or two-minute news segment, it could hardly have taken countless scientists months, years, or even decades to understand it. The highest the best-intentioned, most responsible of broadcasters can aim for is to present a grossly simplified version, and a gross simplification of *anything* is by definition a false, or at least a very incomplete account.** Newspapers can make a better attempt, but they too are trapped in the prison of necessary simplification; for example, they cannot expect their readers to follow (or their typesetters to typeset!) page after page of mathematical equations.

Often TV news shows are keen to present only the most sensational in science, which very often means stepping right outside science into the pseudosciences or the downright fraudulent, while still retaining the "science" label. A further complication is the popular media's corruption of the concept of "balance." Balance in journalism is obviously a good thing: if there is a genuine dispute in science—as there once was, for example, between the Big Bang and Steady State cosmologies—a discussion between proponents of the conflicting views is likely to be enlightening; we can at least come away with the correct impression that there's a debate going on. Where the corruption occurs is in the many instances where there is not

* It's equally inconceivable that Cavuto could have been unaware that, at the same time, states like New Jersey were basking in warm, near-summery sunshine rather than suffering their customary early-January snow and ice.

** A few years ago I received a request to write a children's science book on nuclear power. Part of the briefing was that I should avoid all mention of subatomic particles, or even that the atom could be split, because subatomic particles were "too difficult."

a debate going on, but the producers, always in search of "sexy television," give the viewer the impression there is. Typically, we're presented with, on the one hand, Talking Head A and, on the other, Talking Head B. What is withheld is the crucial datum that Talking Head A is a distinguished scientist, representing the conclusions of every researcher in the relevant field, while Talking Head B hears voices. The two opinions are quite falsely presented as if of equal weight; if challenged, the broadcasters are likely to claim piously that they are "leaving it up to the viewer to judge"—which, of course, they're not, because they've failed to supply the principal basis upon which the viewer could attempt a judgment.

In this looking-glass world we have, for example, Michael Crichton, whose qualifications were in medicine, was offered as an expert on climatology. (In a yet more hilarious example, unqualified demagogue Ann Coulter has been interviewed about her opinions on global warming.) This particular corruption easily translates to political bureaucracies. A feature of the Bush II administration and later the Trump administration has been the filling scientific posts on the basis of political loyalty rather than relevant scientific qualification.

# # #

Not just the mainstream media contribute to the corruption of both science itself and the image of science in the popular mind. As noted, a certain amount of public misunderstanding of scientific matters derives from the efforts of popularizers to explain to laypeople material that simply is not explicable in lay terms. But this is a relatively minor problem alongside the efforts of those who, for political or ideological reasons, deliberately set out to distort science in the public mind, sometimes mounting an outright attack in an attempt to destroy the institution of science itself. There are plenty of examples in these pages.

There are also those who strive to corrupt the public understanding of what science is, portraying it as "the enemy." Their motivation may be part of an ideological agenda, or it may simply be to gain power or financial profit. In *Sleeping with Extra-Terrestrials* (1999), Wendy Kaminer reserves especial venom for the self-styled gurus and authors of pop self-help/spirituality books—often massively bestselling—and their deliberate falsification and denigration of science. It is a weary rhetorical trick (the straw man argument) to misrepresent the arguments of one's debating opponent and then attack what was never claimed in the first place, and one would have thought we the public would have wised up to it long ago; yet, in "spirituality" as in politics, this dishonest device has been used with success for centuries, and seems today to be even more effective than ever, as media institutions gullibly or thoughtlessly promote the dissemination of false ideas. To choose just one of Kaminer's examples, in their introduction to *The Celestine Prophecy: An Experiential Guide* (1995), James Redfield and Carol Adrienne state that "those who take a strictly intellectual approach to this subject will be the last to 'get it,'" and advise readers to "break through the habits of skepticism and denial." As Kaminer summarizes these and many other pop-spiritual authors, "Skepticism they view with contempt, as the refuge of the unenlightened." If you want to read *The Celestine Prophecy*, its own authors advise, you should leave your brain at the door.

Science, in this view, is the enemy of understanding—much as it used to be the claim of bogus spirit mediums that the souls of the departed would fail to materialize should there be anyone in the circle sufficiently skeptical to notice the strings and pulleys. What the gurus are essentially saying is "Only through believing bollocks can you find enlightenment." Of course, the pop gurus don't put it quite like that: those who notice the strings and pulleys are "insufficiently spiritually evolved" or suffer from "closed-mindedness." Those who swallow this stuff whole are, by contrast, the enlightened and open-minded.

One of the leading authors in this field is Dr. Deepak Chopra, whose publications include *Quantum Healing: Exploring the Frontiers of Mind/ Body Medicine* (1989) and *Ageless Body, Timeless Mind: The Quantum Alternative to Growing Old* (1993), in which he seems to promote such ideas as that people can cure themselves of cancer by adjusting their own internal quantum mechanics. (This may be a misrepresentation: it's hard to know exactly what Chopra means except that there's a lot of quantum involved.) Few members of the public and more particularly few of Chopra's eager customers pause to reflect that Chopra himself today looks, well, older than he did when these books were published. Of course, such paradoxes are to be expected from a man whose arguments against Darwinian evolution and in favor of Intelligent Design include that there was no such thing as a self-replicating molecule for billions of years after the Big Bang until DNA appeared on earth, without any apparent awareness that there was no such thing as the earth itself until billions of years after the Big Bang; if Chopra has a means of detecting the lack of self-replicating molecules in the rest of the universe, he should publish details of this technology in *Nature*.

As has often been remarked, the trouble with having an open mind is that people come along and put things in it.

6

THE CORPORATE CORRUPTION
OF SCIENCE

The next generation would be justified in looking back at us and asking, "What were you thinking? Couldn't you hear what the scientists were saying? Couldn't you hear what Mother Nature was screaming at you?"

—Al Gore, introducing *An Inconvenient Sequel*, August 2017

Sometimes there's not enough Prozac to get through the day.

—David Ellenberger, Climate Reality Leadership Corps Training, Denver, March 2017

In July 2017, two organizations, the Bioscience Resource Project and the Center for Media and Democracy, created the Poison Papers website (https://www.poisonpapers.org). This vast online trove of documents, most of them collected using freedom-of-information laws by Carol Van Strum, author of *A Bitter Fog: Herbicides and Human Rights* (2014), reveals horrifying details of the nefarious activities of various chemicals manufacturers and the regulators supposed to monitor their activity for the public good.

The Environmental Protection Agency (EPA), whose name appears often in this chapter, sometimes as hero, sometimes as villain, is frequently to be found in its latter persona as one trolls through the Poison Papers.

One particularly revealing transcript is of a meeting called by the EPA at a Howard Johnson Inn in Arlington, Virginia, on October 3, 1978. Scientists at the Food and Drug Administration (FDA) had spotted that the toxicology testing work being done at Industrial Bio-Test Laboratories

(IBT)—then the biggest lab of its kind in the US, performing over one-third of all toxicology testing done in the country—was often substandard. Since IBT did a lot of the work upon which the EPA relied when determining the safety or otherwise of new commercial chemicals, this was obviously a matter of great concern. IBT scientists were taking shortcuts in some of their testing; they were failing to report negative results; in some instances they were, it seemed, inventing data. The EPA's regulators had been blithely taking the results received from IBT at face value. After the FDA raised the alarm, the EPA's personnel were forced to re-examine IBT's data, and reluctantly concluded that as much as 80 percent of it might be flawed. The testing was so shoddy that scientists at IBT had even begun to refuse to sign their own reports, for fear of repercussions down the line.

So, was this the moment for the EPA to go public and alert the populace to the fact that many of the chemicals it had approved for use were perhaps not as safe as everyone thought?

No.

Instead the EPA called the meeting in Arlington, Virginia. Invitees included not just EPA bigwigs but people from Canada's approximate equivalent, the Health Protection Branch, plus representatives from the chemicals industry.

Were the latter about to get the grim news that swaths of the chemicals the EPA had approved would have to be pulled from the market, at least temporarily, while the toxicology tests were repeated?

Again, no.

Instead, the head of the EPA's Regulatory Analysis and Lab Audits division, Fred Arnold, told them that *none* of the affected chemicals were to be withdrawn, whatever the potential threat to public safety. It was far better for the EPA, for the industry and even—so Arnold claimed—for the public if the boat remained unrocked. After all, he pointed out, much of the work done at IBT *had* been perfectly satisfactory. He was even content for the EPA not to reevaluate those reports the experimenters had left unsigned.

The "much of the work" to which he referred would be, according to the EPA's own best guesstimates, about 20 percent of the total. (A belated EPA investigation, done in 1983, revised this figure downward to 16 percent.)

It seems from the meeting transcript[46] that the spokesman for the Health Protection Branch, David Clegg, was not entirely delighted by Arnold's approach to the problem, but nonetheless went along with it after token protest. The industry representatives, on the other hand, must have been delighted.

In 1981, several key IBT officials were indicted by a grand jury, and in April 1983 there began a criminal trial to determine if the individuals accused had engaged in deliberate scientific fraud. In October that year the court decided they had, and three were sent to prison. There were repeated reports that some of the chemical manufacturers had actively colluded in the fraud, paying people at IBT to produce the spurious data, but this has never been proven in court.

Already, in July 1983, the EPA had belatedly agreed to put 34 IBT-approved pesticides on probation until their manufacturers provided evidence of their safety. Bearing in mind the sheer number of IBT chemical tests that the EPA had accepted, this seemed a very mild response.

At the heart of this appalling story is not just a case of scientific fraud whose human costs have never properly been assessed, but also one of collusion between regulators and regulated that should never have happened. How could it have come to pass?

Reading the Arlington transcript, it's evident the EPA's Fred Arnold is treating his counterparts in the chemical industry as colleagues: there's a cozy, collegiate feel to the whole discussion. Us versus the world. There's no reason, of course, why regulators should necessarily treat their industry counterparts as adversaries—most of the time, anyway—yet it's important, too, that they not be seen as, somehow, allies.

In fact there's a sociological term, "regulatory capture," that describes one of the ways in which government regulators can become corrupted by the industries they're supposed to be policing.

The psychology is fairly obvious. We all tend to like to talk informally about our work—that's why so many groups of office workers totter off to the pub at five-thirty to wind down for a while before going home. Being a regulator in a state or federal monitoring agency can be a lonely business: a lot of what you're dealing with during the day can be too confidential for discussion with workmates in the pub or your spouse at the dinner table.

On the other hand, the people you're chatting with on a daily basis in the industry you're regulating—*they* know what you're talking about, can sympathize with your problems, and can understand your jokes. It's all too easy for the regulators to start believing that they and their industry counterparts are somehow "on the same team"—with results that are potentially catastrophic to the public.

# # #

Insidiously influencing the regulators is one method that industry uses to try to ensure that public science is bent in its favor. Another is to use the law to harass current critics and thereby intimidate potential future ones into silence. In this chapter we'll find plenty of examples of this ethically dubious tactic, such as the threats by chemical manufacturers to sue over Rachel Carson's *Silent Spring* (1962).

Following legal threats received from Monsanto by its authors, in the late 1990s the publisher Vital Health suddenly abandoned its plans to release the book *Against the Grain: Biotechnology and the Corporate Takeover of Your Food* by Marc Lappé and Britt Bailey; the Monsanto claim was that the book misrepresented the company's bestselling pesticide, Roundup (glyphosate). One can understand Vital Health's fears; based in Ridgefield, Connecticut, the company is no HarperCollins-style behemoth of the publishing industry, and could rapidly have been driven out of business by the chemical industry giant. The two authors took their book to the even smaller (and aptly named) Common Courage Press, which published it in 1998.[47]

Also in 1998, in the UK, the longtime printer of the *Ecologist* magazine unilaterally and without notice pulped, at its own expense, all copies of a forthcoming special issue entitled *The Monsanto Files*. The magazine reprinted with a different company. What's interesting here is that Monsanto denied point blank having made any approach to the original printer. In a way, as the magazine's editor, Zac Goldsmith, pointed out, this is even more frightening than if the corporation had actually issued a threat. What have we come to when even the possibility of a threat arriving is enough to intimidate a printing company?

In the spring of 1996, a few weeks after the UK government had acknowledged that at least ten people had so far died from consuming infected beef during an outbreak there of bovine spongiform encephalopathy (BSE, or "mad cow disease"),* Oprah Winfrey was discussing the issue on her show with Howard Lyman of the Humane Society. Lyman** suggested—perfectly correctly—that BSE was likely not just a UK problem but one that could be affecting the herds of any country, including the US. Winfrey responded: "It has just stopped me cold from eating another burger."

---

* By June 2014 the official UK death toll had risen to 177. The outbreak was caused by using sheep and cattle remains in bonemeal as feed for other cattle.

** Whose website now, delightfully, has the URL http://www.madcowboy.com.

The result was a lawsuit from a group of Texas cattle ranchers calling themselves the Cactus Feeders that accused Winfrey and Lyman of "food slander." From February 1998 until August 2002 this legal travesty continued, consuming untold thousands of hours of the defendants' time, not to mention many millions of dollars in legal fees. The original trial was decided in favor of Winfrey and Lyman. So was the appeal that the Cactus Feeders took to the Fifth Circuit Court, which also denied the Cactus Feeders a rehearing. The group then took its case—identical to the one presented to the Fifth Circuit Court—to a Texas state court. whence Lyman, as a nonresident of Texas, removed it to a federal court. The Cactus Feeders appealed the removal to federal court, an appeal that was denied. Finally the federal judge not only threw out the Cactus Feeders' case but did so with prejudice. By now so much time had passed that the statute of limitations had expired. Even had the Cactus Feeders decided to go to the Supreme Court, which seemed a distinct possibility, it was too late.

The Cactus Feeders probably felt their time and their millions had been well spent. Few of us have pockets as deep as Winfrey's. How many of us could afford to take the risk of committing the "crime" of "food slander"? How many other commentators, concerned not just about BSE but about the many abhorrent practices of the US meat industry, were terrorized into silence by this example?

On the other hand—a factor perhaps unconsidered by the Cactus Eaters—how many members of the US public decided, *because of the publicity generated by the lawsuit itself,* never again to knowingly eat a burger sourced from Texas beef?

That's a factor the laws operative in several states (in flagrant disregard of the First Amendment) forbidding criticism of the activities of the agriculture industry can do nothing about.

The German vitamin purveyor Matthias Rath has made a great deal of his fortune in South Africa, where, during the misguided administration (1999–2008) of Thabo Mbeki, he sold huge quantities of his products on the basis that swallowing massive doses of vitamins was a better deterrent against HIV/AIDS than anti-retroviral drugs. In this he had the full support of Mbeki and the administration's equally lethal health minister, Manto Tshabalala-Msimang. In 2008, after Mbeki had resigned from office, the country's legal system cracked down on Rath's activities and he moved his focus to Russia.

One measure of the success of the Mbeki administration's approach to the AIDS problem in South Africa, which rejected "western" medicine in favor of "traditional remedies," is that the rate of infection in the country by 2004 was approximately one in three. In the years 2005 and 2006 alone, some 336,000 South Africans died of the disease. While there's no suggestion Mbeki and Tshabalala-Msimang were actively corrupt in all this, there's little doubt they were useful idiots for a number of entrepreneurs who undoubtedly were.

You can find a good short account of Rath's activities in Chapter 10 of Ben Goldacre's book on medical travesties, *Bad Science . . .* but only in the 2009 paperback and subsequent editions, not in the original 2008 UK hardcover. When that edition was going to press, Goldacre and the newspaper for which he wrote his celebrated *Bad Science* column from 2003 until 2011, the *Guardian*, were defending themselves against a major lawsuit brought by Rath. To judge by other lawsuits brought by the vitamin salesman, this one was designed to intimidate and harass.

In this instance the tactic didn't work. The newspaper dug in its heels, and eventually Rath backed down, leaving himself with a hefty bill for the legal expenses of all parties—and Goldacre was able to write in safety, for later editions of his book, an account of Rath's doings in South Africa. Goldacre has told of the long hours and tedium he put into his defense effort. In his later book *Bad Pharma* (2012), Goldacre warns anyone tempted to sue him that the experience taught him the merits of refusing to knuckle under to malevolent legal threats. Even so, again, we have to question how many of us are made of the same stern stuff as Goldacre and the *Guardian*.

The *Guardian* was involved in another instance where the UK legal system was exploited by pseudoscientists in an effort to silence dissent. In a 2008 article, science writer Simon Singh identified the bogus claims made by some UK chiropractors about the efficacy of their "alternative" or "complementary" therapeutic discipline. Edzard Ernst of Exeter University, with whom Singh had written the book *Trick or Treatment* (2008), had done extensive research into chiropractic and concluded that it could be useful in the temporary relief of back pain, but that claims for a broader utility were unfounded. In his article, Singh complained that the British Chiropractic Association (BCA), the nongovernmental body that is supposed to regulate dubious practices in the profession, was failing to crack down on those of its members who were making exaggerated claims and thereby endangering patients.

Rather than put its house in order, the BCA sued him for libel.

When the case came before Mr. Justice Eady, he ruled that Singh's use of the word "bogus" implied deliberate deception, and decided in favor of the BCA. Later the BBC's Pallab Ghosh pointed out the shortsighted folly of this decision:

> [H]ad Justice Eady's ruling stood, it would have made it difficult for any scientist or science journalist to question claims made by companies or organizations without opening themselves up to a libel action that would be hard to win.[48]

Luckily Singh, despite facing bankruptcy should he lose, took the case to the UK's High Court, where in 2010 he was granted the right to appeal and to base that appeal on the grounds of fair comment, a defense Eady had dismissed.

The short-term consequence was that the BCA abandoned its case. The longer-term consequence was the setting up, by the groups Index on Censorship, Sense About Science and English PEN, of the Libel Reform Campaign. With considerable public and cross-party political support, this succeeded in stirring the UK government into taking action to amend the country's libel laws so that this sort of nonsense could not occur again—so that powerful and deep-pocketed individuals and organizations could no longer use the laws to muzzle criticism. Although it was disparaged for not going far enough, the Defamation Act 2013 is generally accepted as a vast improvement over the previous situation.

It would be useful if the US, at both federal and state levels, could follow the UK's example and enact legislation to ensure that frivolous lawsuits of this kind can be summarily dismissed and thus no longer utilized by charlatans, frauds and corrupt but powerful corporations to suppress scientific (and other) criticism.

# # #

In a "Perspective" published by RAND in 2016,[49] Christopher Paul and Miriam Matthews outline the new model of propaganda utilized in recent years by Russia under Putin to bamboozle foes and allies alike. They call it the Firehose of Falsehood.

Using as wide a diversity of media as possible, from *faux* news TV channels to internet social networking channels to full-time employees hired to troll Twitter and other networking sites 24/7, the Russian propagandists repeat their desired message over and over with only minor variations, so that the recipients come to accept it as the truth—or at least an acceptable version of the truth—even if at the outset they knew objectively it was false. The Russian propaganda outlets "lack commitment to objective reality," as the essayists politely put it. On any issue of the day they aim to get in first with the version of it that the Russians want to project, on the grounds that once a lie starts running the truth will never catch up. Similarly, the propagandists aren't concerned about consistency; indeed, contradicting today what they said yesterday can be portrayed as a strength, in that it implies the speaker has carefully considered the issue and changed his or her mind accordingly.

It's an old maxim that the truth will out. It also appears to be, judging by the extraordinary success of the Russian propagandists, a misguided one.

Although the authors discuss the Firehose of Falsehood as purely a Russian propaganda strategy, it's one we can recognize in domestic use as well. Paul and Matthews were writing the year before terms like "alternative facts," "post-truth," and, most pernicious of all, "fake news"—in its revised sense of "inconvenient truth"—entered public currency. Too many of our public figures are prepared not only to tell lies that are in obvious and outrageous defiance of reality but, when challenged, to double down on those lies, secure in the knowledge that a sizable portion of the public will believe them, and believe them even more when the truth—or "fake news"—is presented as correction. It's the final consequence, perhaps, of a longstanding belief in the mainstream media that "balance" is sacrosanct, "balance" being something that's midway between the truth and a lie.

And the Firehose is a propaganda strategy utilized in the various wars on science. Most but not all* of the twentieth- and twenty-first-century assaults on scientific reason have been mounted by corporate interests that have had the resources to deploy the kinds of methods Paul and Matthews describe: the blanket coverage, the oft-repeated falsehood, the creation of an alternative worldview that bears only occasional connection to reality.

Right down to the trolls. "According to a former paid Russian Internet troll," say Paul and Matthews, "the trolls are on duty 24 hours a day, in 12-hour shifts, and each has a daily quota of 135 posted comments . . ." If ever a science or other blogger posts on the subject of climate change, the trolls arrive swiftly and in droves, all repeating very much the same message—almost as if their action were coordinated. As a result, most bloggers and many news sites block or at least moderate comments on such articles.

---

* There are exceptions, such as anti-vaxxerism, anti-fluoridation campaigns, 9/11 conspiracy theories, worries that your microwave oven is frying your brain . . .

That, too, brings howls of protest. They're "stifling debate," they're "crushing free speech." Except, of course, that it's impossible to have any sort of debate with propagandists who simply repeat, over and over, the same old lies despite their having been debunked a thousand times before . . . or who just hurl abuse. Try debating with a tantrum-throwing toddler and you get the general idea.

Killing the debate is what the trolls are all about, and they can be very good at it. What does the genuine reader do who wants clarification of something the original writer said if, in understandable self-defense, the writer has blocked comments? Trolls are just one end of the long spectrum of misinformation that the corporate corrupters of science, like their Putinist counterparts, have created. The orthodox and the new media are in there too, as for that matter are politicians whose allegiance has been bought through campaign contributions and perks. All contribute to the creation of a false worldview.

Smoking? Hey, maybe it's all just a hoax that it's bad for you. Global warming? Another hoax and, even if it's not, a warmer world grows more food so what're you worried about? Pollinators are dying off because of pesticides? But where would agriculture be without pesticides? You're suffering a twinge of pain? Let me offer you some opioids . . .

We can't really get on top of this war on reason until we rein in the actions of the greedy and corrupt few, and we can't hope to do that until our politicians stop using the exact same techniques against us.

Paul and Matthews conclude that there's no hope of countering the Russian propaganda, no matter how blatantly false it might be, through mere use of the truth.

It's a chilling conclusion.

# # #

The topic of the negation, suppression, misrepresentation and corruption of science for profit, or to protect profits, by large commercial corporations is a huge one. No less than the corporations who pay them, the scientists who lend their names to such endeavors are corrupting science, either deliberately or unconsciously, although the names of the individuals involved in such dishonesty tend to become less celebrated—if that's the word—than those of the Jan Hendrik Schöns and the Hwang Woo Suks (see pages 47 and 59–61).

The issue is considerably clouded by the existence of countless think-tanks (often masquerading as objective scientific institutions) and astro-turf organizations (industry-funded fake grassroots groups).

Take the groups that ExxonMobil has funded in order to promote global warming denial through the 1990s and on into the current century. Here's only a partial list:*

Accuracy in Academia, Accuracy in Media, Advancement of Sound Science Center, Advancement of Sound Science Coalition, Air Quality Standards Coalition, Alexis de Tocqueville Institution, Alliance for Climate Strategies, American Coal Foundation, American Council on Science and Health, American Enterprise Institute for Public Policy Research, American Enterprise Institute–Brookings Joint Center for Regulatory Studies, American Friends of the Institute for Economic Affairs, American Petroleum Institute, Annapolis Center for Science-Based Public Policy, Arizona State University Office of Climatology, Aspen Institute, Association of Concerned Taxpayers, Atlas Economic Research Foundation, Capital Research Center, Cato Institute, Center for Environmental Education Research, Center for the Study of Carbon Dioxide and Global Change, Chemical Education Foundation, Citizens for the Environment and CFE Action Fund, Clean Water Industry Coalition, Committee for a Constructive Tomorrow, Consumer Alert, Cooler Heads Coalition, Council for Solid Waste Solutions, Earthwatch Institute, Environmental Conservation Organization, Foundation for Research on Economics and the Environment, Fraser Institute, George C. Marshall Institute, Global Climate Coalition, Greening Earth Society, Greenwatch, Harvard Center for Risk Analysis, Heartland Institute, Heritage Foundation, Hudson Institute, Independent Commission on Environmental Education, Institute for Biospheric Research, Institute for Energy Research, Institute for Regulatory Science, Institute for the Study of Earth and Man, James Madison Institute, Lexington Institute, Locke Institute, Mackinac Center, National Council for Environmental Balance, National Environmental Policy Institute, National Wetlands Coalition, National Wilderness Institute, Property and Environment Research Center, Public Interest Watch, Reason Foundation, Science and Environmental Policy Project, and Tech Central Science Foundation.

Then there are the journals *Climate Research Journal* and *World Climate Report*, and the websites *Junk Science* and *Watts Up With That*?

---

* I'd have done this as a bulleted list, but my publisher would have thrown a fit over the number of pages it would have filled.

How effective are the astroturf organizations in promoting science disinformation among the public? Really quite effective, if we look at the response to the publication in early 2007 of the latest report from the UN-sponsored Intergovernmental Panel on Climate Change (IPCC). Compiled from the research of hundreds of climate scientists worldwide, this report set out definitively what had already been clear since the late 1990s: the activities of humankind are accelerating global warming, and the point of no return, after which catastrophic climate change will be inevitable, is extremely imminent.

Some parts of the media took seriously the ludicrous industry-generated "talking point" (really an *ad hominem* attack) that all those hundreds of scientists were simply saying climate change existed because they were worried about losing their jobs or grants.

This attack on scientists is only marginally less idiotic than the contention that climate change was a hoax invented by the Chinese as a means of gaining a trade advantage over the US.

Of course, simply being funded by a commercial operation does not necessarily mean one's publicly expressed scientific conclusions are bogus. However, the partial list given here of ExxonMobil fundees (for a more complete one see www.exxonsecrets.org) represents a measurable percentage of the world's total number of scientific global warming deniers, and it's hard to find the papers those individual scientists have published to back up their public pronouncements. In 2004 *Science* did a survey of 928 randomly selected peer-reviewed papers containing the words "global climate change" to find out how many concluded global warming was not largely caused by our burning of fossil fuels. The answer was exactly zero. Thus not only is the "debate" on climate change monstrously lopsided in terms of numbers, but a high proportion of those in the minority are receiving funding from corporations that misguidedly perceive it to be in their own interests to deny human-generated warming.

Far less known is that much of the global warming denial industry is funded not by the oil industry but, as George Monbiot revealed in his book *Heat* (2006) and Naomi Oreskes and Erik M. Conway revealed in their *Merchants of Doubt* (2010), by the tobacco companies.

TOBACCO AND THE CLIMATE

In yet another bogus report, this time by Surgeon General Richard Carmona, he claimes [*sic*] there is no safe level of secondhand smoke and calls for a workplace ban on smoking.

"The scientific evidence is now indisputable: second-hand smoke is not a mere annoyance," Carmona said. "It is a serious health hazard that can lead to disease and premature death in children and nonsmoking adults."
"The scientific evidence is now indisputable," he says.
The truth is every study used by the anti-smoking group on second hand smoke has been proven to be flawed and the data manipulated.

—*Smoking Aloud* website, December 2010[50]

It seems like an odd pairing—denial of the ill effects of smoking and denial of human-generated global warming—but in reality they're inextricably intertwined in that much the same group of willingly mercenary scientists have given both corporate-inspired denial movements whatever scientific credibility they might seem to possess in the eyes of a hurried public. Moreover, the tobacco corporations' campaign very directly served as the template for the later climate change denial campaign funded and guided by fossil fuel giants such as Koch Industries and ExxonMobil.

The tobacco industry was among the first to discover the value of deceptive public relations. Back in the 1920s, society frowned upon women smoking in public, correctly interpreting the erotic connotations that could be read into the act—just look at how suggestively the *femmes fatales* tend to smoke in 1940s film noir. George Washington Hill, head of the American Tobacco Company, called in Edward Bernays—creator of the first PR firm, coiner of the term "public relations," and a nephew of Sigmund Freud—and charged him with bringing about the necessary behavioral and attitudinal change. Bernays bought space in the 1929 New York City Easter Day Parade and hired a bunch of attractive young women as marchers, smoking cigarettes and waving these "torches of freedom." He also supplied the newspapers with press releases and photos, and very soon emancipated women all over the US were making it a point of honor to smoke in public.

Some decades later, in early 1993, the tobacco company Philip Morris was trying to work out the best way of responding to the 1992 publication of the EPA's massive and damning report, *Respiratory Health Effects of Passive Smoking*. They approached a PR company called APCO, which pointed out that the public tended to regard statements on health from tobacco companies with, um, cynicism. APCO proposed, therefore, to set up a *faux*-grassroots organization, a "national coalition intended to educate the media, public officials, and the public about the dangers of 'junk science'" and to assail *in general* the credibility of the science provided by the

US government's scientific agencies, the EPA included; thus the industry-generated claim that there was no connection between passive smoking and respiratory disease in adults and children was camouflaged by a welter of false (and perhaps, who knows, genuine) science, all supposedly derived from the work of scientists unconnected with tobacco. The organization APCO proposed became in due course the deceptively titled The Advancement of Sound Science Coalition (TASSC). Global warming was among the other scientific areas where TASSC proceeded to "cast doubt" upon established, thoroughly peer-reviewed science by pretending there was a difference of opinion in the scientific community.

This "merchandizing of doubt" was one of the many ways in which corporate power sought to corrupt the public understanding of science. In *Doubt Is Their Product* (2008), David Michaels explores the deployment of this tactic in some detail; and in *Merchants of Doubt* Naomi Oreskes and Erik M. Conway discuss its evolution, with particular focus on its application to the war against climate science.

For a long time, an important element creating an illusion that there was scientific debate over the reality of climate change was the website *Junk Science*; in the Orwellian doublespeak characteristic of modern ideological corrupters of science, the website describes much genuine scientific research as "junk science" while promoting its own unsupported claims as "sound science." Along with decrying legitimate climate science, the site's pet peeve seems to be decrying the science showing the harm of passive smoking.

The person who runs *Junk Science* is Steven Milloy, who started it in 1996 (according to the website) while working for APCO; in 1997 he became executive director of TASSC, which by 1998 was funding the site. In addition, for a number of years he wrote a weekly column, also called *Junk Science*, for the website run by Rupert Murdoch's flagship cable TV channel Fox News. Milloy is also apparently responsible for the Free Enterprise Education Institute and the Free Enterprise Action Institute, both of which are funded by ExxonMobil and the latter of which is headed by one Thomas Borelli, previously the executive at Philip Morris who oversaw the funding of TASSC. In addition, Milloy was hired by Donald Trump in 2016 to be part of his presidential transition team.

You can get something of the flavor of *Junk Science* from the description in its August 10, 2017, post[51] of veteran journalist Seth Borenstein as an "AP blowhard and climate bedwetter" for accurately reporting on the garbling of climate science by the Trump administration: "Borenstein isn't a journalist so much as he is a propagandist for the communism-via-climate movement." You get the impression the site has given up on making converts, but today preaches mainly for the benefit of the choir and five-year-olds.

In comparison with *Junk Science*, *Watts Up with That?* can seem quite sciencey at first glance. Although the comments are full of the usual vituperation about libtards and their commie conspiracies, the articles often seem to be not just science-based but actually eye-numbingly overladen with technical detail, almost as if you'd strayed into some Lewis Carrollian version of *Nature* or *Science*; in fact, in July 2010 the *New York Times* columnist Virginia Heffernan was sufficiently deceived by the appearances to recommend readers to the site "for science that's accessible but credible"[52]—a recommendation she rapidly retracted after others pointed out to her what a howler she'd committed.*

As for the site itself, the climate-science qualifications of its "meteorologist" proprietor, Anthony Watts, appear to extend little further than a brief period at Purdue University studying engineering and meteorology, plus work as a TV and radio weather presenter. He has received funding from the Heartland Institute.

Reverting to TASSC's tangled web, one of TASSC's Advisory Board of eight was Frederick Seitz, Chairman of the (Exxon-funded) Science and Environmental Policy Project, but a scientist distinguished enough that he was President of the National Academy of Sciences 1962–69 and President of Rockefeller University 1968–78. His qualifications were not in climatology or any other of the environmental sciences, but in solid-state physics. In 1979, he became a permanent consultant to R.J. Reynolds, and was put in charge of commissioning research from US universities that might help rebut scientific conclusions about the dangers of cigarette smoking. Sometime around 1989 he left Reynolds, whose CEO commented, "Dr. Seitz is quite elderly and not sufficiently rational to offer advice."

In 1984, Seitz co-founded the Exxon-funded George C. Marshall Institute, and in 1994 that organization published his report, "Global Warming and Ozone Hole Controversies: A Challenge to Scientific Judgment." At some stage he became associated with the Oregon Institute of Science and

---

* One couldn't help sympathizing with her a bit, although clearly she'd committed an act of extremely careless journalism. Her daft recommendation came at the end of a column in which she (justifiably) castigated some ex-members of the Science Blogs group for puerile and often vindictive behavior. You can imagine their comments on her egregious error.

Medicine, a "maverick" operation founded in 1980 by Arthur B. Robinson, who started his career as a chemist—under Linus Pauling, no less!—but split off from the mainstream largely, it seems, because his Christian fundamentalism clashed with the findings of real science.

In 1998, the Oregon Institute of Science and Medicine and the George C. Marshall Institute co-published *Research Review of Global Warming Evidence*, written by Robinson and prefaced by Seitz. Promoting the madcap notion that increased atmospheric carbon dioxide would bring an era of lush fecundity to the earth—a new Eden—this was produced in a format exactly mimicking that of *Proceedings of the National Academy of Sciences*, a gambit that mightily confused the media as well as an unknown percentage of the 17,000 graduates (many nonscientists, few qualified to discuss climatology) who signed the accompanying Oregon Petition, which Seitz wrote:

> We urge the United States government to reject the global warming agreement that was written in Kyoto, Japan in December, 1997, and any other similar proposals. The proposed limits on greenhouse gases would harm the environment, hinder the advance of science and technology, and damage the health and welfare of mankind.
>
> There is no convincing scientific evidence that human release of carbon dioxide, methane, or other greenhouse gases is causing or will, in the foreseeable future, cause catastrophic heating of the Earth's atmosphere and disruption of the Earth's climate. Moreover, there is substantial scientific evidence that increases in atmospheric carbon dioxide produce many beneficial effects upon the natural plant and animal environments of the Earth.

Rightly, the Clinton administration ignored the petition and the *Research Review* as pseudoscientific twaddle; the incoming Bush administration, however, was able to use it as part of its excuse for withdrawing from the Kyoto Protocol in 2000. State Department papers released in June 2005 revealed the administration acknowledging Exxon's active involvement in determining its climate-change policy, alongside that of the Global Climate Coalition . . . another spurious entity.

Ron Arnold, a logging consultant, onetime director of the Reverend Sun Myung Moon's Unification Church front group the American Freedom Coalition, and one of the primary scourges of environmental protection in the US, spelled out the tactic of industries funding, and if necessary creating *faux* activist groups in 1980 in *Logging Management Magazine*:

Citizen activist groups, allied to the forest industry, are vital to our future survival. They can speak for us in the public interest where we ourselves cannot. They are not limited by liability, contract law, or ethical codes . . . [I]ndustry must come to support citizen activist groups, providing funds, materials, transportation, and most of all, hard facts.

All of which might seem reasonable enough, except that the "hard facts" on offer are usually anything but, instead being deliberate misrepresentation. Arnold himself perverted the concept of "wise use" (of forests), originally developed by Gifford Pinchot in *A Primer of Forestry* (1903) as the need to balance the demands of nature conservation against commercial interests, to mean the wholesale exploitation of wild areas and to hell with the consequences; we (i.e., someone else, later) could more easily solve problems created by the destruction than we (i.e., industry) could forgo the immediate profits.

The timber and mining industries gleefully funded the "wise users" while their campaigns against environmentalists included death threats and arson. These industries cannot have been unaware that the groups they funded were associating with some extremely sordid allies, including rightwing militias; one organization, the innocuously named National Federal Lands Conference, sought to overturn at county level the federally guaranteed protections of wild lands and consequently, because of its anti-federal venom, attracted militant rightwing extremists.* Only when two men associated with the fringes of the "wise use" movement, Timothy McVeigh and Terry Nichols, bombed the Murrah Federal Building in Oklahoma City in 1995, killing hundreds, did the logging industry tiptoe quietly away from the "movement" it had created. More recently the "wise users" have formed alliances with far-right Christian fundamentalist groups (the ideological link being that many of these groups believe mankind is divinely ordained to exploit all natural resources to the hilt), so we may expect another outrage such as the Oklahoma City bombing in due course. Matters have not been helped by the efforts of people like Pat Robertson and his Christian Coalition to cast environmentalists as the bogeymen the faithful flock should irrationally fear and detest.

The struggle between evangelical fundamentalism and environmentalism continues, with the new focus of dominionists** being climate change.

---

* Antisemitism is, for some reason, another characteristic of the NFLC—and of course another attractant for the rightwing militias.

** So-called because they believe God gave humankind dominion over all the earth, to do with as we wish without concern about consequences.

To the biblical literalist, environmental concerns are all a bit irrelevant anyway, because doesn't it spell out in Revelation that very soon now (as it has been "very soon now" since the time of Christ, if we're to believe the New Testament*) the world will be destroyed amid fire and tribulation and all that other good stuff? But even those evangelicals who have a conscience about small-scale things such as pollution—it's hard to pretend it doesn't matter when children are suffering brain damage from lead or mercury poisoning—often think climate change is an issue too big for human comprehension; it's surely in God's province to deal with that sort of hassle, and He'll surely look after us until the fire and brimstone happens. Besides, everyone at church knows the scientists are all atheists telling us lies straight from the pit of hell, to borrow the famous 2012 words of then-Congressman Paul Broun (R–GA) about evolution and the Big Bang.

What is quite terrifying is that some senior cabinet members of the Trump administration, including those in science positions, have similar beliefs. Ben Carson, Secretary of Housing and Urban Development, is a Seventh-Day Adventist, a biblical literalist and young-earth creationist who expects the End Times imminently. Scott Pruitt, currently busily destroying the EPA, is a deacon at First Baptist Church, Broken Arrow, Oklahoma, which implies he shares Carson's general view: young earth, End Times soon, God will look after all good Christians while condemning everyone else to the Lake of Fire. Sonny Perdue, Secretary of Agriculture, is a member of the Second Baptist Church in Warner Robins, Georgia, another creationist, dominionist outfit; his son, Jim Perdue, is pastor there. The religious views of Sam Clovis,* nominated by Donald Trump to the Department of Agriculture's top scientific post, its undersecretary for research, are less clear, but his outspoken homophobia on rightwing radio speaks for itself. In May 2014 Clovis told Iowa Public Radio: "I have looked at the science and I have enough of a science background to know when I'm being boofed."

---

* There are many verses in the New Testament that state that the end is near. The most concise is I Peter 4:7: "But the end of all things is at hand." The clearest statement that the end is near is likely Mark 9:1: "And he [J.C.] said unto them, Verily I say unto you, That there be some of them that stand here, which shall not taste of death, till they have seen the Kingdom of God come with power." See also Luke 21:25–27, I John 2:18, Matthew 24:49–34, Matthew 16:27–28, Luke 9:27, Mark 13:24–27, 30.

** In November 2017, Clovis withdrew from the nomination for reasons connected with Robert Mueller's investigation of irregularities in the 2016 election.

The entirety of that "enough of a science background"? A degree in political science.

Clearly part of the environmental science-denying attitude of the evangelical right is a consequence of GOP politicians, themselves currying financial favor from the fossil-fuels titans, promoting such ideas among this important sector of their electoral base. According to Philip Schwadel and Erik Johnson in a March 2017 paper,[53] however, this is not the whole story. Biblical literalism is an important factor too, perhaps even a more important one: it's not just that rightwing politicians are persuading evangelicals to go along with the desires of the fossil-fuels corporations, it's that evangelicals are insisting GOP politicians toe the line on their own, theologically derived notions. Although the 35 million or so biblical literalists in the US make up only a little more than ten percent of the population, they thus wield a very great amount of political power. We all recall how Jon Huntsman, Jr's presidential campaign was destroyed by his August 18, 2011 tweet reading: "To be clear. I believe in evolution and trust scientists on global warming. Call me crazy." Rightwing pundits such as Rush Limbaugh decried it as a suicide note, but to the rest of us it was a welcome declaration in favor of rational, intelligent thought.

Despite all this, and despite the poisonous influence of high-profile evangelicals such as Calvin Beisner of the Cornwall Alliance, which fights hard to make sure our politicians promote politics of inaction that will maximize the human suffering incurred by climate change, there is some hope among evangelicals. Some genuinely grassroots movements, such as Young Evangelicals for Climate Action, have sprung up that are campaigning for their churches to take a more rational—and more humane—attitude toward environmental issues, including the biggest environmental issue of them all. After all, doesn't the Bible also talk of our responsibility to act as good stewards of the earth, not destroy it?

#

Our species is responsible, through individual action, industry and agriculture, for the release into the atmosphere of about 30 billion metric tons of CO_2 annually. That works out to about four metric tons per person per year. We should stress, though, that this four-ton figure is an *average*. The per capita figures for different countries vary widely. The worst CO_2-polluting country per capita in the world is Qatar (40 metric tons annually) but, since it has a population of only about 2.5 million, this doesn't affect things much. Second on the list is the USA, at 18 metric tons; since the country has a population in excess of 320 million, this per capita figure has a far more significant effect on the global total.

The per capita figure for China is a mere one-third of the USA's, at about 6 metric tons, a point worth remembering next time someone tells you there's little point in the USA doing anything about its CO_2 emissions until China does likewise. It's because of China's huge population—about 1.4 billion—that the country emits more CO_2 than we do. Furthermore, despite the denialists' sneer, China is working aggressively to bring that per capita figure down, with considerable investment in renewable energy sources at the expense of fossil-fuel ones. (The Chinese government's motivation here is probably an economic one: it's not hard to work out that government money invested in renewables now will soon be paying for itself many times over.)

In 2001, the UK journalist and militant climate-change denier Peter Hitchens—brother of the famed atheist Christopher Hitchens—wrote in the *Mail on Sunday* that "The greenhouse effect probably doesn't exist. There is as yet no evidence for it." Clearly the *Mail on Sunday* boasts the same cavalier attitude toward truth as its daily sister newspaper, because the evidence for the existence of the greenhouse effect is all around us in the form of such everyday things as living creatures and liquid water. Without an atmospheric greenhouse effect, our planet would be very much colder than it is, and almost certainly a dead world.

As early as 1824, the French physicist Joseph Fourier realized the earth's atmosphere must be acting as a blanket. Knowing the distance of our planet from the sun and the amount of energy emitted by the sun, it took only a fairly simple calculation to demonstrate that the earth should be far cooler than it is. Fourier correctly deduced that the atmosphere was responsible for the discrepancy. A few decades later, in the 1850s, the UK physicist John Tyndall demonstrated the existence of the atmospheric greenhouse effect experimentally, although he identified the wrong gas—water vapor—as the primary driver. It took the Swedish physical chemist Svante Arrhenius, in the 1890s, to correctly identify atmospheric carbon dioxide as the main greenhouse gas.*

---

* It's not just CO_2 we have to worry about: there are other greenhouse gases, indeed including water vapor. Most important aside from CO^2 is probably methane. As Carl Pope points out in *Climate of Hope* (2017; with Michael Bloomberg), so much methane leaks from oil and gas wells or pipelines that by 2018 the oil and gas industry is projected to account for 90 percent of U.S. methane emissions. Remember, in the short term, methane is eighty-four times as good at holding on to solar heat as CO_2."

Because of the greenhouse gases, while much of the radiation from the sun can penetrate to the lower atmosphere (the troposphere), once it has bounced around a bit and lost energy it has difficulty escaping back out into the upper atmosphere and hence into space. The difference between incoming energy and outgoing energy fuels fun stuff like evolution, weather, and life; but, now that the balance has become upset because of increased greenhouse gases, there's a dangerous warming of the troposphere.

However Hitchens and his *Mail* readers might choose to delude themselves, the atmospheric greenhouse effect, and CO_2's role in it, has been known about, understood, and measured for well over a century. And by the middle of the last century it was obvious that pumping ever-increasing quantities of greenhouse gases into the atmosphere, primarily from the burning of fossil fuels, was warming the troposphere to levels that would eventually be hazardous.

The fact that there is any public doubt at all about this has nothing to do with uncertainties in climate science. Making climate predictions is an infernally complicated business, and it's entirely understandable that different climate scientists, working along different lines, will disagree on the details; what's remarkable is how very much they all, save a few "contrarians," agree on the general picture. There is a greater degree of agreement among climate scientists that human activities—especially fossil-fuel use—are warming our atmosphere, and dangerously so, than there is among biologists about the theory of evolution.

That otherwise adequately educated people should be so misinformed about this level of certainty in the science is a product of decades of deliberate obfuscation on the part of the fossil-fuel corporations, who learned well the lesson of the campaign mounted earlier by the tobacco companies that, so long as the public were kept unsure about the current state of the science, any necessary action to save human lives that might eat into short-term profits could be delayed. All that was needed was to buy a few malleable scientists and, through campaign contributions, a few corrupt politicians. In fact, perhaps because in large swaths of the country the only local news media available are effectively rightwing propaganda channels, this scheme worked far better in the US than the corporations could have dreamed: they succeeded in recruiting virtually the entirety of the professional Republican Party to their cause, with GOP politicians at all levels terrified of the electoral consequences should they dare to tell their grassroots supporters the very evident truth that climate change is happening and that we're causing it.

As for the fossil-fuels corporations themselves, they've known the truth about global warming, and the role of their products in it, for decades. In November 1959 in New York, to celebrate the one-hundredth anniversary of the US oil industry, the American Petroleum Institute and the Columbia Graduate School of Business organized a symposium called Energy and Man. The symposium was addressed by none other than the physicist Edward Teller, whom we met in connection with hydrogen bombs and the Strategic Defense Initiative. He told the assembled oil barons and others:

> I would . . . like to mention another reason why we probably have to look for additional fuel supplies. And this, strangely, is the question of contaminating the atmosphere . . . Carbon dioxide has a strange property. It transmits visible light but it absorbs the infrared radiation which is emitted from the earth. Its presence in the atmosphere causes a greenhouse effect . . . It has been calculated that a temperature rise corresponding to a 10 percent increase in carbon dioxide will be sufficient to melt the icecap and submerge New York. All the coastal cities would be covered, and since a considerable percentage of the human race lives in coastal regions, I think that this chemical contamination is more serious than most people tend to believe.[54]

In supplementary discussions Teller quantified his prediction. It seems, though, that the reaction of the American Petroleum Institute—or at least the reaction it reported publicly—was to assume that technology would have vehicle emissions well under control long before any untoward global-warming effects could come into effect. In other words, they kicked the can down the road.

Even so, you'll occasionally come across apologists for the fossil-fuels industry deploying the "But who could have known?" argument. Well, every significant figure in the US oil industry knew nearly sixty years ago.

# # #

By the end of this century we can expect, at a minimum, a rise in average global sea levels of about 1m (3.3ft), although some estimates suggest a rise of twice that. But this is before we start taking into account some further events that are not as yet inevitable, but very soon will be if we continue pumping CO_2 into the atmosphere at current rates.

The first of these—and it may already be too late to prevent it—is the melting, in whole or in large part, of the West Antarctic Ice Sheet. This will, on its own, raise sea levels by about 3m (10ft), in a process that might take centuries but might take mere decades. If the prospect of having to abandon many of our coastal cities because of just a 1m or 2m (40in–80in) rise seems grim, imagine the consequences of a 4m–5m (13ft–16.5ft) rise. Less immediate a threat is the melting of the Greenland Ice Sheet, which would add another 3m (10ft) to the oceans' rising waters.

All these figures are dwarfed by the sea-level rise that would be caused if the East Antarctic Ice Sheet melted: about 60m (200ft). Taken along with the contributions of the other great meltings, this would be enough to completely immerse the Statue of Liberty. Sailors of the future, assuming any survived, would be able to look down on the statue's torch some 20m (66ft) below the water's surface.

Sea-level rise is not merely a future problem. It's estimated that about $2 billion of the $65 billion bill for the damage caused in 2012 by Hurricane Sandy was due to sea-level rise. That's on top of the contributions made to the bill by other climate-change-related factors.

One of the states whose administration is in denial, and one of those that will be affected most immediately by its consequences—hurricanes and, in particular, sea-level rise—is Florida. But it's not just at governmental level that this denial exists. The 2017 Miami–Dade Real Estate Study, carried out between June 5 and July 6 of that year and reported by the *Miami Herald*,[55] interviewed one hundred of the top realty professionals in the area—brokers, agents, analysts. Although 59 percent of the agents interviewed expressed personal concern over the impact that changing sea levels and other manifestations of global warming might have on future realty prices locally, almost two-thirds (64 percent) reported that such concerns weren't expressed by any of their clients. This implies that far more than two-thirds—the vast majority, in fact—of Miami–Dade property buyers are oblivious to the fact that climate change could decrease the livability of the home of their dreams, and decimate its value, within an alarmingly few years.

Perhaps this has something to do with the fact that officials of the Florida Department of Environmental Protection have been ordered to avoid the use of terms like "climate change," "global warming" and "sustainability" in any of the department's communiques. According to former department employee Kristina Trotta, cited in the *Miami Herald*,[56] "We were told that we were not allowed to discuss anything that was not a true fact"—in other words, to ignore reality in favor of ideologically based pseudoscience.

The policy, although unwritten, is directly traceable to the personal climate-change denial of Governor Rick Scott.

# # #

Estimating the death toll caused by climate change is as difficult as determining for certain how much of any given weather disaster is directly attributable to climate change. As an example of the confounding factors that aren't immediately apparent, take the matter of suicides.

For a study published in preliminary form online by *Proceedings of the National Academy of Sciences of the United States of America* (PNAS) in July 2017,[57] Tamma A. Carleton of the University of California, Berkeley, studied 47 years' worth of data on suicide rates and climate in India, and concluded:

> This analysis of India, where one fifth of the world's suicides occur, demonstrates that the climate, particularly temperature, has strong influence over a growing suicide epidemic. . . . I show that high temperatures increase suicide rates, but only during India's growing season, when heat also reduces crop yields. My results are consistent with widely cited theories of economic suicide in India. . . . I estimate that warming temperature trends over the last three decades have already been responsible for over 59,000 suicides throughout India.

The true figure is likely to be far higher, Carleton points out, because in India deaths generally go under-reported. And her estimate may very well be conservative even without taking this factor into account: just since 1995 over 300,000 Indian farmers and farm workers have committed suicide.

What isn't open to debate is that drought is a primary driver of suicide among Indian farmers; the exacerbation of drought is one of the consequences of climate change. In 2015, one of India's worst recorded drought years, 12,600 farmers took their own lives.

The Indian government and various of the state governments are making attempts to alleviate the problem, such as introducing public-funded insurance schemes against crop failures, but, until farmers can be educated into changing their practices to take account of the shifting climate, such measures are likely to be little more than band-aids.

# # #

A decade ago, the four main climate-denialist arguments had been demolished by clear scientific demonstration, and very publicly so. Climate
scientists had wasted countless hours that they could have spent doing
valuable research instead demonstrating beyond all rational doubt what
everyone in the community, bar the usual few "mavericks," already knew.

• No, the warming of the atmosphere was not being appreciably affected
by changes in the temperature of the sun.

• No, the measurements of the planet's surface temperatures were not
being corrupted by over-concentration on urban areas (so-called "urban
heat islands"), which are warmer than open spaces because of the concentration there of humans and their activities.

• No, the computer models of past and future temperatures were not
seriously flawed. Here the reasoning is simple: if all sorts of different teams,
adopting all sorts of different approaches, produce results that all point in
a closely similar direction, you can be pretty sure there's nothing seriously
wrong with your models.

• No, the fact that a handful of papers have appeared in respectable peer-
reviewed journals (as opposed to the plenty that have appeared in unrespectable, dodgily if at all peer-reviewed ones) does not undermine what
we might call the standard model of global climate. Without exception,
whatever denialist "authorities" such as Anthony Watts might say, those
few papers have been shown to have flaws that render their conclusions
groundless.[58]

In the decade since then, not one of those denialist arguments, despite
having been comprehensively invalidated, has disappeared from the public
"debate" on climate science—in other words, from the blare of the fossil-fuel companies' noise machine (to appropriate David Brock's phrase). Those
very same arguments are still being drearily regurgitated by bought—and
dismayingly powerful—politicians such as James Inhofe and Lamar Smith,
with a Greek chorus of dishonesty being supplied by virtually the entirety
of Fox News and the other right-leaning media. In this sphere of science
corruption and denial, as in so many others, it appears industry has invested wisely in the maxim that, when it comes to persuading human beings,
facts are almost irrelevant.

There are a slew of other reasons to believe the money was invested well.
It's embarrassing enough that a mere 58 percent of Americans understand

that climate change is real and that human activities are the primary cause. Even worse is that, according to the 2017 results of an annual (since 2008) poll done by the Yale Program on Climate Change Communication and the George Mason University Center for Climate Change Communication,[59] just 13 percent of the US public, when asked what percentage of climate scientists concur on this, chose the option "over 90 percent." Given that the true figure, between 97 and 98 percent, has been actively publicized by the scientific community for well over a decade, it becomes obvious to what extent elements of the supposed news media have distorted—through either malice or dumbing down ("Hey! Here's a new story about Rihanna!"), which can itself be malicious in intent—the information presented to their audience.

# # #

Several completely independent lines of argument related to the notion that the earth might be not warming but cooling have been put forward by climate-science deniers.

The first of these concerns the famous "hiatus" in atmospheric warming. This line of argument admits that for most of the time since the Industrial Revolution the atmosphere has been warming—whatever the cause might be—but insists the *climate scientists' own measurements* show that this warming came to a halt around the end of the twentieth century.

The fallacy is childishly simple. Like most natural phenomena, the warming of the atmosphere does not behave in exact accordance with an idealized picture. In particular, a graph of temperatures from one year to the next shows a zigzagging rise rather than a smooth geometrical curve. It's quite easy to find examples of a year being cooler than the one before it, or even than the several before it—especially since events like El Niño that affect global temperatures don't happen every year.

Because of El Niño, the year 1998 was the hottest on record to that date. There wasn't to be a hotter year after it until 2005. By taking 1998 as their base year, therefore, climate denialists were able to claim that, judging by the next seven years, it was "obvious" global warming had stopped. Of course, it's been a bit more difficult to make that claim since 2005, because after that, almost without exception, each fresh year has set a new high for the hottest on record, but this hasn't stopped the argument—the zombie argument, we could say—from being disinterred at regular intervals, rather like Senator James Inhofe's infantile claim that snowfall in winter disproves global warming.

A quite different denialist assertion is that in the 1970s scientists were predicting there was a new ice age on its way: "The world could be as little as 50 or 60 years away from a disastrous new ice age, a leading atmospheric scientist predicts," is a quote that's often dug out from a 1971 *Washington Post* article.[60] The truth is that, at the time, atmospheric scientists were indeed worried about the cooling effect of all the manmade aerosols being released into the atmosphere. (The damage these were doing to the ozone layer would soon become evident, too.) What the scientists, including the subjects of the *Washington Post* article, S.I. Rasool and Steve Schneider, were always careful to add was an "all other factors being equal" caveat. Of course, all other factors weren't equal: far more important to the equation was the ever-increasing amount of greenhouse gases being pumped into the atmosphere. The aerosols did have the effect of ameliorating the rise of global atmospheric temperatures for a few decades, but they were fighting a losing battle.

Another strand of this argument is that famed atmospheric scientist James Hansen was involved in the "imminent ice age" research. Well, yes . . . to the extent that he lent Rasool and Schneider a piece of software he'd written, and so they acknowledged his help.

One of the pieces of "evidence" produced to back up the claim that 1970s scientists were worried about a new ice age is a 1977 *Time* magazine cover showing a picture of a penguin beneath the headline "How to Survive the Coming Ice Age." The trouble is that the image is a fake, done sometime after 2007. In that year the *Time* cover actually appeared, but with the headline "The Global Warming Survival Guide." Spot the difference.

A more recent ice-age related concern came in the early 2000s, when it was realized global environmental changes were affecting the flow of the Gulf Stream—the warm ocean current that, starting in the Gulf of Mexico, bathes the eastern coast of the US and Canada before dividing into two to cross the Atlantic. The northern of these two streams, the North Atlantic Drift, washes against the western coast of Europe, including the British Isles, keeping that area warmer than it would otherwise be. A disruption of the current could well lead to another "Little Ice Age" in Europe.

The term "Little Ice Age" came into the literature in 1939 to describe the period of unusual cold Europe suffered between the sixteenth (or earlier) and nineteenth centuries, when people could go skating on London's River Thames, and even hold parties on the ice. The "Little Ice Age" wasn't actually an ice age—it wasn't a global phenomenon but seems to have been confined primarily to the northern hemisphere, and not even the entirety of the northern hemisphere. Although no one's really sure of its cause (assuming there was just a single cause), a temporary disruption of the Gulf Stream is among the candidates. Obviously another such disruption, which is certainly still on the cards thanks to overall global warming—no wonder some scientists prefer the term "global weirding"—could quite conceivably give rise to another "Little Ice Age."

Which would mean that temperatures in Europe would rise to intolerable extremes less swiftly than otherwise. This isn't immensely reassuring, if you're a European, because the situation would give rise to new and likely violent weather systems.

Finally, there's the canard that there's good climate science showing the globe is cooling rather than warming, but the IPCC has covered this up.

The "good science" concerned was a pair of papers done by John Christy and Roy Spencer—two genuine but contrarian climate scientists—and published in 1992 in *Journal of Climate*. They looked at sets of satellite measurements of microwave emissions by the atmosphere; these emissions can be used as clues to atmospheric temperatures. According to their analysis, the overall trend wasn't one of warming but of a slow but steady and perceptible cooling.

When one line of approach appears to show one thing and all the other lines of approach show the opposite, there's clearly a conundrum worth investigating, so sure enough other climate scientists investigated the Christy/Spencer claims. They found various errors not in the data but in the analyses Christy and Spencer had done—errors that all pointed in the same direction: toward cooling. It seems the two scientists were seeing what they wanted to see. As an example, one error was to conflate some data from the stratosphere with that for the troposphere. It's known that one result of a greenhouse effect in the lower atmosphere is a cooling of the stratosphere,* so these extraneous data would dilute the evidence of lower-atmosphere warming.

Christy and Spencer have acknowledged these errors, but this hasn't stopped the denialist community from referring to the analyses as if they were still valid.

# # #

---

* While the stratosphere continues to receive the same amount of energy from above—i.e., from the sun—less energy is reaching it from below, because it's trapped in lower atmospheric regions.

Sometime before November 17, 2009, a few weeks before the long-scheduled 2009 United Nations Climate Change Conference—where it was hoped to get international agreement on a new framework for the mitigation of climate change in the years beyond 2012—unidentified hackers gained access to the supposedly secure backup server of the University of East Anglia's Climatic Research Unit (CRU), one of the world's leading climate science research establishments. On that date, the harvest of the hack, a folder containing just over 60 megabytes of e-mails and other private computer files, was posted to various sites around the web—mainly denialist ones like *Watts Up with That?* and *Air Vent*.

Almost immediately, cherry pickings from the thousand or so e-mails flew around the denialist blogosphere, and soon "interpretations" of the selected information were being spoonfed to the world's mainstream media—many of whose outlets, including such supposed redoubts of responsible journalism as the *New York Times* and *Newsweek*, were foolish enough to take this thoroughly spun material at face value. Headlines like "Climategate: The Final Nail in the Coffin of 'Anthropogenic Global Warming'?" proliferated in outlets that really should have known better.*

To this day, no one is certain who the hackers were—in early 2017 there were, inevitably, suspicions that the Russians might have been behind the theft, since the Russians seemed to be busily hacking everything else in sight—but at the time it was assumed, probably correctly, that the project had been underwritten, at least indirectly, by the fossil-fuels industry. Most authorities quickly dismissed the suggestion that the hackers were merely enthusiastic amateurs—some dark version of Anonymous, perhaps. On February 1, 2010, the *Independent* quoted Sir David King, formerly the UK's Chief Scientific Advisor: "[The hack] bore all the hallmarks of a coordinated intelligence operation."

A decade after the event, what stands out about the whole Climategate controversy was how very, very little the "skeptics" were able to find as fuel among that large mine of data. There was definite evidence that Phil Jones, head of the CRU, was actively resisting Freedom of Information Act (FoIA) requests, but it soon became evident that there were good reasons for this "impropriety":

---

* That particular headline was directly above a typically dunderheaded James Delingpole column in the UK's *Daily Telegraph*, a journalist and a newspaper that very much do *not* "know better."

- some of the information being requested was not the CRU's to give

- the CRU was woefully understaffed for the time-consuming business of processing and responding to FoIA requests

- various of the bulk FoIA requests were coming not from genuinely interested individuals and organizations but from denialist outfits using them as a means of harassment

Jones and his colleagues were guilty of further sins. For example, in their personal e-mails they quite often said snarky things about other climate scientists, about science denialists, about administrators and even—oh, shock of shocks—about politicians.

As the denialists didn't bother to point out—and likewise, shamefully, far too many journalists—none of the above malfeasances, if malfeasances they were, had anything to do with the validity of the climate science.

Most of the media attention focused on fewer than a handful of brief extracts, carefully cherry picked for effect. Here's the most famous of them, from a Phil Jones e-mail dated November 16, 1999:

> I've just completed Mike's [Michael E. Mann's] *Nature* trick of adding in the real temps to each series for the last 20 years (i.e. from 1981 onwards) and from 1961 for Keith's [Keith Briffa's] to hide the decline.

Far too often this was reported as something more like:

> I've just completed Mike's *Nature* trick and Keith's to hide the decline.

The denialist community gave a collective gasp. Just look at the words "trick" and "hide the decline"—what better evidence could there be of an intent to deceive, of the rascally climate scientists cooking the books?

The use of the word "trick" is easily disposed of. Most of us recall in our youth mastering the trick of riding a bicycle. Jones was using the word in an analogous sense here, to describe a nifty statistical method that Mann had reported in *Nature* of combining data from two disparate sources.

As for "hide the decline," it had been known for some while that after about 1950 and especially after about 1960, for reasons that are unknown but may well have to do with the shifting composition of the atmosphere, the correlation between tree ring data and actual measured ambient temperatures, hitherto pretty reliable, starts to break down. This had been reported in a letter by Briffa and others to *Nature* in February 1998 and no one had batted an eyelid. "Keith's [trick] to hide the decline" was nothing more than a means of taking account of this discrepancy, the "decline" being a decline in accuracy of the correlation rather than any decline in global temperatures—in fact, as the history books show, around the time of Jones's e-mail world temperatures were at a [then] record high.

It cannot be doubted that the Climategate hysteria threw a spanner into the works of the Copenhagen Summit, and so the perpetrators of the hack and the subsequent coordinated PR campaign doubtless regard their efforts as a success. Of course, if you evaluate the consequence of those efforts—yet further delay in global action to counter imminent climate disaster—in terms of factors such as human death toll, damage to the global economy, and environmental devastation since then and over the next few decades, the word "success" looks Orwellian.

Another consequence of the Climategate fiasco was that climate scientists around the world saw a significant uptick in the amount of hate mail they received from the viciously ignorant, up to and including menaces and death threats. In his books *The Hockey Stick and the Climate Wars* (2012) and *The Madhouse Effect* (2016; with Tom Toles) Michael Mann has recounted the very real sense of fear he had to live with, not to mention the attempts at professional character assassination. By many accounts Phil Jones had it even worse.

And all for what? To increase the profits of a few international corporations and their billionaire owners? Any pretense that the motivation for the hack and the subsequent PR campaign was to show that climate science was just some great Alex Jones-style conspiracy is outright laughable—except, of course, to the under-informed, who seem to have swallowed the false claims whole. And yet, as I pointed out in my book *Denying Science* (2011), when independent authorities have examined the Climategate documents, they've found that the entire denialist case built upon them is a mirage, a deceptive tissue of nothingness:

> Since the opening salvo of the Climategate farrago there have been no fewer than four official vindications* of the CRU scientists and their scientific work: The Parliamentary Science and Technology Committee exonerated Phil Jones (he has "no case to answer" and his "reputation [. . .] remains

---

* A number that has since risen to—depending upon which ones you choose to count—at least six. There have also been nongovernmental analyses by journalistic institutions like the *Guardian*, the Associated Press and FactCheck that again have shown the scientists guiltless of any but the most trivial malfeasance.

intact"), the International Panel set up by the University of East Anglia under Professor Ron Oxburgh exonerated the CRU researchers ("their work has been carried out with integrity"), the Muir Russell Review did likewise, and in September 2010 the Conservative-led UK government stated flatly that: "The government agrees with, and welcomes, the overall assessment of the Science and Technology Committee that the information contained in the illegally-disclosed emails does not provide any evidence to discredit the scientific evidence of anthropogenic climate change."[61]

One odd little irony of the Climategate episode was that Sarah Palin was among the most prominent denialist politicians triumphantly making political capital out of the hack. Aside from the fact that Palin's knowledge of climate change is limited to speculation as to whether she'll still be able to see Russia from her back porch, there was her reaction to her own e-mails being hacked and leaked during her 2008 campaign as John McCain's vice-presidential nominee. Rightly, the hacker was prosecuted and convicted. "Violating the law," wrote Palin sanctimoniously on Facebook, "or simply invading someone's privacy for political gain, has long been repugnant to Americans' sense of fair play." But not, it seems, to Palin's own sense of fair play.

Glaciergate was an even smaller storm in an even smaller teacup than Climategate, but it succeeded temporarily in rallying the denialist faithful.

In the IPCC's Fourth Assessment Report (AR4), released in 2007, there was what was in effect a typo: a prediction of the date when the Himalayan glaciers would disappear was rendered as 2035 rather than 2350 (as correctly given elsewhere in the report). The error was confined to a single section of the report; it was included in neither the technical nor the policy summaries. It was so trivial that all that was required to correct it was an erratum slip in the printed version, an update note in the online version.

Nevertheless, the climate-denialist community went bananas (okay, *more* bananas) with delight over what was soon dubbed Glaciergate by science-denying journalists such as James Delingpole—the man famed for trying to lecture the Nobel prize-winning president of the Royal Society, Sir Paul Nurse, on how science works.* Journalist Johann Hari, in a *Nation* article called "Climategate Claptrap II" (April 15, 2010), discussing the antics of climate-denialist icons such as Christopher Monckton, explored the stupidity of this reaction:

---

* Videos of the encounter can be easily found on YouTube. Eat your heart out, Monty Python.

When it comes to coverage of global warming, we are trapped in the logic of a guerrilla insurgency. The climate scientists have to be right 100 percent of the time, or their 0.01 percent error becomes Glaciergate, and they are frauds. By contrast, the deniers only have to be right 0.01 percent of the time for their narrative—See! The global warming story is falling apart!—to be reinforced by the media.

# # #

And, meanwhile, what of tobacco smoking?

It would be easy to say that the science confirming the harms done by both smoking and passive smoking has been universally accepted, and certainly in the developed world the practice has almost flatlined; those clusters of huddled figures you used to see outside office blocks and restaurants have now largely disappeared. I recently attended a convention where, among the two thousand or so attendees, there seemed to be no objections at all to a ban on smoking that extended not just to the buildings of the academic institution housing the convention but also to the extensive campus: you'd have had to go a mile or more to find somewhere to light up.

In the US, federal, state and civil suits have extracted many millions by way of recompense from the tobacco companies for the dishonest marketing of their products through pretended ignorance of the basic medical science. Numerous countries have introduced strict laws on cigarette packaging, either demanding the packets be covered in prominent and gruesome reminders of the health consequences of smoking or by banning the use of fancy patterns and trademarks. Bans on smoking in public buildings, transport facilities (buses, trains, planes, cabs), restaurants, bars, and workplaces are ubiquitous in developed countries and widespread elsewhere.

The tobacco companies have therefore turned their predatory gaze toward the developing world. There, some governments have, like those of the developed nations, taken up the cudgels against the tobacco industry . . . only on occasion to find themselves sued by an industry that may have larger annual profits than the GDP of the country concerned. So far the industry has lost such lawsuits but, even so, tobacco sales continue to rise worldwide, often through aggressive marketing of cigarettes to children.

But consider what global sales of cigarettes might be today had science— and, often, brave scientists and science-respecting politicians—not stepped up to the plate to combat the false science and the pseudoscience pumped into the media environment, like toxic smoke into a crowded room, by the tobacco titans, by the merchants of doubt. The situation may be bad, but it's not as bad as it could have been.

Even so, figures from the World Health Organization suggest that smoking accounts for five to seven million premature deaths per annum, with an additional 600,000 deaths from secondhand smoking. About 50 percent of smokers are killed by their habit.

THE HOLE IN THE OZONE LAYER

montegoblack: How much did we spend to ban R13 refrigerant from air conditioners and replace with R34 to fix the ozone hole? We now know ozone comes from the Sun and when the Earth is tipped in the winter the ozone can not reach the northern hemisphere. No ozone from the Sun, ergo ozone hole.
—Comment on *American Thinker* website, December 2010[62]

Long before there was a TV series called *Buffy the Vampire Slayer* there was the 1992 movie upon which it was based. Although it seems the movie is almost forgotten now, except by connoisseurs, there's a classic scene that's worth digging out. Buffy and her pals are concerned about the environment, because that's the "interesting" thing to be. Alas, they're also airheads, as we soon discover when they meet to talk over the issues of the day.

"What about the ozone layer?" says one.

"Yeah, we've got to get rid of that," comes the response.

Although satirically intended, it wasn't a wholly inaccurate depiction of common misunderstandings about the ozone layer, the "hole" in it, and the importance of taking resolute and rapid action.

The sun emits electromagnetic radiation at just about every conceivable wavelength, including the visible spectrum and the "almost-visible" wavelengths that are a bit longer and a bit shorter than we can see. Infrared light (longer-wavelength/lower-energy than visible light) plays an important role in global warming; here we're concerned with ultraviolet, or UV (shorter-wavelength/higher-energy than visible light). It's the UV in sunshine that causes sunbathers' skin to tan and burn; it's also the UV that can cause those sunbathers—or anyone else—to develop skin cancer.

Fortunately only a small proportion of the incoming UV radiation from the sun actually reaches the ground, being blocked thanks to ozone in the stratosphere; ozone, unlike oxygen and the other atmospheric gases, has the property that it absorbs light of UV wavelengths.

The reason there's so much ozone present there is because of . . . UV light itself. UV, being very energetic, has the tendency to split up the standard oxygen molecule, O_2, into its two constituent atoms: O + O. Most of those free-wheeling atoms will sooner or later get together with another of their ilk to recombine as O^2 molecules, but a few will latch onto existing O^2 molecules to form a molecule of O_3, or ozone. In other words, the UV light from the sun actually generates the "barrier" that's keeping much of it from penetrating to lower regions of the atmosphere—and eventually to our hot skin.

It's not just skin cancer that we'd have to worry about if the upper atmosphere didn't have enough ozone. If the UV (and other high-energy radiation) from the sun and elsewhere were allowed to bathe the surface of the earth at full dosage, so to speak, in due course our planet would be essentially a dead world.

So there was a great deal of concern in scientific circles when a British Antarctic Survey team led by Joseph Farman, Brian Gardiner and Jonathan Shanklin found there were problems with the ozone layer; they published their discovery in May 1985 in a paper in *Nature*.[63]

In wintertime in the polar regions, because of the tilt of the earth's axis to the plane of its orbit, there are months of darkness: the upper atmosphere isn't just receiving no UV light from the sun, it's receiving no direct sunlight at all. During those months, then, ozone levels in the stratosphere gradually diminish, until the arrival of spring, when the return of sunlight, including UV, starts building up the ozone again. What Farman, Gardiner and Shanklin observed in the Antarctic was that something was interfering with the process of ozone-layer rebuilding—something was breaking down the O_3 molecules faster than the incoming UV could generate them.

Either this could be part of a long-term natural cycle or it could be a consequence of human activity. It was a matter of urgency to find out which in order to see if we could do something to reverse the trend before it was too late.

As long ago as 1970, the Dutch scientist Paul Crutzen, trying to work out why the ozone layer wasn't a whole lot thicker than it actually was, had demonstrated that nitrous oxide (N_2O) from ground-based sources could make its way to the stratosphere, where it reacts with oxygen to produce the gas nitric oxide (NO), in the process "using up" extra oxygen atoms. Obviously this would affect the amount of ozone being created by the incoming UV. N_2O is produced by such sources as soil bacteria and supersonic aircraft. Focusing on the former, Crutzen theorized that the increasing use of fertilizers in agriculture—due to the celebrated Green Revolution—and the consequent increase in soil bacteria might well be a cause of ozone depletion.

Crutzen was right in principle, although we now know the contribution to ozone depletion from N_2O is relatively minor.

A few years later, in 1974, two scientists at the University of California, Irvine, Frank Rowland and Mario Molina, correctly fingered a group of chemicals known as the chlorofluorocarbons, or CFCs, as the culprits. The maverick UK environmentalist James Lovelock had earlier discovered that these chemicals—widely used in aerosol cans, fridges, air-conditioners, and many other applications—did not easily break down in the atmosphere. Sherwood and Molina (correctly) theorized that some of these CFCs were making their way up into the stratosphere, and worked out the mechanism whereby they were interfering with the formation of ozone.*

Crutzen, Rowland and Molina deservedly shared the 1995 Nobel Chemistry Prize. Industry was less complimentary to their work. The trade magazine *Aerosol Age*, instead of tackling the science, accused Sherwood and Molina of being KGB agents intent on destroying US industry.

Although no one knew about the ozone hole over the Antarctic at the time the three were discovering these mechanisms, the message was clear enough. If we carried on using CFCs in the foolhardy way we had been doing, we were sooner or later—probably sooner—going to run into real problems, increased rates of skin cancer being just the beginning. Clearly something had to be done, and fast.

Not according to the chemicals manufacturers, however. According to the industry, while the science underpinning the conclusions of Rowland and Molina might be interesting in a speculative sort of way, it was nothing more than that. The main protagonist in the industry's attempt to muzzle and/or discredit the science was DuPont, which deployed all the tactics the tobacco companies had used in their attempts to discredit medical science and the fossil-fuel companies were already beginning to use in their attempts to stifle climate science and, indeed, plain straightforward physics.

---

* UV is efficient at breaking down not just O_2 molecules but also CFC molecules. Two of the constituents of the latter, chlorine and fluorine, are efficient catalysts in the breakdown of O_3 to O_2. The chlorine and fluorine are merely catalysts—they don't participate in the reaction—and so the levels of these two elements don't naturally diminish much. The more we continue pumping CFCs into the lower atmosphere, the more of them reach the stratosphere, and the greater the depletion of O_3 to O_2.

In the same way that the deniers of climate science trot out alternative "explanations," such as variations in solar output and long-term natural cycles of global temperatures, so the chemicals industry and its useful idiots suggested that a major cause of ozone depletion could be naturally occurring chlorine, produced by either erupting volcanoes (Ross Island's active volcano Mount Erebus was a target of real and simulated suspicion) or ocean spray. Yes, there is naturally occurring chlorine in the lower atmosphere, and some of it does indeed make its way to the stratosphere, but it forms only a small proportion of the chlorine present there.

Even if this were not the case, even if it were true that naturally occurring chlorine was a major contributor to ozone depletion, the argument was, just like its climate-science-denying analog, a spurious one. CFCs were the contributor to the depletion of the ozone that we could do something about. It didn't really matter if there were other factors involved.

The other very familiar argument produced by the chemicals industry was that a curtailment of CFC use would hurt the economy and destroy jobs. It's odd how in such instances, when science has eventually prevailed, these industry-predicted economic crashes have never happened—to the contrary, the required new technology to replace the hazardous old one has generally provided an economic boon—and yet still the argument keeps reappearing like an embarrassing rash.

Founded and funded by DuPont, the astroturf organization the Alliance for Responsible CFC Policy promoted through the media and argued in Congress that the science on CFCs was "not yet settled," even as more and more evidence came flooding in that CFC use was indeed damaging the planet's ozone defenses. The pretense was that DuPont was only too eager to act according to the science:

> Should reputable evidence show that some fluorocarbons cause a health hazard through depletion of the ozone layer, we are prepared to stop production of the offending compounds

read a June 1975 DuPont ad in the *New York Times*. Of course, DuPont was willing to do no such thing. A couple of weeks later DuPont's chairman, Irving S. Shapiro, was quoted in Chemical Week as saying that the science was just "a science fiction tale . . . a load of rubbish . . . utter nonsense."[64] The world disagreed with Shapiro, especially so after the discovery of the ozone "hole" over the Antarctic. Moving with, for once, commendable speed, the international community drew up and agreed upon the Montreal Protocol on Substances that Deplete the Ozone Layer, requiring the phasing out of CFCs; in 1987 the protocol was ratified by every member nation of the UN (although a few have been slow to sign the numerous revisions that have subsequently appeared).

Thanks to the Montreal Protocol, the rate of depletion of the ozone layer has been dramatically reversed. We can't really say as yet that the problem has been entirely solved because, as noted, chlorine and fluorine last a long time in the stratosphere. The situation is nowhere near as bad as with the buildup of greenhouse gases in the lower atmosphere, where the recovery time is reckoned to be on the scale of hundreds if not thousands of years, but it's reckoned that—assuming we don't start pumping out any further ozone-depleting gases—by 2050 or so the ozone layer may fully recover.

And so our story ends with the Montreal Protocol. Science and politics managed to forestall a major global crisis. We can all relax.

Well, not quite.

DuPont wasn't giving up that easily. Months after the signing of the Protocol, three US senators—Max Baucus (D–MT), David Durenberger (R–MN) and Robert T. Stafford (R–VT)—wrote in February 1988 to DuPont chairman Richard E. Heckert suggesting it was now about time his company stopped marketing CFCs. This was, you'll doubtless have calculated, a mere baker's dozen years since the DuPont claim that "Should reputable evidence show that . . . fluorocarbons cause a health hazard . . . we are prepared to stop production . . ." Heckert responded:

> DuPont stands by its 1975 commitment to stop production of fully halo-genated chlorofluorocarbons if their use poses a threat to health. This is consistent with DuPont's long established policy that we will not produce a product unless it can be made, used, handled and disposed of safely and consistent with appropriate safety, health and environmental quality criteria. At the moment, scientific evidence does not point to the need for dramatic CFC emission reductions. There is no available measure of the contribution of CFCs to any observed ozone change . . .[65]

Two weeks later NASA's Ozone Trends Panel reported the latest evidence confirming that CFCs were indeed depleting the ozone layer. Seeing as a DuPont scientist had been part of that panel, there was little the corporation could do except change its tune, pronto. A few days after that, DuPont began to phase out production of CFCs. And thereafter, to the company's credit, it became one of the pioneers in developing replacements less detrimental to the environment and to human well-being.

No, that *still* isn't quite the end of the story. As late as 1994 Frederick Seitz, whom we met earlier as a denier of the harmful effects of smoking and of anthropogenic global warming, published a paper at the George C. Marshall Institute called "Global Warming and Ozone Hole Controversies: A Challenge to Scientific Judgment" that claimed exactly what you might expect. And if you grub around on sites like Steven Milloy's *Junk Science* you can even today find proudly counterfactual "proofs" that the science of ozone depletion is hogwash.

The Lead Wars

Prior to April 2014, the public water supply of the city of Flint, Michigan came from Lake Huron and the Detroit River via the Detroit Water and Sewerage Department, which had carried out the necessary processing and purification. The new water supply came from the Flint River and, crucially, for reasons that seem to have been primarily economic, the new processing failed to include corrosion inhibitors—chemicals added to reduce the amount by which the water corrodes metals with which it comes in contact. This switchover occurred after Governor Rick Snyder (R) appointed an emergency manager for the city in September 2013.

Since Flint was a relatively poor city, decades of under-investment in infrastructure meant most homes and businesses hadn't had their plumbing replaced in many years, if ever—in other words, most of the pipes were lead and, because of the lack of corrosion inhibitors, the lead content of the water coming out of residents' faucets soared.

Lead is a neurotoxin—a poison that affects the nervous system, including the brain. Although its symptoms include such calamities as loss of fertility, what it's known for primarily is its effects on behavior and learning ability. Children are, for various reasons, especially susceptible to it—about five times as vulnerable as adults, according to current estimates. The reason lead's such an effective poison is that, once ingested, it mimics calcium, one of the most important minerals in the body. Although we're probably most familiar with calcium's role in bone formation, calcium also plays an important part in various aspects of brain and nervous-system function. When calcium ions are replaced with lead ions, things slowly start to go wrong—horribly wrong.

It's the slowness of the changes that helps make lead poisoning so insidious. Anyone who's lived with a family member in the throes of physical or mental decline will know how difficult it is to notice that their imperceptible daily deteriorations are adding up to something grim. Similarly, with sufferers from lead poisoning, it's hard to recognize that their intellectual abilities have sagged a bit over the past few years—especially if one's suffering the same decline oneself. (A further complicating factor is that, as my daughter's fond of pointing out to me, people do tend to get a bit stupider *anyway* with advancing years.) The net result is that by the time someone's in the advanced stages of lead poisoning it's too late to do anything about it.

In Flint, the problem was spotted relatively quickly. As early as August 2014, just a few months after the shift in supply, the city issued a warning that residents should boil any water they might want to drink or use for cooking; of course, boiling has no effect on lead content, but that's politicians for you.*

In January 2015, despite a firestorm of local protests over foul water, the city's officers decided, on grounds of expense, against a proposal to start using the Detroit supply again. Detection by the EPA in February of dangerously high lead levels in Flint's drinking water led to the city's councilors changing their tune in March . . . but their resolution to readopt the Detroit supply was vetoed by Flint's emergency manager, Jerry Ambrose.

The saga now entered a phase where politicians on both side of the partisan divide went into "Problem? What problem?" mode. Even as the EPA was announcing more and more experimental results that showed there was a real crisis with Flint's water, the city's Democrat mayor, Dayne Walling, pulled the TV stunt of drinking a glass of it to show how safe it was—as if, in the context of lead poisoning, a single glass of water was going to make much difference one way or the other. All through 2015 the Michigan Department of Environmental Quality (MDEQ), under its director, Dan Wyant, was telling residents there was nothing to worry about.

Eventually, though, it was impossible to deny the crisis any longer. On January 5, 2016, Michigan's governor, Rick Snyder, declared a state of emergency for the city, and a few days later, on January 16, President Barack Obama followed suit. The consequent millions in state and federal relief have been spent on supplying Flint residents with bottled water for all uses, on replacing lead water pipes with modern ones—a process that's expected to be complete by 2019—and on effecting the shift back to the supply offered by the Detroit Water and Sewerage Department. The total eventual

---

* To be fair, there were concerns, too, over contamination by coliform bacteria in the water. There were also concerns over chlorine levels. Obviously the whole affair was a mess.

cost will likely run in excess of $200 million[66] and perhaps to twice that.[67] Other estimates run even higher. According to the *Detroit Free Press*[68] and *SciTech Connect,*[69] the cost of adding the anti-corrosive treatment to the Flint River water would have been $50,000 per annum.

At the time of writing (July 2017), Flint residents are still having to use bottled water. There's a whole slew of criminal lawsuits under way over the authorities' handling of the crisis. The courts have dismissed numerous civil actions, usually on the grounds of sovereign immunity and official immunity, two doctrines that protect elected and appointed state officials from the legal consequences of their actions, no matter how heinous.

Environmentalists have suggested the Flint water crisis represents only the tip of a nightmarish iceberg. In early 2017, the similar plight of the residents of East Chicago, Indiana, came to national prominence. The problem in fact dated back to 2009, when parts of the city were found to be suffering significant lead and arsenic contamination. In 2016, EPA testing revealed alarmingly high levels of lead contamination in the city's West Calumet Housing Project, and there were calls for East Chicago to be declared a state emergency. Indiana's then-governor, Mike Pence (R), rejected these calls, but his successor, Eric Holcomb (R), acted on them. In February 2017, federal authorities greenlighted mandatory evacuation and relocation for the housing project's residents, declaring it a Superfund area.

While the source of the lead that's contaminating the project is mainly a relic of the industrial facilities that once operated there, there's a more current problem: air pollution from the Indiana Harbor Coke Co. factory, which bakes coal to make coke for use in steel mills. The plant has a long track record of air pollution violations with the EPA, one of the main pollutants being lead. So, when the residents of the West Calumet Housing Project weren't drinking the stuff, they were likely breathing it.

For reasons unknown, the EPA, despite all of the Indiana Harbor Coke Co.'s violations, has failed to bring legal action against the company to force it to clean up its act, and there are fears now that it may never do so. The *Chicago Tribune* quoted Eric Schaeffer, head of the Environmental Integrity Project and a one-time top enforcement official at the EPA:

> This case should be a slam-dunk . . . [But] I've been told by career staff at the agency that everybody is kind of frozen since Pruitt arrived. Nobody is willing to pull the trigger to enforce the law.[70]

Or, as Emma Foehringer Merchant summarized sarcastically in *Grist*:

The company that owns the facility told the *Tribune* that it's "exploring a number of projects" to deal with the continued pollution. With proposed budget cuts of 31 percent at the EPA, it may be up to companies to monitor pollution and clean up after themselves. That'll work.[71]

As noted, children are especially susceptible to the effects of lead, a fact that the EPA seems to have insufficiently recognized. Currently, the EPA reckons a safe level of lead in drinking water is 15ppb (parts per billion). The American Academy of Pediatrics, in contrast, sets the safe limit at a mere 1ppb. The EPA's figure seems rooted in historical underestimates of the dangers of environmental lead. We'll come back to those historical underestimates in a moment.

The drinking water at Caroline Elementary School, Ithaca, New York, was found in 2016 to contain 100ppb of lead—fully one hundred times the American Academy of Pediatrics recommended maximum. That same year, testing showed that a Montessori school in Cleveland, Ohio, had 1,560ppb of lead in its drinking water, and Leicester Memorial Elementary in Leicester, Massachusetts, had one drinking-water outlet that tested at an astonishing 22,400ppb.

And this isn't just a case of a few outliers. In 2016, a Massachusetts survey revealed that just under 50 percent of the survey's 40,000 tests showed lead contamination over 1ppb. In New Jersey in the spring of 2017, the organization Environment New Jersey, in response to lethargic state testing, compiled the results of tests done in 47 of the 70+ school districts in the state's Bergen County, which apparently was quicker to act than most. The figures roughly matched those in Massachusetts the previous year, although slightly worse, with about 55 percent of the drinking-water outlets testing positive for lead.[72] Most people involved, including state and local politicians, pointed toward parsimony in infrastructure spending as the culprit. While states and the federal government have taken steps to eliminate the use of lead in new plumbing, politicians and environmental authorities have been far slower to take action on the expensive process of replacing older, lead-containing water systems.

They know these old systems utilizing lead are dangerous; they know children are especially vulnerable to the effects of lead; and yet politicians and voters have over decades chosen to save pennies rather than address the problem.

The experience in the testing of New York City schools seems to show the willingness, perhaps unconscious, of environmental authorities to attempt to preempt negative results. During testing in summer 2016, water sampling was routinely preceded by a two-hour flushing-out of the water system. Of course, this assured minimal measures of lead in the subsequent sampling, since the water being tested had been barely in contact with the plumbing. After trenchant criticism from environmental scientists, the city agreed to rerun the tests without the prior flushing-out. Preliminary results, reported in early February 2017 after about one-third of the testing had been done, revealed that about nine times the previously reported number of samples contained lead at levels not just over 1ppb but over 15ppb.

At least the states and cities mentioned above are actually doing some testing. Many others don't bother.

As long ago as 2009, the EPA agreed to update its standards on lead contamination, which were horrendously outdated. By August 2016, no new standards had even been proposed, far less instituted. Accordingly, a consortium of environmental groups—including the Sierra Club, United Parents Against Lead National, California Communities Against Toxics, New York City Coalition to End Lead Poisoning and New Jersey Citizen Action—sued the EPA for "unreasonable delay." It took until June 2017 for the Ninth Circuit Court of Appeals to start hearing oral arguments in this "unreasonable delay" lawsuit. On December 27 the court ruled—alarmingly, only by majority decision—in favor of the plaintiffs and ordered the EPA to update its lead standards within 90 days.

# # #

Lead pollution as a result of human activity is nothing new. We know this thanks to research done during the twentieth century by Clair Cameron Patterson—"The most important scientist you've never heard of," as Lucas Reilly describes him in the title of his long 2017 essay about the man and his work (see Bibliography). The title slightly hyperbolizes—Patterson was the main subject of an episode, "The Clean Room" (aired April 2014), of the widely watched TV series *Cosmos: A Spacetime Odyssey*—but its gist is accurate.

During the latter part of World War II, Patterson and his wife, Lorna McCleary Patterson, both played a peripheral role in the Manhattan Project. Soon after the end of the war, Clair began working for a PhD at the University of Chicago under the geochemist Harrison Brown. (Lorna got a job as an infrared spectroscopist and at the time was the family breadwinner.) Brown was keen to investigate the properties of meteorites, in particular the abundances they contained of the trace elements uranium and lead; because uranium decays into lead at a known rate, it should be possible to use the relative abundances of these two elements to determine an age for the meteorites, and hence for the solar system. From there it should be just a small step to determining an accurate age for the earth.

Brown set Patterson the task of determining, through mass spectrometry, the lead content of the meteorites, and charged Brown's fellow postgraduate student George R. Tilton* with measuring the uranium abundances.

Almost as soon as Tilton and Patterson started work, however, they each realized there was something seriously askew with their results. The problem, as Tilton swiftly concluded, was that the lab in which he was working was contaminated with uranium because of the work other researchers had done there before him. He moved to a different lab and, sure enough, his results now made sense.

But this didn't explain Patterson's false readings for lead, unless—and here he made a conceptual leap that was to change countless millions of lives—the lab, his equipment, his samples, his clothing, the air, himself and his colleagues . . . unless *everything* was contaminated with lead.

It took Patterson until 1953, by which time he and Brown were working at Caltech, to build himself a lab that was free enough of lead contamination for his purposes—the "Clean Room" of that *Cosmos* episode's title. The end result of his research, using his own measurements and Brown's math, came in 1956, when he was able to publish the first accurate estimate of the age of the earth, 4.5 billion years, far higher than the preceding accepted estimate of around 3.3 billion years.

That feat alone should have been enough to make Patterson's a household name, but his true importance lies in what he did afterward. By examining cores of ocean sediments and ice from the glaciers of Greenland and eventually Antarctica, he was able to show that human civilization and pervasive environmental lead contamination went hand-in-hand, and had done so ever since the Romans utilized lead pipes in their clever plumbing systems.** He was even able to "map" the drop-off in lead pollution corresponding to the Black Death, the plague that swept Europe in the

---

* Who decades later would write a very useful memoir of Patterson—see Bibliography.

** Ironically, Roman scholars warned of the dangers of lead, but since the people who were suffering the most obvious ravages of lead poisoning were those involved in mining and smelting the metal—i.e., slaves—no one much cared. So the Romans happily carried on poisoning themselves.

mid-fourteenth century, wiping out, by most estimates, between thirty and sixty percent of the population. But by far the heaviest lead pollution belonged to the modern, post-Industrial Revolution era, and in particular to the twentieth century. This latter increase was far, far greater than could be accounted for by the world's steadily growing population and the spread of technology.

With one exception. It was an exception that it didn't take long for Clair Patterson to identify.

In 1916 the inventors Charles Kettering and Thomas Midgley teamed up to try to find an additive for gasoline that would eliminate engine knock, a problem that had plagued cars since Kettering had devised the pioneering crankless model, the 1912 Cadillac. Their first successful additive was tellurium, in 1921; unfortunately, while it eliminated the engine knock, it stank to high heaven. Later that same year they hit upon tetraethyl lead, which not only got rid of knock in a pleasingly odorless manner but improved engine performance. The first leaded gasoline—marketed as Ethyl—went on sale in 1923, and soon was very much in demand. However, some of the workers employed by the newly founded Ethyl Gasoline Corporation, including Midgley himself, started to display serious side effects, both physical and psychological. Some died, by suicide or otherwise.

In 1924, Midgley—one of the lucky ones who recovered—and Kettering called in Robert A. Kehoe, an instructor in the University of Cincinnati Department of Physiology, to tackle the health issues at the Ethyl Gasoline Corporation; the following year Kehoe became the company's Chief Medical Adviser.

In October 1924, four of the tetraethyl lead workers at the Standard Oil Company's Bayway plant in Elizabeth, New Jersey, were driven to delirium by the lead fumes amid which they worked, and a fifth died as a consequence of them. Kehoe's remedy was a simple one: install fans in all factories where Ethyl was being produced. This seemed to work, and production resumed.

Not everyone was as convinced as Kehoe and his bosses that the tetraethyl lead problem really had been solved. Yes, the factories seemed now to be safe, but what about all the tetraethyl lead that was being spewed into the atmosphere from car exhausts? What effect would breathing this stuff have on the general public?

One such critic was Yandell Henderson, Director of Yale's Laboratory of Applied Physiology. He predicted that, as more and more cars came onto the roads, most if not all of them using the new lead-laced fuel, they'd present an ever-increasing hazard to public health. Most dangerously of all, the effect on health wouldn't be something individuals would immediately notice; instead, the sickness would creep up on them undetected over years and decades. Henderson's prediction was dismissed by Standard Oil and its science mouthpiece, Kehoe, as scaremongering nonsense. The federal government—following a 1925 conference on the controversy called by the Surgeon General, at which Henderson was joined by the pioneering industrial toxicologist and occupational health researcher Alice Hamilton in raising the alarm—sided with the oil company.

Benjamin Ross and Steven Amter, in their book *The Polluters* (2010), skewer Kehoe's pro-industry contention:

> At the 1925 national conference [Kehoe] propounded a position which synthetic chemical manufacturers would adopt in many subsequent controversies. Tetraethyl lead should be banned if an "actual hazard" was demonstrated—and only then. In the face of scientific uncertainty, society should err on the side of utility; demonstrable economic benefits should always outweigh unproven risks. Chemicals, in short, were innocent until proven guilty.

In other words, Kehoe's stance matched exactly that of every other corporate plea against good science that demonstrates the dangers of pollution, whether the pollutants be asbestos, CFCs or greenhouse gases: it's better to risk human lives and welfare than corporate profits. Moreover, as Ross and Amter continued, "The research that might discover an actual hazard from tetraethyl lead was in Kehoe's hands." And, unsurprisingly, the research done by Kehoe and his colleagues at the University of Cincinnati's Kettering Laboratory of Applied Physiology, which he founded with oil and chemicals industry sponsorship in 1930, never did discover such a hazard—at least publicly. Surprising? Kehoe's industry sponsors vetted the results of any researches he did and decided whether or not they should be published.

This wasn't the only way in which Kehoe offered what we might regard as a textbook example of the corrupted scientist. There was also the fact that the only people equipped to peer review the Kettering Laboratory scientists' work were other Kettering Laboratory scientists, because no one else was doing any relevant research. Preconceptions, right or wrong, thus solidified into accepted knowledge without any proper critical analysis; basic scientific errors slipped through.

So Kehoe and his colleagues said the use of tetraethyl lead in gasoline presented no hazard to the public health. Kehoe also presented as established fact what we now know to be pseudoscience: the notion that there's a *natural* level of lead in the bloodstream and, above that, there's a safe level. Kehoe's estimates of the "natural" level of bodily lead were based on the levels he found in experimental subjects. These bogus claims weren't challenged until the publication of Clair Patterson's groundbreaking paper "Contaminated and Natural Lead Environments of Man" in *Archives of Environmental Health* in September 1965.

What Patterson had realized, as a consequence of all the research he'd done in the years since clearing his lab of lead, was that the commonplace lead concentrations found in people's bloodstreams weren't natural at all—they were a result of omnipresent lead pollution.

There were various causes. Lead in water pipes: check. Lead in the paints used to cover walls in homes, offices, and factories: check. Lead used in the paints in children's paintboxes: check.* Lead employed in soldering in the canning industry: check. Lead in pesticides: check. Newspaper ink: check.

Lots of other sources of lead contamination could be fingered—including, trivially, a few natural ones, such as volcanic eruptions—but there was one that stood out head and shoulders above the rest: the tetraethyl lead in gasoline.

It comes as a surprise to learn that Kehoe had peer reviewed Patterson's *Archives of Environmental Health* paper. It seems his view was that, if he recommended against its publication, this would (rightly) be seen as an act of suppression when, as would inevitably soon be the case, Patterson's paper was published somewhere else. Far better to let the *Archives* publication go ahead and then afterward pan the paper as the rantings of a crank.

And this is exactly what the toxicological community did—after all, the only research on bodily lead upon which the toxicologists could draw was that being done at the Kettering by Kehoe and his cronies. The American petroleum industry—which, thanks to Harrison Brown's wiles, had actually been conned into funding quite a lot of Patterson's work over the years—not only withdrew all promises of future monies but joined in the chorus of condemnation.

The problem with such tactics is that sooner or later people see through them. When Senator Ed Muskie (D) called on Kehoe to appear before his Subcommittee on Air and Water Pollution for a hearing on the dangers of lead in 1966, Kehoe, who had sailed through congressional hearings easily enough before, his expertise being taken for granted, suddenly found him-

---

* A particular cause for concern because what child doesn't habitually suck their paintbrush?

self facing more scrutiny. And one of the other expert witnesses Muskie called was Clair Patterson, who demolished not only Kehoe's (pseudo)scientific claims but also politicians' failure to insist on independent rather than industry-commissioned data. In the words of George Tilton in his memoir of Patterson:

> He furthermore believed it was wrong for public health agencies to work so closely with lead industries, whom he considered often biased in matters concerning public health.
>
> His views drew support from some of the public (e.g., Ralph Nader), but were once again strongly opposed by others, notably by R.A. Kehoe, the highly regarded authority on industrial poisoning. A battle line was drawn that was to last about two decades.

Patterson was, however, still a relatively lonely figure. Although 1970's trailblazing Clean Air Act, which Muskie heavily influenced, did recognize that gasoline additives might represent an environmental threat, little attention was given to lead. When the EPA that same year commissioned the National Academy of Sciences to evaluate the hazards of ethyl gasoline, Kehoe was one of the experts consulted, Patterson wasn't. No cigar for guessing the conclusions of the NAS report. Yet the EPA itself must have sensed there was something in Patterson's claims, because in 1972 the agency proposed, although not an elimination of tetraethyl lead as a gasoline additive, at least a phased reduction in its use.

The Ethyl Gasoline Corporation, using the same economic arguments that Kehoe had advanced robotically at every hearing for decades ("until the science is settled . . ."), persuaded the EPA to delay the implementation of the phase-down. Then, in 1973, it sued the EPA, using the same arguments, and actually won the case in 1974 at the federal appeals court level. In 1976, however, a US Court of Appeals reversed the decision and denied the petitioners the right to proceed to the Supreme Court.

Patterson, all this while, had been doing more research. He compared lead levels in the bones of Egyptian and South American mummies with those in the bones of modern Americans, and discovered that the levels in the latter were not just a few times higher than those in the former but a whopping 500+ times higher. Others, notably the Harvard University Medical School pediatrician and child psychiatrist Herbert Needleman,*

---

* Whose much later review paper, "Clair Patterson and Robert Kehoe: Two Views of Lead Toxicity" (*Environmental Research*, August 1998), offers an interesting historical account.

had begun to publish their own conclusions about the hazards of lead in the bloodstream, conclusions that directly contradicted the ones Kehoe had been pronouncing without rebuttal since the 1920s.

And then, in the wake of the EPA's delayed phasing down of the amount of lead in gasoline, it was found that levels of lead in the bloodstreams of US children were dropping. By 1986, Patterson's case was regarded as more or less proven and Kehoe's reputation lay in tatters. In that year the EPA decreed that the use of tetraethyl lead as a gasoline additive be phased out entirely by the end of 1995. That final eradication, when it came, had been delayed for decades thanks to the corruption of the scientific narrative by the gasoline industry, the tetraethyl industry, and their scientific mouthpieces.

It was delayed by a matter of mere weeks too long for Clair Cameron Patterson, who died on December 5 of that year.

# # #

Contrary to many dystopian predictions and to perceptions fueled by the internet and our nightly dose of TV news, rates of violent crime in the US have been gently falling during the twenty-first century. This seems counterintuitive, since the economic inequalities initiated during the Reagan years and exacerbated during all subsequent administrations have generated rapidly rising poverty levels. Throughout history there's been a fairly explicit relationship between levels of poverty and levels of crime, and it's starkly improbable that this relationship could suddenly have reversed. Clearly there must be some other factor at work.

No one knows for sure what that other factor might be, but plenty of people, bearing in mind the known psychological symptoms of lead poisoning, have pointed to the declining levels of environmental lead contamination and the corresponding drop in the serum lead levels of the average citizen, and have proposed what seems a very obvious explanation: there's less violent crime because fewer of us are dangerously nuts due to lead exposure.*

---

* See for example "Environmental Policy as Social Policy? The Impact of Childhood Lead Exposure on Crime" by Jessica Wolpaw Reyes, *B.E. Journal of Economic Analysis & Policy*, vol 7 no 1, 2007. For a popular roundup see "Lead: America's Real Criminal Element" by Kevin Drum, *Mother Jones*, February 11, 2016. A later piece by Drum that's of relevant interest is "New Zealand Study Provides More Support for Lead-Crime Hypothesis," *Mother Jones*, December 27, 2017.

"Obvious" doesn't mean "right," of course. Yet we remember the widespread lead pollution in the Roman Empire and the horrific levels of sadistic violence the Romans promoted in their penal system and their circuses. Some archaeologists, too, go so far as to suggest that lead poisoning, and its consequent mental debilitation, may have contributed to the Empire's collapse.

It's a sobering thought.

THE VILIFICATION OF RACHEL CARSON

If man were to follow the teachings of Miss Carson, we would return to the Dark Ages, and the insects and diseases and vermin would once again inherit the earth.

—Robert H. White-Stevens, Assistant Director, Agricultural Research Division, American Cyanamid, interviewed on *CBS Reports*, April 3, 1963

Following a major three-part serialization in the *New Yorker* during the summer of 1962, Rachel Carson's fourth and final book, *Silent Spring*, published on September 27, 1962, was an immediate bestseller: it had advance sales by publication day of 40,000 copies, while its selection as a main choice for the Book of the Month Club accounted for another 150,000. Its sales to date run into the millions. Its message was that chemical pesticides and herbicides, which seemed to be such a boon to agriculture, were entering the food chain and inexorably climbing it, with results that were often disastrous. Although not objecting to these pesticides *per se*, Carson was concerned that their continued massive overuse—abuse, in fact—could lead to a future where the springs were silent because bird and other populations had largely been annihilated. She spotlighted what many members of the public and politicians did not fully appreciate and, sadly, often forget today—that our health as a species depends upon the health of the environment in which we live and of which we're a part.

Silent Spring was a bit of a departure for Carson, a mild-mannered marine biologist who'd worked for over a decade at the US Bureau of Fisheries until leaving in the mid-1950s to become a full-time writer. Her three earlier books had been lyrical accounts of marine biology: *Under the Sea-Wind* (1941), *The Sea Around Us* (1951), and *The Edge of the Sea* (1955). When it came to the dangers of synthetic pesticides, a topic she knew was important and should be brought to public attention, she didn't feel she was the right person to write it, and looked around for others.* She may also have been nervous about facing the firestorm she knew the chemicals industry would unleash in response to the book. In the end, though, she wrote it.

One reason *Silent Spring* so captured the imagination of the US public was a growing awareness that the testing of nuclear bombs—largely regarded by the average American as something that happened a long way from home—was not, after all, entirely without consequences in the homeland. In November 1961 the preliminary results had been made public of the so-called Baby Tooth Survey, a project undertaken to check the levels of strontium-90 in children's teeth; the results showed that, the more recently a child had been born, the higher the strontium-90 level—a clear indication that human beings were picking up man-made radioactive contamination from the environment.**

Another reason was the Thalidomide tragedy. The drug Thalidomide had been introduced in 1957 and widely prescribed—far more in Europe than in the US, in fact—to help alleviate pregnant mothers' morning sickness. Initially no one knew why an increasing number of babies were being born with only vestigial limbs. It wasn't until 1961 that William McBride put two and two together to conclude that Thalidomide was the cause of the birth deformities (see page 39).

So the public was primed to accept the idea that invisible components of the environment could cause harm, and that apparently beneficial products of science could have horrendous side effects.

Critics of Rachel Carson have often zeroed in on her claim that artificial pesticides were born in the quest by the US for chemical weapons during World War II. *** Allied to this has been the criticism that she failed to stress that what may be poisonous in a large dose is often (in fact, usually) innocuous if the dose is small enough. In fact, Carson was perfectly correct: World War II chemical-warfare endeavors were indeed the crucible

---

* She briefly collaborated with an individual called Edwin Diamond, but they parted company on apparently unfriendly terms. Years later Diamond mounted a bitter personal attack on her in the form of a book review in the *Saturday Evening Post*. Although his charges that she essentially faked a lot of her evidence in *Silent Spring* may have seemed convincing to *Post* readers at the time, they're looked at with stark incredulity today.

** The results of the Baby Tooth Survey were a factor in John F. Kennedy's zeal to negotiate the 1963 Partial Test Ban Treaty.

*** To a great extent the attacks mounted by the chemicals industry and, more recently, by free-enterprise ideologues have focused on the person rather than the book.

for what would, soon after the end of hostilities, become the commercially produced artificial pesticides. And she was certainly not so stupid as to be ignorant of the question of dosage. Quite the contrary. She didn't encourage a blanket ban of dichlorodiphenyltrichloroethane (DDT) or the other artificial pesticides. "I think [these] chemicals do have a place," she testified to a Senate subcommittee on the matter in June 1963. And in *Silent Spring* she spelled this out:

> No responsible person contends that insect-borne disease should be ignored. The question that has now urgently presented itself is whether it is either wise or responsible to attack the problem by methods that are rapidly making it worse.

Even so, just the publication of her book, and the size of its audience, was enough to impact the profits of the pesticide and herbicide manufacturers; the industry trade group, the National Agricultural Chemicals Association, spent over $250,000 (over $2 million in today's terms) on efforts to counter the book, mainly through attacking—indeed, often vilifying—its message and its author. *Silent Spring* was also seen as a threat to those working in the sciences, the white-coated experts whose work underpinned the advance of technology. Then as now, many of the experts are largely funded by industry, so it can be a little difficult to distinguish the motivations of individual attacks, and attackers.

Leading the field was Dr. William J. Darby, who took not just a hatchet but the entire edged contents of his toolshed to the book in his review of it for *Chemical & Engineering News*,[73] dismissively headlined "Silence, Miss Carson."

It's clear, reading Darby's essay today, that he regarded Carson's book as a threat to everything he held dear, a threat that must be extirpated at any cost; he even criticized the unorthodox layout of her sources list! Not to mention her readership:

> Those consumers will include the organic gardeners, the antifluoride leaguers, the worshipers of "natural foods," those who cling to the philosophy of a vital principle, and pseudo-scientists and faddists.

Like the lead industry had earlier and the asbestos industry would later, Darby sought to confuse the issue by saying that the possibility of illness through overexposure to toxic chemicals is really just a hazard for those handling the materials every working day, not for members of the general public. The fact that exposure to relatively low levels of toxins over a long period can be* just as damaging as exposure to high levels over a shorter one is a possibility that he signally ignored, presumably hoping his readers would do likewise.

Although insisting, "It is doubtful that many readers can bear to wade through the book's high-pitched sequences of anxieties," Darby was sure this was a dangerous work:

> Such [an] attitude . . . means the end of all human progress, reversion to a passive social state devoid of technology, scientific medicine, agriculture, sanitation, and education. It means disease, epidemics, starvation, misery, and suffering incomparable and intolerable to modern man. Indeed, social, educational, and scientific development is prefaced on the conviction that man's lot will be and is being improved by a greater understanding of and thereby the increased ability to control or mold those forces responsible for man's suffering, misery, and deprivation.

Oddly enough, these are words with which Carson might have largely agreed. She wasn't against the use of science and technology to mold the world to make it a better place to live in. All she was questioning, as we saw, was "whether it is either wise or responsible to attack the problem by methods that are rapidly making it worse."

Darby concluded:

> The responsible scientist should read this book to understand the ignorance of those writing on the subject and the educational task which lies ahead.

The main scientific attack dogs championed by industry were two scientists working at the chemicals company American Cyanamid, Robert H. White-Stevens and Thomas H. Jukes. They focused on the annual bird count released by the Audubon Society—a highly vocal supporter of Carson—and used it in their attempts to undermine her case. White-Stevens wrote: "Thus robins, over which Miss Carson despairingly cries requiem as they approach extinction, show an increase of near 1,200% over the past two decades." Because of his genial, avuncular manner, White-Stevens was frequently hauled out as spokesman for the anti-Carson faction in TV

---

* I stress the "can be." Obviously there are plenty of instances where low dosages do us no harm—like the tiny amount of naturally occurring arsenic we ingest every day.

broadcasts on the issue, including the April 3, 1963, edition of *CBS Reports* cited at the head of this section.

In May 1977, the Court of Appeals for the Second Circuit reconsidered a libel case won in the lower courts by White-Stevens, Jukes and entomologist J. Gordon Edwards against Robert S. Arbib, Jr., and Roland C. Clement of the Audubon Society and the *New York Times*. In the April 1972 issue of the society's publication, *American Birds*, Arbib had written:

> We are well aware that segments of the pesticide industry and certain paid "scientist-spokesmen" are citing Christmas Bird Count totals (and other data in *American Birds*) as proving that the bird life of North America is thriving, and that many species are actually increasing despite the widespread and condemned use of DDT and other non-degradable hydrocarbon pesticides.
>
> This, quite obviously, is false and misleading, a distortion of the facts for the most self-serving of reasons. The truth is that many species high on the food chain, such as most bird-eating raptors and fisheaters, are suffering serious declines in numbers as a direct result of pesticide contamination; there is now abundant evidence to prove this. In addition, with the constant diminution of natural habitat, especially salt- and freshwater marshes, it is self-evident that species frequenting these habitats are less common than formerly.
>
> The apparent increases in numbers of species and individuals on the Christmas Bird Counts have, in most cases, nothing to do with real population dynamics. They are the result of ever-increasing numbers of birders in the field, better access to the Count areas, better knowledge of where to find the birds within each area, and increasing sophistication in identification.

The *New York Times*' supposed sin was to have reported ("accurately," as the judges tartly noted) on Arbib's editorial while also mentioning the names of the three scientists as examples of those who were continuing, despite having been personally informed by the Audubon Society of the true explanation for the bird statistics, to use the "bird count" argument.* The question under consideration was whether the scientists had explicitly been called "paid liars" by the Audubon Society officers, and the judges concluded that they hadn't. They had been accused merely of repeating information that they knew to be false. (I confess I find the distinction tricky.) Although sympathizing with the trio, the judges found against them.

---

* Two others mentioned in the *New York Times* piece, Norman E. Borlaug and Donald A. Spencer, didn't join the action.

One of the other scientists mentioned in the 1972 *New York Times* piece, Norman E. Borlaug, father of the Green Revolution and the 1970 Nobel Peace laureate, had predicted that, without DDT and the other chemical pesticides, modern agriculture would collapse. This statement was grasped eagerly by the Montrose Chemical Company, DDT's main US manufacturer, who used it as the centerpiece of an anti-*Silent Spring* publicity campaign.

Many years later, in the Summer 1992 issue of the magazine *21st Century Science and Technology*, J. Gordon Edwards, one of the three appellants in the doomed 1977 libel case, published an oft-cited, point-by-point attack called "The Lies of Rachel Carson." Sure enough, he was continuing to use the "bird count" argument.

That same magazine in its Summer 2002 issue reprinted Edwards' piece while also mounting one of the more vicious latter-day attacks on Carson, an editorial called "Bring Back DDT, and Science With It!" by Marjorie Mazel Hecht:

> The 1972 US ban on DDT is responsible for a genocide 10 times larger than that for which we sent Nazis to the gallows at Nuremberg. It is also responsible for a menticide which has already condemned one entire generation to a dark age of anti-science ignorance, and is now infecting a new one.
>
> The lies and hysteria spread to defend the DDT ban are typical of the irrationalist, anti-science wave which has virtually destroyed rational forms of discourse in our society. If you want to save science—and human lives—the fight to bring back DDT, now being championed by that very electable candidate for the Democratic Presidential nomination, Lyndon H. LaRouche, Jr., had better be at the top of your agenda.
>
> Sixty million people have died needlessly of malaria, since the imposition of the 1972 ban on DDT, and hundreds of millions more have suffered from this debilitating disease. The majority of those affected are children.

A subhead later in Hecht's editorial reads: "Banned to Kill People." She then cites the 1990 comment of Alexander King, a founder of the Malthusian Club of Rome, to the effect that DDT was a curse because it kept the world's human population unnecessarily high.

As you probably guessed from the glowing endorsement in the above extract, *21st Century Science and Technology* is a magazine much in awe of perennial fringe outsider Lyndon LaRouche, and has the same sort of scientific status as the journals produced by organizations like the Church of Scientology and the Discovery Institute. As well as supporting LaRouche,

21st Century Science and Technology opposes established climate science and, perhaps most bizarrely, supports the atomic theory developed in his dotage by Robert Moon, a genuinely significant nuclear physicist who—like Linus Pauling with homeopathy and Vitamin C in his later years, like Alfred Russel Wallace with spiritualism in his later years, and so on—started to back scientific losers. In Moon's instance it was the notion that the protons and neutrons in the atomic nucleus were arranged according to the regular convex polyhedra—the Platonic solids. His model wasn't unlike the one that the great cosmologist Johannes Kepler developed in the folly of his early career to describe the orbits of the planets. Like Kepler's, Moon's model was hopelessly wrong. At least Kepler had the excuse of being born four centuries earlier.

Reverting to the 1960s, among the chemical companies that sought to denigrate Carson and her work was Monsanto, later to become world-renowned for the "aggressive" promotion of its genetically engineered organisms and in particular the pesticide glyphosate, which it markets as Roundup.

The company's president attempted to characterize Carson as a "fanatic defender of the cult of the balance of nature." The company also seized upon *Silent Spring*'s first chapter, "A Fable for Tomorrow," in which Carson depicted a future world largely devoid of wildlife; Monsanto commissioned and published a parody, "The Desolate Year," in which the sufferings of a future world arose from the lack of chemical pesticides.

The lawyers for Velsicol, manufacturer of chlordane, one of the several pesticides indicted by Carson along with DDT, sent publishers Houghton Mifflin a letter promising a lawsuit for massive libel damages should they go ahead with the book. Houghton Mifflin, recognizing a blowhard threat when they saw one, replied robustly. When the same company made similar threats to *Audubon Magazine*, which was planning to publish extracts from the book, the magazine's editor, John Vosburgh, responded even more robustly. Not only did he go ahead and publish the extracts, alongside them he published his own editorial outlining the company's threats and describing the industry's behavior in strongly unflattering terms.

After it had published the second installment of its serialization of *Silent Spring*, the *New Yorker* received this (possibly satirical) letter from a reader:

> Miss Rachel Carson's reference to the selfishness of insecticide manufacturers probably reflects her Communist sympathies, like a lot of our writers

these days. We can live without birds and animals, but, as the current market slump shows, we cannot live without business. As for insects, isn't it just like a woman to be scared to death of a few little bugs! As long as we have the H-bomb everything will be O.K. P.S. She's probably a peace-nut too.

It's a comment that could have easily appeared today on any of countless anti-science websites, where popular targets of abuse include any area of established science that might be seen as inhibiting the writer's liberty to behave like an antisocial jerk at the expense of the rest of us. Thus environmentalism in general gets flayed, and Rachel Carson in particular.

In 2012, the fiftieth anniversary of the publication of *Silent Spring* brought with it not just countless reverential editorials, articles, TV pieces and the like but also a new wave of vilification. It would be nice to say that many of the articles critical of *Silent Spring* were sober debates on scientific grounds, but I have yet to find them; even if supposedly arguing the science, the authors seemed unable to resist gratuitous insults.

Take, for example, "Rachel Carson's Deadly Legacy," by two representatives of industry-favoring thinktanks, Henry I. Miller of the Hoover Institution and Gregory Conko of the Competitive Enterprise Institute (CEI). This piece was initially published by *Forbes* on September 27, 2016, but no longer appears on that site.* Describing the book as "an emotionally charged but deeply flawed denunciation of the widespread spraying of chemical pesticides for the control of insects," which is fair enough as criticism, whether valid or not, the pair soon move into hyperbolic mode:

> But the fears she raised were based on gross misrepresentations and scholarship so atrocious that, if Carson were an academic, she would be guilty of egregious academic misconduct. . . . It is extraordinary that anyone in the mainstream scientific community could continue to embrace the sentimental claptrap of *Silent Spring*, so we were surprised to see the commentary, "In Retrospect: *Silent Spring*," in the scientific journal *Nature* in May by evolutionary biologist Rob Dunn. Science is, after all, evidence-based, but Dunn's puff piece is a flawed and repugnant whitewash of Carson's failure to present actual evidence to support her assertions, and of the carnage that she caused. It also demonstrates that Dunn knows little about the history or toxicology of DDT. . . . The legacy of Rachel Carson is that tens of millions of human lives—mostly children in poor, tropical countries—have been traded for the possibility of slightly improved fertility in raptors.

---

* I have no idea of the history here. However, I was able to find large extracts of the piece posted at the blog *The Atheist Conservative*: http://theatheistconservative.com/tag/professor-robert-h-white-stevens/.

The CEI, to which Conko owes allegiance, is a frequent attacker of Carson. Posing as a serious scientific thinktank, this organization, today largely funded by the Koch brothers, is almost as notorious for its science refusal as the Heritage Foundation. Others of its campaigns have claimed that smoking is safe (or at least, in the usual corporate-denialist phrase, "the science is still out"), and that fossil-fuel CO_2 emissions are actually good for you: "They Call It Pollution, We Call It Life!" was the slogan for that one. The CEI also played a big part in publicizing the Climategate *faux*-scandal.

It was the CEI that created the website RachelWasWrong.org, "aimed solely," as Michael Mann puts it in his *The Hockey Stick and the Climate Wars* (2012), "at discrediting Carson's legacy." That site has now become (or been subsumed into) the doublespeakishly titled SafeChemicalPolicy. org which, complete with pictures of lovable kids, promotes unsafe chemicals. Here's a sample article-opener from when I checked the site in June 2017:

> If you believe the headlines, honeybees may soon be endangered, pesticides are to blame, and regulations offer an easy solution. Allegedly, the problem stems largely from our naive trust in agro-technologies, particularly pesticides. Specifically, Greenpeace and myriad others blame a class of pesticides called neonicotinoids, claiming that these chemicals "might just be the prime culprit in the honeybee plague known as Colony Collapse Disorder (CCD)."

The article goes on to tell us that colony collapse disorder isn't really important to bees at all—that there are far more serious threats. In a sense this is arguably true, just as extreme old age is far more of a threat to health than a dose of flu; on the other hand, if you catch the dose of flu when you're in extreme old age . . . And, too, while the guilt of neonicotinoid pesticides may not have been conclusively proven, they remain, at the time of writing, very strong suspects.

As we've seen, Carson's modern vilifiers are fond of levelling the charge that the abandonment of DDT by the World Health Organization in its fight against malaria, in consequence of *Silent Spring*, has cost millions of lives—more, indeed, than Hitler's activities. Another to make this accusation was Michael Crichton in his daft conspiracy-mongering novel *State of Fear* (2004). The reality is that the WHO stopped using DDT against malaria-carrying mosquitoes because the insects had developed resistance to it, thanks to the overuse of the chemical in agriculture—much as today the overuse of antibiotics in agriculture has led to the development of antibiotic-resistant superbugs.

Carson's attackers also frequently imply—perhaps assuming their readers won't know any better—that Carson was the originator of the scientific content of *Silent Spring*, and that the conclusions therein were exclusively hers. As Elena Conis points out in her long essay "Beyond Silent Spring: An Alternate History of DDT,"[74] public and governmental acceptance of DDT was never as wholehearted as we've been led to believe, and there were plenty of relevantly qualified scientists who expressed skepticism about DDT's overall benefits—particularly on the grounds that, at the same time as it was killing off insect pests, it was also killing off beneficial insects. It's not helpful to wipe out pollinators if you want to increase agricultural production.

In 1944, the Office of Scientific Research and Development released a report that emphasized that DDT should be used with caution, and kept away from livestock, drinking water, fish, and more. At the end of the report the FDA pharmacologist Herbert O. Calvery pointed out that far too little was known about the safety levels of DDT: animal testing had shown that some species were far more tolerant of it than others, with no clear way of predicting which were the tolerant species. Buildup of DDT in the body was another of Calvery's concerns—and is still a concern today. Aside from worrisome evidence that DDT may be carcinogenic, there are demonstrable links between DDT, a known endocrine disrupter, and various reproductive disorders, such as miscarriages and reduced semen quality.

So Carson's fears were grounded in pre-existing science, and they've been emphatically confirmed by more recent science.

# # #

John F. Kennedy was one of the countless readers of *Silent Spring* who worried about what it had to say. Not equipped himself to evaluate Carson's assertions, he instructed the Life Sciences Panel of the President's Science Advisory Committee to look into the matter. The committee issued its report in May 1963, generally favoring Carson's book. The following day Senator Abraham Ribicoff (D–CT) welcomed her to a Senate subcommittee hearing on pesticides where, although she was by now in very frail health because of the breast cancer that would soon kill her, she gave effective testimony.

The environmental movement gained ever-increasing traction through the 1960s, with the result that in 1970 the Nixon administration created the EPA. In 1972 Congress gave responsibility to the EPA for ensuring that pesticides and their usage were safe for the environment and for human health. A further consequence of Carson's work was the Toxic Substances Control Act of 1976, which charged the EPA with regulating the manufacture, sale, and use of chemicals in order to protect us from "unreasonable risk of injury to health or the environment."

By 1975, use of all of the chemical toxins identified in *Silent Spring* had been much restricted if not outright banned in the US.

In Europe, although the book was still a bestseller in several countries, the impact of *Silent Spring* was more muted. Even so, the environmentalist revolution reached Europe fairly quickly.

A refreshing piece of news about possibly changing industry attitudes came on October 26, 2012, in the fiftieth anniversary year of *Silent Spring's* publication, when the American Chemical Society erected a plaque at Chatham University, Pittsburgh, as a National Historic Chemical Landmark to commemorate the book's legacy. The plaque reads:

> Rachel Carson's *Silent Spring*, published in 1962, was a landmark in the development of the modern environmental movement. Carson's scientific perspective and rigor created a work of substantial depth and credibility that sparked widespread debate within the scientific community and the broader public about the effect of pesticides on the natural world. These discussions led to new policies that protect our air, our water, and, ultimately, our health and safety. Carson's book promoted a paradigm shift in how chemists practice their discipline and helped to establish a new role for chemists in investigating the impact of human activity on the environment. The legacy of *Silent Spring* continues today in the chemistry community's increased focus on green chemistry practices and the public's heightened support for sustainability in all areas of our lives.

With the advent of such new insect-borne diseases as Zika, there's a good case to be made for the controlled reintroduction of the use of DDT—this time with the full awareness that overuse of this chemical can seriously damage the environment, the wildlife within it, the pollinators we rely on for our agriculture and, yes, human beings. This is not to be confused with the thundering by Carson's detractors that DDT is utterly harmless and all the scientific concerns about its safety utter lies. Whether we could learn to use DDT wisely and in moderation, or whether we'd soon revert to a pre-*Silent Spring* world—and all the nightmare futures that its environmental abuses might engender—is another question.

The Beatification of Sugar

Back in the 1960s and 1970s, there was a nutritionist who said it all—well, a lot of it, anyway. In his book *Pure, White and Deadly* (1972), the English academic John Yudkin—who since 1954 had been the Professor of Nutrition at Queen Elizabeth College, London—set out for the general public what he had for years been spelling out in studies published in the science journals: the overconsumption of sucrose was causing an epidemic of a group of related diseases including obesity, coronary heart disease (CHD), Type 2 diabetes, and stroke—what we'd today call the metabolic syndrome.

Yudkin wasn't the only one to have raised this alarm—others included Walter Metz, head of the DOA's Carbohydrate Nutrition Laboratory, Jean Mayer, of Harvard's School of Public Health, and another UK nutritionist, Thomas L. Cleave*—but he was the most prominent. He was therefore the primary target for the counterattack mounted against science by the sugar industry.

The sugar industry set the tone early. A June 1976 article by Jean Mayer in *New York Times Magazine*, "The Bitter Truth about Sugar," describing sugar as highly addictive for children and fingering it for obesity and Type 2 diabetes, caused consternation in the sugar industry, which of course was in the habit of portraying sugar as healthy and wholesome. *Reader's Digest* planned to run an extract from the article. John Tatem, then president of the Sugar Association, later boasted about how his organization

---

* Cleave was an interesting character. A Royal Navy surgeon, he was, by the time he retired in 1962, Director of Medical Research at the RN's Medical School. Unlike Yudkin, Cleave reckoned refined grains were also a culprit.

Cleave's deductions were based upon an hypothesis of his own: that the more the foods in our diet digress from the natural state—i.e., the more they're refined or processed—the worse for our health they're likely to be. It was this, rather than any knowledge of how the body processes carbohydrates, that led to his inference that eating white flour, and baked goods in general, was likely harmful. As with refined sugar, white flour was a relatively recent appearance on the dietary scene and our bodies hadn't had the evolutionary time to adapt to either. People coming from non-western to western cultures were particularly susceptible to the metabolic syndrome diseases because, obviously, there had been no evolutionary time at all for their bodies to adapt to these foods. Moreover, it was easy to eat far more of these foods when they were available in refined form than it was of them in their natural state: it's easy to put a heaped spoonful of sugar in your tea, far more work to get that sugar through eating three or four apples.

had managed to pressure *Reader's Digest* into scrapping that plan. He'd told the magazine's editors that the article was a "scientific farce" and a "journalistic disgrace." As we now know, Tatem was lying: Mayer's science was, at least for its day, impeccable.

The counterattack was enormously successful. For decades conventional medical science attributed the metabolic syndrome not to sugar but to the overconsumption of fats leading to excessive cholesterol in the bloodstream. The chances are your primary physician will still tell you this; when I recently spent a spell in the hospital I received, on my departure, a sheet of dietary guidelines based almost entirely on this principle.

In recent years, Yudkin's baton has been taken up by such figures as science journalist Gary Taubes and endocrinologist Robert H. Lustig. The former's *The Case Against Sugar* (2016) is the definitive account of the science, pseudoscience, politics and industry corruption during this whole saga. Lustig's 2009 lecture, *Sugar: The Bitter Truth*, screened by University of California Television, became a viral hit on YouTube; when last I checked (June 2017) it had been viewed there over seven million times. In the lecture he recounts how he had already come independently to the conclusion that sugar rather than fat and cholesterol was the villain in the metabolic syndrome epidemic when someone at a conference asked him if he'd read John Yudkin's *Pure, White and Deadly*; on his finally laying hands on a copy of the book,* he discovered the nutritionist had reached the same conclusions decades before him, but had been effectively written out of history by the medical community under the influence of the food industry.

Another physician to recognize fairly early that the fat/cholesterol hypothesis was flawed was the cardiologist Robert Atkins, whose Atkins Diet, as promulgated through a series of bestselling books beginning with *Dr. Atkins' Diet Revolution* (1972), espoused the extreme reduction of carbohydrate intake in favor of getting one's calories from proteins and fats. There's some controversy over how effective the Atkins diet is, mainly because Atkins himself never did any proper scientific research on it, instead just citing his countless case files. Certainly Atkins's own proposed theoretical underpinnings for his theory were weak.

In 2003, Atkins died as a consequence of slipping on an icy sidewalk and hitting his head. Instantly the rumors circulated that he'd died of a heart attack and had been grossly obese at the time of death—ho ho, so much for

---

* In 2012 it was reissued with a new introduction by Lustig.

the Atkins diet, eh?—but in fact the cardiac ailment he'd been suffering for some years beforehand was, according to his cardiologist, viral rather than congestive in origin. While it's true that he weighed 117kg (258 pounds) at death, 29kg (63 pounds) of this—about one-quarter—was due to weight gain through fluid retention during the nine days he lay in a coma before dying.

# # #

Why did we become so convinced heart disease was a product of high levels of cholesterol in the blood due to too much fat in our diet?

This was the theory of Ancel Keys, founder of the University of Minnesota's Laboratory of Physiological Hygiene, and developer of the famous K-rations used by US troops during World War II. His other wartime contribution was to perform an infamous (to use the adjective employed in the National Lipid Association's obituary of him) study of the mental and physical consequences of starvation, using conscientious objectors as experimental subjects. This research was the basis of his *The Biology of Human Starvation* (1950); the whole episode is explored in Todd Tucker's *The Great Starvation Experiment* (2008).

Sugar was a major component of the K-ration package, which contained both hard candy and a bar of chocolate, and it's plain Keys regarded sugar as a useful energy source—unsurprisingly, perhaps, since some of his other research had been funded by the Sugar Research Foundation (which later became the Sugar Association).

Largely thanks to the efforts of Keys and the Sugar Association, in the 1960s and 1970s there was a decided transatlantic split in nutritional opinion on the sugar issue. While UK nutritionists weren't unanimous in their opinions, the predominant view in the medical literature was that there was a relationship between sugar consumption and the metabolic syndrome diseases. This view barely made an appearance in the medical literature in the USA. As early as 1974, the editors of the UK medical journal *Lancet* were wary of the growing push to restrict dietary intake of fat, whose importance to health they recognized. "The cure," they said, "should not be worse than the disease."

Although we can read Keys's violent rejection of the work of Yudkin and the others as motivated by the fact that they were threatening his own cholesterol theory, it's tempting also to point at his funding links with the sugar industry. One of the bases for Keys's attacks on Yudkin was that Yudkin's work was based on inadequate experimental evidence:

It is clear that Yudkin has no theoretical basis or experimental evidence to support his claim for a major influence of dietary sucrose in the etiology of CHD; his claim that men who have CHD are excessive sugar eaters is nowhere confirmed but is disproved by many studies superior in methodology and/or magnitude to his own; and his "evidence" from population statistics and time trends will not bear up under the most elementary critical examination.[75]

What Keys chose to ignore was that exactly the same could be said of his own research. Take, for example, the famed Seven Countries Study. From 1956 Keys and a group of colleagues examined data on cohorts of middle-aged male subjects, totaling over 12,500, from seven different countries—Finland, Greece, Italy, Japan, the Netherlands, and Yugoslavia, in addition to the US—over a period of decades, and proved to their own satisfaction that there was a good correlation between dietary fat, blood cholesterol, and what we'd now recognize as primary components of the metabolic syndrome: obesity, CHD and stroke.

But wait a moment. Isn't that rather an odd collection of countries to choose? If you're selecting European countries to include in a survey like this, what about France? Surely France is a bit more "mainstream Europe" than far-north Finland?

The truth was that Keys knew the French diet contained saturated fats galore, yet French rates of heart disease were relatively low. The same could be said for West Germany, another curious omission from Keys' list. Similarly, Switzerland, Sweden . . . In his book *Fat Chance* (2013), Robert Lustig explains:

> The Seven Countries study started out as the Twenty-two Countries study. . . . For these seven, the relationship between dietary fat and heart disease looked pretty convincing. But when all twenty-two countries were plotted . . . the correlation became a lot less convincing. He also chose to leave out "indigenous tribes," such as the Inuit (North America), Tokelau (Oceania), and Maasai and Rendille (both Africa), who ate only animal fat and have among the lowest prevalence of heart disease on the planet.

In other words, presumably aware that some national diets flouted his preconceived ideas about the relationship between saturated fats and the metabolic syndrome, Keys shamelessly cherry picked which countries to study—and, sure enough, got the results he'd hoped for.

There are plenty of other problems with the Seven Countries Study, such as the methodology with which the data were analyzed and that the

European countries were at the time recovering from World War II and, in Italy's case, decades of fascism—in other words, in recent times the customary diets of the subjects there had been significantly disrupted, for good or ill.

This isn't to say that the results Keys and his colleagues produced were entirely worthless. In particular, it was evident from the Seven Countries Study that people in southern Europe were far less affected by the metabolic syndrome than were their northern and US counterparts. It was this that drew the enthusiasm of Keys, his colleagues and in due course the US government for the so-called Mediterranean Diet—plenty of fish, olive oil, pulses, fruit, raw vegetables, with minimal consumption of red meat. Over the years, Keys and his wife Margaret would together write three successful recipe books based on his dietary ideas: *Eat Well and Stay Well* (1959), *The Benevolent Bean* (1967), and, of special note here, *How to Eat Well and Stay Well the Mediterranean Way* (1975). That the Mediterranean Diet was conducive to good health was clearly not just some cockeyed notion; as a single example among millions, Keys himself lived to be 100.

Twenty-five years after his initial work alongside Keys and the rest on the Seven Countries Study, Italian researcher Alessandro Menotti reanalyzed the data that Keys had used and found that the best correlation for fatal heart attacks was not with saturated fats or dietary cholesterol but with the consumption of sweet foods—pies, pastries, cakes, and the like. This would have been an even more staggering match were it not for the fact that, following the classification system of the original, Menotti didn't include ice cream, chocolate, and sodas as sweet foods. To turn this around, had Keys and his colleagues chosen to look for a correlation between sugar and the metabolic syndrome, they'd have found it—with ease!—but Keys had so steadfastly set his face against the "sugar hypothesis," or any other "hypothesis" save his own fat/cholesterol one, that he didn't even look.

From all accounts, this refusal to recognize the mote—or beam—in his own eye was very much in keeping with Keys's character. One of the reasons his fat/cholesterol hypothesis became so much the established rubric among US nutritionists was that he was, to put it bluntly, an intellectual bully: if you didn't agree with him, you were the enemy, and he treated you accordingly. Since he was a towering figure in the nutritional world, few US nutritionists were prepared to take him on.

He certainly behaved this way toward Yudkin, and in the end, even though Yudkin had no academic reasons to fear the man, Keys triumphed. By the time of Yudkin's death in 1995, his ideas on the dangers of sugar had largely been forgotten. In part this was because some of the scientific work done in the 1970s did appear to support Keys's cholesterol theory—there definitely seemed to be a link between the levels of cholesterol in the bloodstream, the levels of cholesterol in the diet, and a person's susceptibility to the metabolic syndrome diseases. The scientists involved couldn't know at the time about "good" cholesterol and "bad" cholesterol, nor could they know that dietary cholesterol has little effect on bloodstream cholesterol levels.*

If dietary cholesterol were the problem, we'd expect to find high levels of obesity, diabetes, hypertension, and CHD in populations with high meat intakes, and populations in most of the wealthier countries of the world do indeed show this symptom. Further, when poorer populations start to enjoy the "benefits" of western civilization, their rates of metabolic syndrome skyrocket.

However, increased meat is not the only dietary change that comes about with the altered circumstances. As we've noted, some cultures have very high meat-eating rates yet lack the correspondingly high levels of metabolic syndrome.

When members of such cultures enter western society—through coming to the big city in search of jobs, perhaps—their metabolic syndrome rates start climbing, but meat and dairy produce can't be the culprits; rather, it's increased sugar consumption via processed foods, sodas, and all the rest.

A further clue is that the French diet, which is typically high in saturated fats but far lower in sugars than the norm for anglophone societies, is likewise marked by a relatively low rate of metabolic syndrome diseases.

# # #

The publication in 1972 of Yudkin's *Pure, White and Deadly* in the US (as *Sweet and Dangerous*) sparked the interest of Senator George McGovern (D–SD), who convened a hearing before the Senate Select Committee on Nutrition and Human Needs. McGovern's political career had been marked by a genuine interest in nutrition and food politics, so his hearing was a genuine effort rather than the kind of puerile play-acting we've come to expect today from the likes of Lamar Smith (R–TX). One of McGovern's

---

* Alarmingly, as recently as 2014 a survey done by Credit Suisse revealed that over half (54 percent) of US doctors *still* don't know this.

witnesses was Yudkin; the rest were likewise highly respected nutritionists with relevant expertise. Although, like good scientists, they held back
from saying with 100 percent certainty that sugar was responsible for the
metabolic syndrome diseases, they did confirm that it was very strongly
implicated and that, in general, overconsumption of sugar was likely to
have a catastrophic effect on individual and public health.

The International Sugar Research Foundation (ISRF)* responded with
what has become a time-honored tactic of industries fighting back against
science they don't like: the mounting of a phony science conference a few
weeks later. At "Is the Risk of Becoming Diabetic Affected by Sugar Consumption?" a bunch of tamed nutritionists expressed the collective answer
that, no, it wasn't.

McGovern's committee didn't feel able to break free of the prevailing
dietary orthodoxy promoted by Ancel Keys and his allies. Their report,
when it came in 1977, *Dietary Goals for the United States*, was written by a
journalist, Nick Mottern, who toed the establishment line: it stressed the
importance of a reduction in consumption of red meat, butter, milk, eggs
and the other usual suspects.

Dietary Goals was the seed for what became, in 1980, *Dietary Guidelines
for Americans*, today compiled every five years by the Center for Nutrition
Policy and Promotion and released through the Department of Agriculture and the Department of Health and Human Services' Office of Disease
Prevention and Health Promotion. The *Guidelines* echoed the prejudices of
the *Goals*, and to a great extent still do, advocating restricted consumption
of saturated fats and cholesterol and substitution with increased carbohydrates.

The net consequence has been a skyrocketing in obesity rates. In 1980,
the rate in the US was 15 per cent; by 2004 it had reached nearly 35 percent,
and today two-thirds of US citizens are considered overweight or obese.**
Is it because people are ignoring the *Guidelines*? Far from it. The US gov-

---

* The creation of the ISRF in 1968 by the Sugar Association represented the formation of
the kind of spurious "research institute" so beloved by science-denying industries in all
fields. To be sure, there was genuine science done by the ISRF, but it was directed toward
conclusions that could support the claim that sugar was a safe constituent of the diet. The
ISRF was in reality just the Sugar Association's existing research department, spun off and
renamed to make the intimate relationship less obvious to consumers and politicians.

** In the UK (the home of Yudkin!), which mimicked the US's *Guidelines* in 1983, just six
percent of the population suffered obesity at the time; that figure today matches its US
equivalent.

ernment's own figures show that since 1980 we've reduced our saturated fat intake by 14 percent, with the female population being both particularly assiduous about reducing fat intake and particularly susceptible to obesity. There's very clearly something wrong with the low-fat model because what's happened with the US population since 1980 is in effect a very loosely conducted but vast trial, involving hundreds of millions of subjects, and it has shown the exact opposite of what the nutritionists predicted it would.

Another consideration is that saturated fats have been an important part of our diet since long before agriculture began; as a mark of their integral role in our species's evolutionary history, they're a significant constituent of human breast milk. About 10,000 years ago, with the introduction of agriculture, carbohydrates became an increasing factor in our nutrition. It's only within the past three centuries or so that sugar has become a staple, and only within the past few decades that our consumption of it has mushroomed.

A fairly useful principle of scientific investigation is that, when seeking an explanation for a recent change, of whatever kind, you look first at recent innovations that might have affected the status quo rather than at preexisting conditions. While the *Guidelines* have been right to note that our intakes of red meat and the like have increased with increasing national affluence, there's no justification for their ignoring the sugar question. As late as the most recent (December 2015) edition, the *Guidelines* continued to promote the fat/cholesterol hypothesis at the expense of attention to the deleterious effects of sugar. Cited in the *Washington Post*, David McCarron of the University of California, Davis, Department of Nutrition summarized: "There's a lot of stuff in the [*Guidelines*] that was right forty years ago but that's been disproved."

# # #

In 1975, the Sugar Association and its corporate supporters decided the ISRF was displaying too much independent scientific integrity and withdrew funding, decreeing that in future the Association would do its own research. It also recruited the PR company Carl Byoir & Associates to mount a campaign defending sugar from inconvenient science—and this campaign was successful enough that it went on to win an industry award. Part of it was the setting up of a bogus scientific thinktank, the Food & Nutrition Advisory Committee (FNAC), comprising nutritionists prepared to speak out in favor of the benefits of including plenty of sugar in the diet.

For the most part the scientists appointed to FNAC were perfectly reputable physiologists, nutritionists and so on; they just happened to be on the "pro-sugar" side of the sugar vs. cholesterol debate. But we can look more askance at the career of Fredrick J. Stare, founder of and professor at the Harvard School of Public Health's Department of Nutrition. Stare was notorious for the amount of research funding he managed to raise from the food industry, especially the manufacturers of sugar-laden foods. Perhaps unsurprisingly, the research he conducted seemed to show that even very large quantities of sugar in the diet had no ill effects—were highly beneficial, in fact. Stare was more than willing to say whatever the Sugar Association wanted him to say. Since he also solicited—and received—departmental funding from the Tobacco Research Council to do studies showing cigarettes had no link to CHD, and later in life sought funding from Philip Morris Inc. for the American Council on Science and Health, which he chaired, we can see that he made a habit of this sort of thing.

Perhaps Stare's most significant achievement on behalf of the Sugar Research Foundation (as the Sugar Association was then known) came in 1967. This was a paper, done with fellow Harvard scientists D. Mark Hegsted and Robert B. McGandy and published in the August 3, 1967, issue of the *New England Journal of Medicine*: "Dietary Fats, Carbohydrates and Etherosclerotic Vascular Disease." This heavily promoted the theory that dietary cholesterol was the prime culprit in CHD and, because of the undoubted prestige of the *NEJM*, was regarded as definitive. What the *NEJM* did not do was require the authors to declare any possible conflicts of interest. Had the journal done so, and had the authors been fully honest, that disclosure would have revealed they'd been paid $6,500 (equivalent to nearly $50,000 today) in order to conduct research that would let them come to the conclusions they did.

Documents unearthed half a century later by Cristin E. Kearns of the University of California, San Francisco, revealed the whole sordid story of how the three scientists worked hand-in-glove with the Sugar Research Associations' John Hickson in order to produce a review that would favor the sugar industry and cast dietary cholesterol as the villain; they even consulted Hickson on drafts of the paper to make sure it satisfied his specifications. Kearns, with coauthors Laura A. Schmidt and Stanton A. Glantz, blew the whistle on the guilty scientists in the paper "Sugar Industry and Coronary Heart Disease Research: A Historical Analysis of Internal Industry Documents," which appeared in September 2016 in *JAMA* (*Journal of the American Medical Association*).

The Sugar Association, confronted by this revelation, commented that "We acknowledge that the Sugar Research Foundation should have exercised greater transparency in all of its research activities," but then went on the offensive: "We're disappointed to see a journal of *JAMA*'s stature [deploying] headline-baiting articles to trump quality scientific research" because, after all, "sugar does not have a unique role in heart disease."

The questions remain: How could such a filthy little secret be kept for almost a full half-century without someone, somewhere having the moral integrity to blow the whistle? And how many thousands—more likely, hundreds of thousands—of people died before their time because of the bogus medical evidence promulgated by these three men and paid for by an equally corrupt industry organization?

Another influential publication produced on behalf of the sugar industry, a ninety-page group of papers by FNAC scientists, edited by Stare, appeared in August 1975 in *World Review of Nutrition and Dietetics*: "Sugar in the Diet of Man." As soon as it had done so, the association issued a press release breathlessly summarizing the supposed findings. Yudkin's theories were a primary target:

> Sugar is not recognized as a causative factor in cardiovascular disease, according to the scientists. Dr. Francisco Grande [a frequent collaborator with and long-term colleague of Ancel Keys], of the School of Public Health at the University of Minnesota and the Jay Phillips Laboratory at Mount Sinai Hospital in Minneapolis, shows how unreliable, incomplete and out-of-date are Dr. John Yudkin's arguments that sugar consumption per capita and the death rate ascribed to coronary heart disease are related. "The extensive work dealing with the production of experimental atherosclerosis in animals currently available does not provide any proof of the atherogenic effect of sugar," Dr. Grande said.

Similarly, sugar wasn't directly implicated in obesity or Type 2 diabetes—those were a matter of overconsumption of calories in general—and was only one of several factors in dental caries:

> The results of the sugar and dental review suggest there should be an extension of efforts by the dental profession to promote better dental hygiene, plaque control and more extensive use of fluorides and fluoride mouthwashes.

In other words, don't blame the sugar, blame the dentists!

This wasn't the first time Big Sugar had attempted to shift the focus away from the role of their product in caries. In the mid-1960s, at the instigation of President Lyndon Johnson, the National Institute of Dental Research (NIDR) announced an initiative to combat childhood dental cavities. The NIDR correctly fingered sugar and sugar products as the guilty parties in the formation of caries: sugar encourages bacteria to stick to the outer surfaces of the teeth, and the bacteria do the rest. The solution to the problem was obviously a mixture of water fluoridation, education in dental hygiene, and a reduction in sugar consumption.

Big Sugar, in the form of the ISRF, was obviously unhappy about that last recommendation. By 1969 it had subverted the NIDR to the extent that, out of nine members of the NIDR's committee to investigate caries, just one was unaffiliated with Big Sugar. A more fruitful approach, the committee now maintained, than recommending kids cut down on their sugar intake was the search for an anti-caries vaccine. By 1971, when the National Caries Program was born, the idea that kids should eat less sugar had been largely discarded.

Returning to 1975: "Sugar in the Diet of Man" didn't just pollute medical science, in due course it heavily influenced the FDA's assessment of the safety of dietary sugar. Under the administration of Richard M. Nixon, who became concerned about the foodstuffs the US population was consuming, the FDA commissioned the Federation of American Societies of Experimental Biology to set up a committee to investigate GRAS (Generally Recognized As Safe) foods. When it came to sugar, the committee quite clearly relied upon "Sugar in the Diet of Man" as a primary source. Through naivety, sloth or some other factor, the committee additionally asked the Sugar Association for consultative assistance. (That help is acknowledged in the report.)

The Sugar Association must hardly have been able to believe its luck. A supposedly objective review was relying heavily upon a collection of papers that had been produced according to association preferences and was now asking the association to help make the report even *more* favorable to sugar! No wonder the association had a field day after the release of the FDA's GRAS report. In his *The Case Against Sugar,* Gary Taubes cites one of the association's ads:

SUGAR IS SAFE!

Sugar does not cause death-dealing diseases . . . There is no substantiated scientific evidence indicating that sugar causes diabetes, heart disease or any other malady . . . The next time you hear a promoter attacking sugar, beware the ripoff . . . Ask yourself what he's promoting or what he is seeking to cover up. If you get a chance, ask him about the GRAS Review Report. Odds are you won't get an answer. Nothing stings a nutritional liar like scientific facts.

The specifics may have changed, but the rhetorical style is instantly recognizable to anyone who has followed the political and corporate war on climate science. Likewise the tactic: cherry pick some dodgy science, which is what the Sugar Association had effectively done, and feed it to some supposedly responsible body; then claim that it's the supposedly responsible body that has independently come to the conclusion.

# # #

Allied to its offensive against the science that condemned its product, and its largely successful campaign to promote what we now know to be at best iffy nutritional science, the sugar industry has over the decades waged a series of wars against artificial sweeteners, which it has sought to portray as dangerous to human health, even when the actual evidence for this has been . . . scanty.

The classic example concerned sodium cyclamate. This was a cheap sweetener commonly used in processed foods and sodas in a 10:1 mixture with saccharine because the slightly unpleasant side-taste of each of the two sweeteners tended to mask that of the other. It was a boon to dieters, to people concerned about the adverse health effects of sugar, and of course to diabetics. Cyclamate was manufactured by Abbott Laboratories, and its success certainly made a dent in the sugar market.

Abbott Laboratories itself sponsored a 1969 study by New York's Food & Drug Research Laboratories into the effects of cyclamate on rats. After two years of consuming a daily amount of the cyclamate/sugar mixture equivalent to over five hundred cans of diet soda—an amount somewhat in excess of the capabilities of even the most determined teenager—it was found that a small proportion (about 3 percent) of the subject rats were showing signs of bladder cancer. As John Yudkin, who discussed this experiment in later editions of *Pure, White and Deadly*, pointed out, two years is "a very long time in the life of a rat." Later experimentation was unable to replicate the result, and it's now generally accepted that the bladder cancers had nothing to do with the cyclamate.

This didn't stop the FDA from banning cyclamate for human consumption, partially in 1969 and then entirely in 1970. Some other countries followed suit, but later, on consideration of the scientific evidence, reversed their ban. The FDA, by contrast, in complete denial of the best available science, still imposes the ban in the US. It is, of course, impossible to establish to what extent this is due to Big Sugar's influence.

A major conspiracy theory, boosted of course by the internet, centers on another popular sweetener, aspartame, which supposedly, even in minor doses, causes all sorts of medical problems, from lupus to multiple sclerosis . . . and we'd know about all this, including the death toll, were it not for an unholy connivance between the manufacturer, G.D. Searle, and the FDA. In fact, there's been a huge amount of scientific investigation into the safety of aspartame and it has given the sweetener a clean bill of health. Despite the science, a number of food producers and retail chains around the world have instituted voluntary bans of aspartame.

This is not to say that we can as yet give all the artificial sweeteners a thumbs-up. A meta-analysis published in July 2017 in the *Canadian Medical Association Journal (CMAJ)*[76] found that the benefits of artificial sweeteners are seemingly mixed, and that their use by no means guarantees weight loss—in fact, in some individual cases it was possible they could cause weight gain.

There were some caveats, however. As the study's authors were at pains to point out, much of the research they were reviewing focused on the presence of artificial sweeteners people added to their beverages and ignored ingestion through eating processed foods. Moreover, most of the trials' subjects were people who were trying to lose weight because they were obese, and it's hard to draw health conclusions when subjects have a pre-existing condition. As University of Manitoba epidemiologist Meghan Azad, lead author on the study, summarized, "There's no clear evidence for benefit from the artificial sweeteners, and there is a potential that they have a negative impact, but we need more research to figure it out for sure."

# # #

Various countries have introduced taxes on sugary beverages in an attempt to reduce consumption, and in 2012 New York City's mayor, Michael Bloomberg, introduced a ban on the sale by fast-food joints, delis, eateries, movie theaters, sports venues, etc., of sodas in containers larger than a pint (473ml). Although fought by the beverage industry with a million-dollar lobbying campaign, the super-sized soda ban was approved by the state's Board of Health and scheduled to go into effect in March 2013. At the last minute, however, it was overturned by State Supreme Court Justice Milton Tingling, who declared that the Board of Health was entitled to issue such rulings only in instances where there was an imminent danger of disease: "That has not been demonstrated herein," wrote Tingling in his decision, apparently unaware that obesity, diabetes, and other metabolic syndrome ills are diseases and that we're suffering an epidemic of them.

The ruling was hailed by the beverage industry and the ignorant as an example of an overreach by the nanny state being reversed. Presumably these advocates of freedom will now seek to abolish the laws barring the sale of alcohol to minors—after all, as Robert Lustig is fond of pointing out, there's little difference, in digestive terms, between what happens to a can of beer after we swallow it and a can of sugary soda: both do equal damage in almost identical ways.

There's also some alarming evidence that the fizzy-beverage industry is in bed with academia. As reported in April 2017 by Paul Thacker for the BMJ,[77] for a number of years following 2011 the sweetened-drinks company Coca-Cola funded a group based at Colorado University, the Global Energy Balance Network, to run annual conferences for journalists on obesity; these conferences promoted the line that it's the amount of exercise people take rather than the quantities of "liquid candy" they swallow that governs their fatness. The prime mover at the university was pediatrician James Hill, and the BMJ uncovered congratulatory correspondence between Hill and the people at Coke describing how successful and cost-effective the conferences were from a PR point of view.

Also involved in the organization of the conferences was the National Press Foundation (NPF); it seems it was only when journalist Kristin Jones complained to the foundation after the 2014 conference that the NPF learned Coke was supplying most of the funding. The 2016 conference was largely underwritten instead by the Mayo Clinic.

In 2013, the Centers for Disease Control and Prevention (CDC), which had hitherto accepted funding from the Coca-Cola Company, realized there was a dissonance between attempting to tackle diseases such as childhood obesity and receiving financial contributions from a company whose business is largely concerned with marketing sugary drinks to kids.

That decision may be about to be reversed, to judge by early signals given out by the newly appointed head of the CDC, Brenda Fitzgerald, in July 2017. Back in 2013, when she was Georgia's Health Commissioner, the state ran a scheme called Georgia Shape. Part of it was an admirable program called Power Up for 30, which encouraged schools to increase children's exercise time by a daily thirty minutes. Fitzgerald deserves credit for being a mainspring of this program. What's less creditworthy, though, is that she accepted funding for almost the entirety of it from Coke, whose public claim is that it's not sugar but lack of exercise that's behind obesity. E-mails from that period, obtained in 2017 by the organization US Right to Know, reveal a very friendly—some might say too friendly—relationship between Fitzgerald and the folks at Coke.

Asked by the *New York Times* after her 2017 appointment to the CDC directorship if she'd be willing for the CDC to accept Coke's sponsorship in the future, Fitzgerald depressingly answered that she'd certainly be open to the idea. (But see page 390.)

# # #

Today, the main holdouts against the realization that dietary sugar intake rather than fat intake bears principal responsibility for the diseases of the metabolic syndrome are nutritionists. As Ben Goldacre joyously points out in his *Bad Science* (2009), the term "nutritionist" has been widely abused—anyone in the world can call themselves a professional nutritionist even if they lack relevant qualifications entirely—but here we're talking about what we might loosely call the academic nutrition establishment. To be sure, some of them will be in hock to Big Sugar or some other branch of the food industry, but even those may be speaking with absolute integrity when they continue to declare that the root dietary cause of the metabolic syndrome is the overconsumption of saturated fats.

This flies in the face, though, of what can now be regarded as established science.

As Ian Leslie recounts in his long article "The Sugar Conspiracy" (2016),[78] a stack of trials, surveys and reviews have demonstrated that the fat/cholesterol hypothesis is false:

• A huge 1993 controlled trial, the Women's Health Initiative, run on behalf of the National Heart, Lung and Blood Institute, was anticipated to show that—just as with men, so with women—a diet low in saturated fats would help resist CHD (and cancer). The results showed no difference in outcome between the control group and the low-fat dieters. This was so in conflict with expectations that the study was rejected as hopelessly flawed. Now, of course, we know better.

• A 2008 survey done by Oxford University researchers covering the countries of Europe found there was actually an *inverse* correlation between dietary fat and heart disease. "France, the country with the highest intake of saturated fat, has the lowest rate of heart disease," reports Leslie, and "Ukraine, the country with the lowest intake of saturated fat, has the highest."

• A 2008 systematic review done for the Food and Agriculture Organization of the UN, covering all previous studies of the effect on disease of fat consumption, found no evidence of any correlation with heart disease and cancer.

• That was the conclusion, too, of a major 2010 review done for the American Society for Nutrition.

The *British Journal of Sports Medicine*, part of the *BMJ* stable, in 2016 carried an editorial with a long and self-explanatory title: "Saturated Fat Does Not Clog The Arteries: Coronary Heart Disease is a Chronic Inflammatory Condition, the Risk of Which Can Be Effectively Reduced from Healthy Lifestyle Interventions."[79] It referred to a recent systematic review that had appeared in the *BMJ* that

> showed no association between saturated fat consumption and (1) all-cause mortality, (2) coronary heart disease (CHD), (3) CHD mortality, (4) ischemic stroke or (5) type 2 diabetes in healthy adults. Similarly in the secondary prevention of CHD there is no benefit from reduced fat, including saturated fat, on myocardial infarction, cardiovascular or all-cause mortality.

The authors stress that current "unblocking the pipe" approaches to the management of heart disease are not completely useless but are in essence profoundly misguided: "[S]tenting significantly obstructive stable lesions fail[s] to prevent myocardial infarction or to reduce mortality." The menace of LDL ("bad") cholesterol has been grossly overstated, they concluded—in fact, it's perfectly possible to significantly increase your risk of a heart attack through significantly reducing your LDL levels. Ancel Keys would have been gratified to learn that the authors recommend the Mediterranean diet, but horrified to discover that they counter-recommend the low-fat version, and in particular frown on the wholesale substitution of vegetable fats for animal ones.

A study by the international organization Prospective Urban Rural Epidemiology (PURE), presented at the European Society of Cardiology's 2017 meeting in Barcelona and published in *Lancet* in late August that year,[80] backed these conclusions. The team followed the eating habits of 135,000 adults in eighteen countries of varying affluence for on average over seven years. Rates for cardiovascular events didn't show any significant difference between high-, low- and medium-fat consumers, but stroke rates did: the high-fat eaters were less likely to suffer strokes. The highest rates of mortality were recorded not among the high-fat eaters, as conventional nutritional wisdom would suggest, but among those whose diets were low in fats. Mortality rates were highest among those who got most of their calories from carbohydrates.

Taking all this together, it's clear that the basis for decades of attacks on Yudkin's proposals, not to mention decades of instruction and prohibition doled out by the medical profession—the notion that there was no need to bring sugar into the equation because the diseases of the metabolic syndrome were proven to be rooted in the overconsumption of animal fats and the consequent accumulation of serum cholesterol—has been shown to be so much hogwash.

The horrifying thing is that this is not fresh news.

Between 1968 and 1973 Dr. Ivan Frantz of the University of Minnesota and his colleagues conducted a study on the effects of diet on the health of no fewer than 9,423 people, aged between 20 and 97. The data he used were of particular value because of their accuracy—the people concerned lived in state mental hospitals and nursing homes, so not only could their diet be strictly controlled (some getting a standard US diet for the late 1960s and early 1970s, with lots of animal fats, etc., while others got what we'd think of as more modern fare), accurate records could be kept of exactly what they ate.

The study, unearthed thanks to the diligence of Christopher Ramsden of the National Institutes of Health, with the logistical help of Ivan Frantz's son Robert, was never published, presumably because its results flew in the face of accepted nutritional wisdom at the time.

What it showed was that those patients consuming the diet where animal fats were largely replaced by vegetable oils did indeed show reduced cholesterol. They also, though, had just as many heart attacks, were afflicted by just as much atherosclerosis and in fact suffered higher mortality rates.

Even older was an unpublished study done in Sydney, Australia, in 1966–73, the subjects being 458 men aged between 30 and 59 who'd recently undergone a coronary event. Again the comparison was between participants whose diet was high in animal fats and others where these had largely been replaced by vegetable oils. Again the results from the "improved" diet were not beneficial.

Nevertheless, we're *still* often advised to follow dietary recommendations concerning fats that have never been supported by much evidence and now seem to have been debunked by plenty of very strong evidence.

But as for sugar? It seems to be a vital weapon in the ongoing campaign our elected politicians have mounted to kill us.

According to the American Heart Association in 2009, the average American daily consumes the equivalent of some thirty teaspoonsful of sugar, mainly in the form of sucrose and corn sugars. That's about 1.4kg (3 pounds) of sugar a week, much of it swallowed not in an obvious way—sweetening for cereal or coffee, soda, donuts, etc.—but in processed foods.

What sort of mortality rates are we looking at through overconsumption of sugar? It's obviously almost impossible to arrive at a figure, but to give some kind of an indication we can look at "Health Effects of Overweight and Obesity in 195 Countries over 25 Years," a study published in June 2017 in the *New England Journal of Medicine* by a group calling itself the GBD 2015 Obesity Collaborators. According to the researchers, about four million people a year, worldwide, die of weight-related disease. Contrary to the meme that's recently grown popular, "obesity's bad for you, but merely being overweight is neutral or even beneficial," fully 40 percent of those four million fall into the lesser "overweight" category.*

As a further measure of the damage done by dietary sugar, the WHO's 2016 *Global Report on Diabetes*[81] informed us that the number of diabetics in the world rose from 108 million in 1980 to 422 million in 2014—a near quadrupling during a period when the global population as a whole has less than doubled.

# # #

In May 2016, apparently as a result of a suggestion by Michelle Obama, the FDA announced that, as from July 2018, it would be rolling out a revised version of the Nutrition Facts labels you find on packaged foods. The

---

* That still means the risk factor for the overweight is quite a lot less, since far more people are overweight than full-out obese. For a measure of this, about 24 percent of adults in the UK are obese, but nearly 40 percent of the adult population is overweight but not to the point of obesity.

most important change was probably the inclusion of a new category to go alongside sodium, fats, cholesterol, etc.: Added Sugars. That way consumers could see what proportion of the stated sugar content was inherent in the foodstuff and what proportion had been added by the manufacturer.

It was an eminently sensible plan. What simpler—and cheaper—way could there be to help people reduce their risk of incurring the sugar-induced metabolic syndrome diseases?

In June 2017 President Trump's FDA announced it was postponing the measure indefinitely.

"We Repudiate the Term 'Asbestos Poisoning'"

April 28, 2006. It's International Workers' Memorial Day, and accordingly a memorial's being unveiled in a public park in the center of Rochdale, UK, watched by a crowd of local people, trade union representatives, and politicians from all the major parties. The memorial's plaque offers a forceful message:

> ASBESTOS
> ONCE A 'MAGIC MINERAL' BUT ALWAYS A KILLER DUST
> CURRENTLY 100,000 DEATHS PER YEAR FROM ASBESTOS
> IN TOTAL, 5,000,000 MAY DIE FROM ASBESTOS EXPOSURE
> REMEMBER THE DEAD—FIGHT FOR THE LIVING
>
> PEOPLE BEFORE PROFIT

Why Rochdale? On March 14, 1924, Nellie Kershaw, aged 33, a mother of two young children and an employee of the Rochdale textile factory Turner Brothers Asbestos (later Turner & Newall), became the world's first recorded victim of asbestosis, a lung disease consequent upon inhalation of asbestos fibers and for which there was then and still is no cure. Kershaw became too ill to work in late 1922, and her employers refused to give her a single penny in sick pay. After her eventual, agonized death, Turner & Newall declined to contribute toward her funeral expenses. She was buried in an unmarked pauper's grave.

The official response of Turner & Newall, on being confronted by the medical evidence explaining Kershaw's cause of death, was:

> We repudiate the term "asbestos poisoning." Asbestos is not poisonous and no definition or knowledge of such a disease exists.

Although, in the later decades of the twentieth century and the opening years of the twenty-first, the use of asbestos was banned in almost all of the world's developed nations, nothing much has changed about the corporate response to the science surrounding the fatal hazards of asbestos use.

An Air that Kills (2004; expanded as *An Air that Still Kills* in 2016) by Andrew Schneider and David McCumber tells a devastating tale of the corruption of science by both the asbestos industry and the US government agencies responsible for regulating it.

The two authors' involvement began in 1999 when Schneider, an investigative journalist with the *Seattle Post-Intelligencer*, where McCumber was his editor, and later the *St. Louis Post-Dispatch*, was tipped off that scores of people were dying of lung diseases in the small town of Libby, Montana, in consequence of the vermiculite mine that, though now closed down, had for long been the town's economic mainstay. Later that same year the EPA's Paul Peronard, alerted by Schneider's stories, arrived in the town and began to examine the remains of the mine workings. As was obvious immediately to anyone visiting the spot, although the mine's final owners, W.R. Grace & Company, had been recognized in the industry for their magnificent job of restoring the area of the mine to its natural state, in reality it had been left just as it was on the day it closed down, with loose ore—including a plentitude of tremolite, the dangerous asbestos with which the mine's vermiculite had been heavily contaminated—open to the elements. It was that tremolite which was killing so many people, not just those who had worked in the mine but—because the tiny tremolite fibers got everywhere, clinging to clothing, hair and skin—their families and acquaintances.

Peronard and his allies within and outside the EPA—allies who soon included Schneider and his journalist colleagues—found themselves facing an uphill struggle, first to gain acknowledgment from Peronard's EPA bosses that the people of Libby were indeed being killed by the tremolite and then to ensure the survivors got some just compensation for their suffering.

For decades officers of not just the EPA but other supposed public- and worker-safety organizations including OSHA, MSHA, ATSDR and CPSC* had been in effect conniving with the asbestos industry in the pretense that asbestos presented no health hazard. The EPA actually had in its files reports done in 1982 and 1985 of the damage the tremolite contamination

---

* Occupational Safety and Health Administration, Mine Safety and Health Administration, Agency for Toxic Substances and Disease Registry, Consumer Product Safety Commission.

in mined vermiculite could do, yet had deliberately ignored them. Those reports had concluded that the number of US garden-products users alone facing health risks from the mineral was in the tens of millions.

The technology being used by government scientists to detect asbestos, polarized-light microscopy (PLM), was woefully outmoded—and quite likely intentionally so, for it recorded far lower levels than were actually present, especially of tremolite, the tiny-fibered form that was causing such havoc in Libby. Even when Peronard and his colleagues demonstrated that more sophisticated technology such as transmission electron microscopy (TEM) was required for an accurate count of tremolite concentrations, the agencies *continued* relying on completely pointless PLM techniques, because "them's the rules."

# # #

As early as 1918, according to a report published in that year by the Commerce Department, insurance companies in the US and Canada had the established practice of declining coverage to applicants who worked in the asbestos industry, because of the known dangers of the mineral. In 1932, the US Bureau of Mines declared, "It is now known that asbestos dust is one of the most dangerous dusts to which man is exposed." In 1933, nearly 30 percent of the workers in the Johns-Manville asbestos company were suffering from asbestosis, yet the following year the company and its insurer were—successfully!—intimidating the editor of the trade journal *Asbestos Magazine* into suppressing all information about the health dangers posed by asbestos.

And so the pattern continued into the 1940s. In the asbestos industry, the frequent lethality of exposure to asbestos was an open secret. The only people who weren't in on it were the asbestos workers themselves and the general public.

The arrival of World War II was a tremendous boon to the asbestos industry, which in 1942 persuaded the War Production Board to proclaim the mineral "a critical war material." The mineral had countless applications in the war effort, from fireproofing tanks and submarines to lining the brakes of armored cars. According to Schneider and McCumber, the US government's reckoning is that some 4.5 million longshoremen were involved in the transportation of the mineral, untold numbers of whom must have died as a result. As for members of the US armed forces, it's thought by the military that the fatality rate from exposure to asbestos nearly matched that inflicted by the enemy.

This was far from the first time that the US government, justifying its actions by citing the exigencies of war, had deceived its people into sacrificing their lives by working with lethal materials they'd been told were safe. During World War I, the United States Radium Corporation (USRC) of Orange, New Jersey, enticed hundreds of young women into contributing to the war effort by painting radium onto clock and other dials to make them luminous. The approved method of bringing the hairs of one's paintbrush to a point was to squeeze it between one's lips.

The pay was good, and an added perk was that by the end of the shift their clothes would, like the clock hands, glow in the dark. (Some of the women even painted their teeth with the stuff—what a novelty for parties!) As for safety, the women were assured there was nothing to worry about.

Until the end of the war and for the first few years after it, none of the "ghost girls"—the nickname they earned through their trick of glowing in the dark at parties—showed any overt signs of radiation sickness, but in the early 1920s all that changed, as one by one they began to sicken and die.

USRC denied all responsibility, even after an independent report the company commissioned demonstrated the clear link between the radium exposure and the sickness. So the company commissioned *another* report, this one tailored to show the women's illnesses had nothing to do with their handling of radium. The women themselves were smeared as would-be parasites, trying to get money out of the company under false pretenses.

In 1924, the pathologist Harrison Stanford Martland decided to investigate, and he was soon able to demonstrate that it was indeed the radium they'd swallowed that was killing the women. It had lodged hither and thither throughout their bodies, and its radioactivity was destroying bone and soft tissue.

Martland's work on radium poisoning is now internationally recognized as a landmark in our relationship with radioactivity, but at the time it was pooh-poohed by USRC. Luckily some of the dying women were made of stern stuff, and in 1927 they brought the company to court. The resulting publicity alerted radium workers throughout the US to the dangers under which they were laboring, but many of the companies involved followed USRC's policy of simply lying about the hazards, claiming radium was perfectly safe. Perhaps the worst of all was Radium Dial of Illinois, which reportedly even stole body parts from mortuaries on the eve of autopsies that would have revealed the role of radium in a victim's death.

It was not until the late 1930s that the matter was finally settled and corporate America was forced to admit that radium was harmful, and that work in radium-handling factories had killed thousands.*

Returning to asbestos:

When peacetime came after World War II, the asbestos industry's conspiracy of silence continued. What was remarkable was that the medical profession, many of whose practitioners must have been highly aware of the suffering and death that exposure to asbestos was causing, failed to rock the industry's boat.

This situation persisted until the 1960s.

One of the clients of New Jersey physician Irving J. Selikoff was the Asbestos Workers Union (AWU). Selikoff noticed that the rate of the lung disease mesothelioma was abnormally high among the AWU members he treated, and it didn't take him long to connect this to the asbestos with which they were working. As a physician who had been a co-recipient of the prestigious Albert Lasker Award for Clinical Medical Research in 1955, he was not someone whose testimony could be easily brushed aside and he was resolute that it wouldn't be. The culmination of his efforts to make sure the public became fully aware of the hazards of asbestos came with the international conference that he organized, "Biological Effects of Asbestos," whose proceedings were published in 1965 in the *Annals of the New York Academy of Sciences*. OSHA duly took note, and introduced measures to limit workers' exposure to asbestos. In practice those measures were widely ignored within the industry and by the underinformed workers themselves, as well as even within OSHA.

By the time Peronard and his colleagues, not to mention journalist Schneider, came onto the scene in the late 1990s, government complacency over the decades had resulted in asbestos—which most people thought had been banned from commercial use years ago (it was, briefly, but then some helpful New York judges lifted the ban)—having become widespread at dangerous levels in all sorts of products throughout the nation, not just in fireproofing and insulation but in brake linings, sheet rock, and even children's crayons.

W.R. Grace, the onetime owner of the Libby mine, was the company mainly responsible for the asbestos-rich fireproofing and insulation that was in use all over the US. Even vermiculite mined elsewhere than Libby showed tremolite contamination, and internal memos that the reporters managed to get out of the company via a Freedom of Information Act or-

---

* Kate Moore's *The Radium Girls* (2017) offers a comprehensive account of the whole saga.

der demonstrated that Grace knew both this and that the tremolite was lethal. Bizarrely, for many years the company had insisted on conducting chest X-rays of its workers, yet had declined to release the details of those X-rays—many of which showed evidence of lung disease—to the men concerned.

In 1981, W.R. Grace's scientists tested the running tracks at Libby's high school, in the construction of which vermiculite tailings had been used. The scientists discovered that the shoes of the runners were kicking up tremolite at levels far above those that OSHA deemed safe. The response of the company was to keep the information secret . . . and the uninformed schools carried on using the running tracks. Who can tell how many Libby schoolkids sickened and died in their later years because of Grace's cost-cutting act of suppression.

In 1982, W.R. Grace's CEO, J. Peter Grace, was appointed by the new US president, Ronald Reagan, to head his Private Sector Survey on Cost Control (PSSCC) or, as it soon came to be called, the Grace Commission. The stated task of the commission was to find economies that could be made in the federal budget. In reality, formed as it was of a platoon of plutocrats and corporate heads, the commission, in the report that it eventually produced in 1984, recommended a passel of industry deregulations that promoted cost efficiency, and thus profits, over worker and public safety. By strange coincidence, one of the beneficiaries of the proposed relaxations was the asbestos industry. Luckily, Congress saw through this ploy and threw the report out. Even so, action on the hazards of asbestos (and equivalents in other industries) was delayed for a few more years.

In 2005 the Grace company and seven of its executives were indicted by a federal grand jury over the Libby, Montana, disaster. Through 2006 the charges were progressively watered down. When the case finally came to court in 2009, the company was acquitted of all substantive charges. The EPA's Superfund cleanup of Libby continues, at huge and largely public cost, to this day.

# # #

Schneider, alongside Peronard and others, investigated a number of commonly purchased items whose asbestos content threatened the public's health. Car brakes—pads and linings—were one. In one instance, employees of an auto parts store told Schneider firmly that the relevant items no longer contained asbestos; when he pointed out that some of those on the shelf had CONTAINS ASBESTOS printed on the packet, he was shown the door.

Tests carried out on the CONTAINS ASBESTOS brands showed, reasonably enough, that they contained asbestos. But so did some of the tests on unmarked brands, and even on those marked ASBESTOS-FREE—and sometimes the levels of unacknowledged asbestos were higher than those from the more honest competitors.

The tragedy is that this need not have been—there are other materials you can use in your car brakes—and would not have been had the asbestos industry not yet again put greed ahead of human life. In the early 1990s the EPA succeeded in getting every automotive-vehicle manufacturer in the US to voluntarily support an asbestos ban in their vehicles. The agreement was scheduled to be signed in late 1993 and to come into effect in late 1994.

At that stage, though, Bob Pigg of the Asbestos Information Association (AIA) intervened. As you might imagine, the "information" the AIA purveyed was the falsehood that the use of asbestos in brakes was perfectly safe—for workers, home auto-repair enthusiasts, and the general public: all people had to do was follow the safety instructions. More or less on the eve of the agreement's signing, the asbestos industry informed all the intended signatories that, if they participated, they'd be sued.

Today, asbestos has been banned in brake linings in almost all developed countries. In the US, its use has been largely phased out by the manufacturers.

Another alarming occurrence of asbestos problems that Peronard and the reporters found concerned some of the major brands of children's crayons. The manufacturers seemed genuinely not to know that asbestos was in their products, and in the main took steps, on receiving the information, to remove both the toxic ingredient and recall the offending crayons. The culprit was the talc used in the crayons, talc supplied by the R.T. Vanderbilt Company.

That Vanderbilt's talc contained contaminating asbestos had been known for decades to federal scientists and regulators. What no one had reckoned with was the company's ability and willingness to buy politicians—and even government scientists. In the 1970s, the newly formed OSHA discovered the asbestos contamination in many Vanderbilt mining products, the talc included, but OSHA, under the Nixon and then the Ford administrations, was in a business-friendly mode, those administrations realizing that corporations could play a great part in electing Republicans if suitably cajoled. Vanderbilt's protests that the company was being unfairly hampered by safety regulations, that the abundant tremolite present in its products was in fact *tremolite of a perfectly safe variety*, thus fell on fertile ground, and—even though this "safe" variety of tremolite was unknown to orthodox science—OSHA did nothing.

Even so, in 1980 OSHA's research wing, the National Institute for Occupational Safety and Health (NIOSH), conducted an investigation that demonstrated conclusively that the asbestos in Vanderbilt's talc was indeed killing the company's workers. The company's reaction was in 1984 to make discreet contact with two of the scientists working within NIOSH, John Gamble and Bob Glenn. These two covertly conducted their own investigation, which just happened to disagree with the 1980 report on almost every major detail. Through Glenn's position as head of the Respiratory Disease Division, Gamble's report was, in response to a carefully orchestrated "inquiry" from Vanderbilt, made public. Any attempts to debunk it could thereafter be painted as a matter of establishment toadies trying to suppress inconvenient science—a frequent tactic of science deniers and a reversal, of course, of the truth. Even though OSHA became completely aware in the early 1990s of the falsehood of Gamble's report, that tactic was terrifyingly effective: regulation of Vanderbilt's talc was abandoned.

Just to be on the safe side, Vanderbilt now claims that its talc—whose use is widespread in many domestic products—contains no asbestos. In a 2014 case before the Delaware Supreme Court, for example, the company "conceded that the industrial talc contained asbestiform minerals but denied that the talc contained actual asbestos or caused mesothelioma."[82] "Asbestiform minerals"? Right.

# # #

On September 11, 2001, two commercial passenger airplanes, piloted by hijackers from the terrorist group Al-Qaeda, crashed into the World Trade Center in downtown New York. Of the almost 3,000 people who died on that day as a result of Al-Qaeda's attacks, not just in New York but at the Pentagon and aboard United Airlines Flight 93 (where the hijackers were thwarted by the passengers), some 2,600 lost their lives in and around the Twin Towers.

That's the death toll for those who died within those horrific initial hours. The true death toll in New York City arising from the September 11 attacks may never be known, because it's still, to this day, increasing. As the two towers and later in the day the adjacent 7 World Trade Center building collapsed into the surrounding streets, they emitted huge clouds of dust. That dust, which deposited itself in a thick layer all over downtown

Manhattan, contained lethal levels of asbestos from the buildings' insulation and fireproofing.

The early responders who probed the wreckage for survivors did so amid an environment thick with the dust, as did New York's firefighters. Those who lived and worked in lower Manhattan discovered, on returning to their homes and offices, that the dust was everywhere. The authorities issued warnings that it might be a good idea not to breathe too much of it, so many landlords brought in cleanup teams. Armed typically with cheap face-masks, mops and buckets, these usually untrained men and women did their best to clear away the mess; once they'd gone, the occupants generally still had remnants of the dust to cope with. Many people, not fortunate enough to have the benefit of a cleanup team, did the whole job themselves.

Peronard and his colleagues, armed with all the experience they'd gained in the efforts to analyze and clear up the tragic affair at Libby, plus the specialist knowledge they had of asbestos, offered to come to Manhattan to help in the rescue operations. Their help was declined. The New York and New Jersey branches of the EPA reckoned they knew everything there was to know on the subject and had no need of a bunch of hicks from flyover country.

But there's a more significant reason why people in New York, including those heroic first responders, are still dying from the asbestos poisoning they received in the days, weeks and months after the 9/11 attacks.

The George W. Bush administration, in between concocting reasons why the attack justified an invasion of Iraq—a country whose leader, vile as he was, was actually an ally in the fight against Al-Qaeda—reckoned that the priority in the US recovery after 9/11 was the economy, and that meant the imperative in New York was to get Wall Street back up and running as soon as possible.

This would inevitably be delayed if the cleanup of downtown Manhattan was unnecessarily complicated by efforts to remove the very last traces of toxic waste.

The White House therefore made the conscious decision to downplay the dangers of the dust, much of which remained airborne over a surprisingly long period. The head of the EPA, Christine Todd Whitman—despite a surprisingly good record in the post on other counts—lied to the public that they had nothing to fear from breathing the city's air. The New York City authorities, headed by Mayor Rudy Giuliani, lied to the public that they had nothing to fear from breathing the city's air. President George W.

Bush himself lied to the public that they had nothing to fear from breathing the city's air.

Asbestos? Yes, there's some asbestos there, but nothing to worry about.

This was, of course, in the teeth of the science. Circumventing the EPA, OSHA, and the other official organizations, Peronard and his allies had infiltrated the city with scientists of relevant expertise, equipped with technology that was capable of detecting and measuring tremolite levels, and the results they obtained were terrifying. When they passed those results along to the bosses at the EPA and OSHA, though, the only response was silence.

The one building in lower Manhattan that *did* get a proper cleanup under the auspices of the EPA was . . . the New York offices of the EPA.

Our own politicians created far more domestic casualties as a consequence of the 9/11 attacks than Al-Qaeda did, and paid no penalty for that crime—indeed, they benefited politically from it, with both Bush and Giuliani being praised for the "firm leadership" they offered the people of New York and the nation in the catastrophe's aftermath.

In 2007, an underground steam pipe exploded in midtown Manhattan, sending up a forty-story-high geyser of steam and mud that lasted for two hours or more.

It's enlightening to compare the reactions of the New York City authorities, under Michael Bloomberg, to those of their 2001 counterparts. As the pipe was some eighty years old and would have been clad in asbestos, every precaution was taken to ensure New Yorkers were safe from the fibers—even though the science soon confirmed that the risks were almost nonexistent. In part this caution on the part of the politicians was because by now the dishonesty of their predecessors was well known and the subject of widespread and—more than justifiably—passionate condemnation. But in part it was the sign of a city administration operating under different and more humane priorities.

# # #

Andrew Schneider, who had a long and distinguished journalistic career investigating matters of public health and safety—he received the Pulitzer in 1986 and again in 1987—died in February 2017 of heart failure while undergoing treatment for lung disease. I haven't been able to ascertain if his investigations of asbestos and other lung hazards played any role in this. Based on the book he wrote with David McCumber, *An Air that*

Kills, Drury Gunn Carr and Doug Hawes-Davis directed and produced the two-hour documentary feature *Libby, Montana* (2004); it was later aired on PBS, in 2007.

Despite regulations and voluntary bans, asbestos is currently estimated to kill about 10,000 Americans every year. That figure's small alongside such statistics as the number of Americans who die annually from tobacco (approaching half a million, according to the CDC in 2017; worldwide the figure is about 7 million) and alcohol (88,000 according to a 2010 CDC estimate), but it's still a lot of human beings who're dead because corporations and their individual executives corrupted the awareness of science for the sake of profits.

BIG BAD PHARMA

I fear that we are heading blindly in the direction of Pharmageddon. Pharmageddon is a gold-standard paradox: individually we benefit from some wonderful medicines while, collectively, we are losing sight and sense of health. By analogy, think of the relationship between a car journey and climate change—they are inextricably linked, but probably not remotely connected in the driver's mind. Just as climate change seems inconceivable as a journey outcome, so the notion of Pharmageddon is flatly contradicted by most personal experience of medicines.

—Charles Medawar, retiring director of Social Audit Ltd, February 2007

The corruption of scientific medicine by big business over the past several decades—primarily by the pharmaceuticals industry but also by equipment manufacturers and not least, in the US, by a healthcare system tailor-made to benefit and protect corporate interests to the detriment of patients—is so considerable and so extensive that it's difficult even to adequately summarize it in a single chapter, far less in a segment of a chapter. To tackle the topic satisfactorily would require an entire book.

And that's why there are some very good books on the subject, such as Ben Goldacre's *Bad Pharma* (2012) and David Healy's *Pharmageddon* (2012), both of which have proven invaluable in compiling the brief discussion below.

A major reason there's so much corruption in the medical industry—specifically in the regulation of the pharmaceutical industry—is that fully 90 percent of published clinical trials are funded by the pharmaceutical industry. In order to save taxpayer dollars, we're having the foxes guard the henhouse—and to hell with what happens to the unfortunate hens.

In an ideal world these trials would be mounted and funded by nations or even international bodies, to ensure the results were independent and unbiased. We don't have that world because our politicians would rather campaign on a platform of cutting taxes than one of reducing the chances that the medicines our doctors, in all innocence, prescribe will actually do the job they're supposed to, and won't kill us. We can't blame just the politicians for this, however: ask any roomful of people if they'd like lower taxes and the chorus of "you bet!" will drown out any more nuanced argument. We vote to keep the system broken.

How much of a problem is this? In 2003, two systematic reviews appeared, one in *BMJ* and the other in *JAMA*, that analyzed published trials of pharmaceuticals to determine what effect industry sponsorship had on the results. Their conclusion was that trials sponsored by a drug's manufacturer were four times more likely to lead to a favorable conclusion than were independent trials. A further systematic review published in 2008 in *Contemporary Clinical Trials* updated the two earlier ones and found that fully 18 of the 20 drug trials that corporations had sponsored to that time gave favorable results for the corporations' products.[83]

Similarly, governments should be funding the continuing education of doctors—some of whom may still be in practice fifty or more years after leaving medical school, where they learned much that is now entirely obsolete. Good doctors do their best to keep up with things, but doing so in any comprehensive manner is both theoretically and practicably impossible—besides, they have patients to treat. But governments are reluctant to spend any money on this continuing education ("Save the taxpayer's dime!") and so, guess what, it's left to the pharma companies to do it. And, again guess what, it seems their own products just happen to be the drugs of choice for all manner of ailments.

As an aside on trials, an interesting and often unrecognized point about randomly controlled testing of drugs is that, the bigger the trial—the more experimental subjects involved—the better it is at finding *dis*advantages. Some of the biggest trials of all, those done to examine the claim that vaccination is the root cause of autism, have been so enormous precisely to find if there's any supportive evidence *at all*. (Universally, of course, they've found none.) If you're testing to see if your new product, Wundadrug™, does something significantly useful, you need only a relatively small number of guinea pigs before the benefits become obvious. Where you might want to run a far larger trial is to find out if Wundadrug™ has any nasty but relatively rare side effects.

So, when you hear pharmaceutical companies boast that their new product has been shown to be effective through randomly controlled trials involving many thousands of respondents, scrutinize the claim hard. It's likely the trial had to be so huge in order to detect any value at all the drug might possess—in other words, only one in hundreds or thousands of patients may benefit from taking it. Of course, you'll have the reassurance that at least Wundadrug™ is unlikely to kill you, because the huge trial would have shown deleterious effects, too . . . unless, of course, the relevant results on this have been buried somewhere in the small print by the experimenters or, more likely, their ghostwriters.

# # #

To stress, the system that has evolved in the medical sciences has become corrupted from root to topmost twig simply because of the way it's funded. To take a few examples:

• Drug companies own the results of the trials they fund. If a trial indicates Wundadrug™ is a stinker, they can simply embargo those results. More to the point, if *five* trials indicate Wundadrug™ is a stinker and one shows it may have some marginal benefit, the company can let the single trial be published . . . thus giving a grossly misleading impression of the truth.

• As has been shown time and time again, teams whose research has been funded by the manufacturers of Wundadrug™ (and of all the other Wundadrugs™) are more likely to find positives when they analyze their results. This generally isn't a question of deliberate corruption and it may not even be conscious, just a matter of looking on the bright side, so to speak. And such matters of funding often go undeclared, or are tucked away in the tiny print, whatever the ethics guidelines might say.

• By the nature of things, Wundadrug™, if determined to be effective and therefore licensed, is going to be used to help sick people in a particular demographic—elderly sufferers from high blood pressure, shall we say. Yet the recruits who sign up as subjects for medical trials need not belong to this demographic at all—often they're young people in need of money. The fact that Wundadrug™ lowers their blood pressure might be encouraging, but it doesn't really tell us much about how the drug will act on the elderly.

• Very often Wundadrug™ will be tested in trials against a placebo. It's always exciting that a new drug works better than nothing at all, but that isn't really what we want the trials to find out. The useful knowledge is how WundadrugTM does against its competitors—not just the obvious matter of whether it's as good as or better than they are in the matter of, for instance, reducing blood pressure, but also in what the side effects are. In real life, doctors prescribe drugs not to trial subjects but to their patients, so that for a particular patient a doctor may prefer to use a less effective drug because, once you take its other effects into account, it's better for that patient.

• Looking beyond trials, the manufacturers of Wundadrug™ buy ads in medical journals. The editors of those journals should be immune to such considerations as where their ad revenue is coming from, and I'm sure many of them indeed are, but . . .

The overall picture is profoundly bleak. No matter how good or assiduous medical practitioners might be, they're often essentially prescribing blind, because the knowledge they should have is either hidden (those unpublished negative trial results) or has been hopelessly distorted by the time it reaches them.

Another problem is that the authors whose names are given at the end of an article in even the best-respected peer-reviewed medical journals may not be the actual authors. They may well have done the research work to produce the raw data, but thereafter the job may have been taken over by a ghostwriter, whose task it is, aside from producing a professional piece, to give to those data whatever interpretation the customer may dictate—the customer generally being a pharmaceutical company. The United States Senate Committee on Finance held a hearing in June 2010 to examine this problem, and in his minority report, "Ghost Writing in Medical Literature," ranking member Chuck Grassley (R–IA) disclosed a rather frightening detail:

> In addition, an editor-in-chief of a medical specialty journal . . . informed Committee staff that at least one-third of the papers submitted to his journal were written by science writers hired by an agency and paid for by a pharmaceutical company. . . . [I]n some cases, it was clear to him that the academic expert had limited input in the writing of the article and . . . [may not have] evaluated the implications of what he was submitting for publication. That editor was also concerned that medical literature "has become inundated with repetitive promotional articles."[84]

There are of course sections of published scientific papers devoted to declaring conflicts of interest, including monetary ones, but these quite legitimately may omit the cost of ghostwriting.

The UK psychiatrist David Healy in February 2002 told the *Guardian*'s Sarah Boseley[85] about a couple of his own experiences with the drug companies' use of ghostwriters. In one, the company Pierre Fabre asked him to speak at a meeting of the European College of Neuropsychopharmacology about the company's antidepressant Milnacipran; his accompanying paper would be published, along with others from the meeting, in a supplement to *International Journal of Psychiatry in Clinical Practice*. Even before he gave his speech, the company had paid an agency ghostwriter to produce "his" paper for him. He refused to put his name to their article, instead writing his own. In the end the journal's supplement appeared with both papers in it, Healy's own and the ghostwritten piece, which made a couple of commercial points Pierre Fabre was keen to promote and which now bore the name of Dr. Siegfried Kasper, the journal's joint editor.

In the second incident Healy received a request to address a conference at the University of Toronto's Center for Addiction and Mental Health; his subject was the antidepressant Effexor, manufactured by Wyeth (now a part of Pfizer). The company's claim was that, unlike other antidepressants, Effexor could cure depression rather than merely ameliorate its symptoms. Once again, the drug company produced a ghostwritten article which it proposed be published under Healy's name.

This time Healy did not object to the article's content as a whole, although he made two significant changes: he added the information that there was evidence antidepressants of Effexor's class could in some cases make patients suicidal, and he denied Wyeth's claim of evidence that Effexor could actually cure depressives. The next time he saw the paper, after its submission to *Journal of Psychiatry and Neuroscience*, he discovered that both of his corrections had been excised. By then his only option was to remove his name from the paper, which he duly did.

Although the ghostwriting situation is reportedly improving, editors of scientific journals from all parts of the world have admitted that, while in theory they're supposed to police submissions for ghostwriting and the introduction of spin favorable to commercial sponsors, in practice this is extremely difficult—and, to do properly, prohibitively time-consuming. They and their staffs can nibble at the edges of the problem, but to a great extent they must rely on the integrity of the researchers themselves.

#

It used to be argued that the reason prescription drugs cost so much, even when their production cost per pill (or whatever) was often extremely small, was that the pharmaceutical company had somehow to recoup the R&D costs not just of that particular drug but of all the other drugs the company produced. Drugs are phenomenally expensive to develop, the reasoning went, and then there are enormous costs involved in running randomized controlled trials to see if the drug actually does what everyone thinks it should. Some drugs fail at the trials stage—they prove to be useless, or at least of not much use—and, of the others, some fail in the marketplace. So when you fill your prescription at the pharmacy you're paying for a lot more than just the drug your doctor prescribed.

That used to be a reasonable argument. In the past few decades, though, things have changed. Where it used to be that the bill for R&D was vastly larger than the marketing budget, now it's the other way round. The marketing budget of the pharmaceuticals industry as a whole for the US is currently about $30 billion per year, and heading northward.

As a bonus, some of the drugs that in previous eras would have been regarded as failures at the trials stage, because they offered no improvement over existing drugs or were actually not as good, can now be brought successfully to market. Enough people told in glossy TV ads to ask their doctor about the purple pill—or the green pill, or the magenta pill, or any other color of pill—will do so, and enough doctors will obediently comply by producing a prescription, to make the marketing exercise a success even if the pill in question doesn't do much, or treats a condition that's little more than a figment of some marketer's imagination.

So nowadays what you're paying for through the inflated price of your prescription drug is not R&D at all—or, to be fair, relatively little—but the *marketing* of the drug.

If even that. In September 2015, Martin Shkreli,* then CEO of the grandiosely titled Turing Pharmaceuticals, bought the manufacturing license for the drug Daraprim—a drug upon which some patients depend for their lives—and promptly increased its price from $13.50 to $750 per dose. In the face of outrage, Turing toyed with patients' hopes and expectations,

---

* Shkreli was found guilty by a jury of securities fraud in August 2017, and in March 2018 was sentenced to seven years in federal prison.

announcing (then withdrawing) price reductions, outlining special pricing schemes, and so on. It all became a tad academic in October of that year when Imprimis Pharmaceuticals announced it was bringing to market a substitute for Daraprim that it claimed was actually better . . . and cost $1 per dose.

There's also been a shift in the nature of the drug companies' marketing. It used to be the case that new pharmaceuticals were brought out to help doctors deal with the ailments they encountered among their patients. In the latter part of the last century, in a process that was particularly encouraged by the legalization of direct advertising of prescription drugs to patients, the drug companies devised a radical new marketing technique, which was in effect the invention of diseases to which their new pharmaceutical molecule just happened to be applicable.

Take the case of bipolar disorder, to use one of Healy's examples. Before the mid-1990s, this condition was essentially unknown. There was manic depression, sure, but manic depression was an extremely rare condition that generally required far more serious therapy than merely popping a pill twice a day—not much to be made there for the pharmaceuticals industry. But then, with the launch of Abbott Laboratories' Depakote, all that changed. The term "manic-depressive illness" dropped out of use, to be replaced by "bipolar disorder."

Further, bipolar disorder was far more common than manic depression had ever been—hundreds if not thousands of times more common, in fact. And, to its shame, the medical profession, not for the first or last time, went along with this marketing-department invention. Within a decade this hitherto unknown term, thanks to generous ad-department funding, had spawned peer-reviewed scientific journals and international scholarly societies. Schools were vetting their students for early signs of bipolar disorder, so treatments could start before it had come into full flower. Patients went along with it too: it's kinda cool to be diagnosed with a serious-sounding illness that requires you to take a pill with a serious-sounding name rather than be told you're suffering from anxiety and/or depression accompanied by the recommendation to get a bit more fresh air and exercise. Just in the same way that the placebo effect is at its most powerful when the "treatment" is expensive and inconvenient—an injection you pay for is more likely to "cure" you than a free pill—so the new, expensive drugs and the serious-sounding new term for the condition, and even the somber face of the physician prescribing the drug for you, did actually make patients feel better.

The trouble obviously is that, before being told they had—roll of drums—bipolar disorder, most of the patients concerned probably *weren't feeling all that bad to begin with*. They may have been anxious or depressed over disruptions at home or at work—who never is?—and come to their doctor in search of something to help them through the night. Instead they were told they had a lifelong "condition."

And all because of a marketing ploy. Rather in the same way that children who were once merely a habitual pain in the ass are now "known" to be suffering from attention deficit hyperactivity disorder (ADHD), a condition treatable using snake oil—I mean, rigorously trialed pharmaceuticals.

A further popular marketing tactic among the pharmaceutical companies is the repackaging of inexpensive old drugs as bright shiny new ones that cost several times as much. One of the most famous examples of a drug company patenting and making huge profits from what was essentially a pre-existing drug came in 2002, when AstraZeneca announced with much fanfare its new antacid proton-pump inhibitor, Nexium—the Purple Pill famed from the company's ubiquitous TV ads. What was less publicized was that Nexium was merely one of the two isomers (mirror-image molecules) that together made up the company's already existing—and far cheaper—proton-pump inhibitor Prilosec, whose patent was approaching expiration. AstraZeneca's argument was that the "purified" molecule acted more rapidly than the mixture that comprised Prilosec, but there was no especially compelling scientific evidence to support this. In essence, then, the patent authorities had colluded with the company to give the drug a second life . . . and grossly overcharge consumers into the bargain. Alas, Nexium is far from the sole example of this process.

And what about vaccines? Does the industry exhibit the same corruption of marketing techniques in this region of its activities? Does it pull strings to cover up mediocre or negative trial results? Does it jack up prices in an attempt at price-gouging?

The short answer is, by and large, no.

The reason's pretty simple. The customers for vaccines are typically governments rather than private insurers and individuals, which means prices per dose are kept low; where the drug company makes its profits on vaccines comes from bulk sales rather than high per-unit price. Because vaccines are administered not just to a few thousand experimental subjects in trials but to millions of individuals in real life, there's no place for an ineffective vaccine to hide, and certainly no way a vaccine might have adverse side effects that could go undetected for long. So much for the claims of the anti-vaxxers that the MMR vaccine causes autism, the polio vaccine causes sterility or homosexuality, and all the rest of the noise out of Conspiracy Central.

There's a lesson to be learned here, perhaps, about how the relationship between modern public health and the pharmaceuticals industry might be cleaned up, but it seems many are unwilling to learn it.

# # #

A further issue that adds hugely to both the expense of the US health-care system and the profits of Big Pharma concerns the stated expiration dates of drugs. At face value it seems like a good idea to label drugs thus—typically the drugs you pick up from the pharmacist are good for about three years after you get them—because no one wants people to sicken through using old drugs. The trouble is that in almost every single case the deadline given by which a prescription drug should be used is not just too short, but ridiculously so.

And expensively so. A 2017 estimate published by *ProPublica*[86] suggests US hospital pharmacists alone throw out about $800 million annually in the form of drugs that have "expired." Add in the contributions from retail pharmacists, old people's homes and other care facilities, reps' samples and normal consumers like you and me, and you can get some idea of the staggering scale of the waste.

The tests that have to be done to estimate the shelf-life of any new drug are time consuming and expensive, and so the FDA puts the onus for them onto the pharmaceutical companies. As in so many other instances, what seems like a saving for the public purse contributes to a huge drain on it. It's quite clearly in the pharmaceutical companies' interests to keep the stated shelf-life as short as possible; however, if it's too short, professional consumers are likely to search for a different drug, one that (apparently) lasts longer. So the companies run tests to ensure the drug retains its properties for a reasonable period and get their accreditation from the FDA. Thereafter there's absolutely no incentive for the companies to invest further to find out how much longer the drug might remain viable . . . and a great deal of financial *dis*incentive.

Healthcare professionals in developing countries have long been aware of such disparities. While it may to us seem a shockingly dangerous dereliction of duty to give patients drugs that are past their sell-by dates, the physicians concerned point out, first, that the success rate of such use is nigh indistinguishable from that of within-expiration-date prescriptions and, second, in deprived areas patients essentially have the choice between these drugs and no drugs at all.

But it's not just healthcare professionals in developing countries who've realized that stated expiration periods for most prescription pharmaceuticals are artificially short. The CDC, the Department of Veterans Affairs and the US Air Force are among the others. As long ago as 1986, the Department of Defense and the FDA, at the instigation of the USAF, set up the Shelf Life Extension Program. This program tests the viability of drugs stockpiled by the CDC and others for use in case of national emergency. The program has found that, although it's not the case for all drugs, the great majority retain their active properties for years after their stated expiration dates.

In 2000, the American Medical Association (AMA), recognizing the benefits that could accrue not just to the US military but to Joe Public if expiration dates were more rationally assessed, agitated with the FDA, the Pharmacopeial Convention and the pharmaceutical industry's trade association, Pharmaceutical Research and Manufacturers of America (PhRMA), for something to be done. The AMA's initiative went nowhere, presumably because there were too many vested interests—not just Big Pharma's—in the status quo.

Marshall Allen[87] cites the discovery of a stash of prescription drugs in a long-forgotten retail pharmacy closet; the drugs were 28–40 years past the expiration dates on their (faded!) labels. The drugs were analyzed and in due course the results were published in *Archives of Internal Medicine*.[88] Of the fourteen drugs tested, twelve had retained their potency for at least the lesser period of 28 years and eight of those were still viable after the full 40 years.

What sort of savings could be made if drugs' potential shelf-lives were properly evaluated? In 2016, the annual cost of running the Shelf Life Extension Program ran to $3.1 million and the savings, in terms of replacement costs for drugs that were in fact still viable, whatever their stated expiration dates, was $2.1 *billion*—a staggering 677 times as much. Now try translating those figures to the entire US healthcare system . . .

# # #

Everybody's keen that government agencies should be as efficient as possible. The difficulty comes when you try to decide what you mean by "efficiency."

One of the less wise definitions has been applied to the FDA since the administration of Ronald Reagan: people there are judged to be doing their jobs properly based on the number of drugs they've managed to approve for the market in a particular calendar year. No account is taken, in this measure of efficiency, of how many *suitable* drugs might have been on offer in that calendar year, nor how simple or complicated it might be to evaluate the merits, demerits or simple uselessness of a particular offering. All that counts is the actual number approved.

Which is of course crazy, and represents a significant potential hazard to the public. Imagine you applied an analogous requirement to the work of, say, a brain surgeon. There'd be patients dying in droves as the luckless surgeon struggled to keep the numbers up.

With the FDA—and other, similar bodies around the world—it gets worse. Just as cops are more likely to pull you over for trivial or even imaginary offenses at the end of the month, to meet their monthly tickets quota, so the poor, overworked civil servants at the FDA display what is known as the December Effect. Fully half of the year's requisite total of approvals are likely to fall in the final month, as the workers strive to meet their quota. Inevitably this means some applications aren't properly vetted—in other words, some useful drugs that could save a lot of misery for the public may be turned down while some duds may go through.

The situation is complicated yet further when the public is applying pressure for a drug to be approved swiftly. You'll regularly see newspaper or internet articles bewailing that a drug exists that could be helpful for some horrendous medical condition, but the FDA is dragging its feet in granting the drug approval. In some instances this perception can even lead to the setting up of public advocacy groups, such as ACT UP, founded in 1987 to speed the approval of anti-HIV therapies. You can see the activists' point: if a disease is painfully and swiftly lethal, who wouldn't opt for the chance of a cure even if it came with some risk of even *swifter* death?

In a less litigious society, the solution to the dilemma would seem obvious: patients could sign a form declaring they recognized the hazards of accepting an incompletely evaluated treatment—that they were going into this with eyes open, and that physicians, drug manufacturers and FDA officials were blameless if things went horribly wrong. The trouble is that, even then, there's no guarantee patients or their survivors wouldn't sue anyway.

The result of these various pressures is that drug-approval times are dropping in all the developed countries. In the UK, for example, between the end of the 1980s and the end of the 1990s the average time it took for a drug to be approved by the regulators dropped from 154 to 44 working days—by more than two-thirds, in other words. This is either a good thing or a very, very dangerous thing, and there's no easy way to decide which.

# # #

Like most governments around the world, the US government has an appalling record of persecuting whistleblowers. So it's not really surprising that, when Big Pharma reacts vindictively against those who expose its malpractices, pharmaceutical whistleblowers face the threat of retaliation by the regulatory agencies that are supposed to be policing the industry.

Take the case of Rosemary Johann-Liang, deputy director of the FDA's Division of Drug Risk Evaluation in 2006 when, on reviewing the medical evidence, she realized GlaxoSmithKline's Avandia, a popular drug for diabetes, carried with it an increased risk of congestive heart failure, and that this should be reflected in a strong warning label. Her FDA bosses reprimanded her for rocking the boat, and gave the Avandia docket to someone else. In May the following year, however, the *New England Journal of Medicine* published a study[89] effectively echoing Johann-Liang's concerns. Avandia's popularity plunged, Johann-Liang was fully vindicated, and the drug now bears a warning label, just as she proposed. It seems, though, that the climate at the FDA became intolerable for her. She now practices as a pediatrician affiliated to the University of Maryland Medical Center.

Then there was the instance of Shiv Chopra, Margaret Haydon and Gerard Lambert of Health Canada's Bureau of Veterinary Drugs, who went public in 1998–99 with their concerns that they were being pushed by their bosses into greenlighting drugs that had safety worries. One was the antibiotic Baytril; the scientists emphasized that using Baytril for livestock increased the risk of breeding antibiotic-resistant bacteria. In 2004, Health Canada responded by firing the three scientists, although naturally this was not the stated reason for their dismissal.

Many whistleblowers are luckier. In the mid-1990s, Apotex threatened researcher Nancy Fern Olivieri with a punitive lawsuit—on the grounds she'd be violating a confidentiality agreement—if she told participants in a trial she was running, partly funded by Apotex, of the company's Deferiprone drug that there were concerns it could cause fibrosis of the liver. In a courageous act of defiance, Olivieri not only told the patients but co-authored a 1998 *New England Journal of Medicine* paper on the drug's dangers. In 2009, the American Association for the Advancement of Science presented Olivieri its Award for Scientific Freedom and Responsibility for this and for her work in general alerting academia and the public to the dangers of corporate dominance of the medical industry.

It's thanks to whistleblowers that some of the worst malpractices of the pharmaceutical industry have come to light, with some of the biggest legal penalties being exacted. In the 2000s, AstraZeneca rep Jim Wetta exposed the company's fraudulent marketing of the antipsychotic drug Seroquel for conditions for which the drug hadn't been approved.* The Department of Justice accordingly hit AstraZeneca with a $520 million fine. In 2009, Eli Lilly was fined $1.415 billion after a group of its sales reps revealed off-label marketing practices connected with the antipsychotic drug Zyprexa. GlaxoSmithKline found itself with a $750 million fine in 2010 for marketing contaminated products after whistleblower Cheryl Eckard exposed hygiene concerns in its factories. Eckard herself was awarded $96 million in compensation for her eight-year ordeal in helping bring the company to justice. Microbiologist David Franklin blew the whistle in 1996 on his new employer Parke–Davis for off-label promotion and other dishonest marketing techniques related to the epilepsy drug Neurontin. That cost Pfizer, by now the owner of Parke–Davis, $430 million when the legal case was settled in 2004; Franklin himself was awarded a share of $24.6 million.

The largest relevant penalty to date handed down by the US courts has been the $3 billion fine imposed in July 2012 on the pharmaceutical giant GlaxoSmithKline after the Justice Department acted on information supplied by whistleblowers. Some of the multiple offenses concerned the encouragement of off-label use of various GSK drugs (including Advair, Flovent, Imitrex, Lamictal, Lotronex, Paxil, Valtrex, Wellbutrin and Zofran). The company had been giving kickbacks and rewards such as luxury vacations to physicians who prescribed and promoted their products. It had misreported safety concerns about some of those products, notably the diabetes drug Avandia, which it represented as conferring cardiovascular benefits when in fact it presented cardiovascular hazards.

The list of crimes goes on and on. In addition to the $3 billion fine, GSK was forced to sign a five-year Corporate Integrity Agreement—in essence,

---

* This practice, known as off-label marketing, is fairly common in the industry. Even though Wundadrug™ has been approved only for use to tackle Disease A, which was the subject of the trials that were run, the company strongly implies to physicians, or even states outright, that it is effective too for Disease B, Disease C, Disease D . . . thereby hugely expanding the drug's potential market. This is of course a dangerous practice, since there's typically no evidence the drug is helpful at all in these instances.

a blueprint designed to ensure that at least a certain level of integrity was instituted at every level within the company, from drug reps on up.

For his book *Bad Pharma*, Ben Goldacre thought it might be useful to examine just how the careers of those GSK officers primarily responsible for the corporation's pre-2012 unbridled white-collar gangsterism might have fared in the pharmaceutical industry in the aftermath of the Justice Department's condemnation—in other words, to ascertain quite how willing the pharmaceutical industry was to do its own policing.

The results weren't encouraging, as Goldacre reported:

> Chris Viehbacher of GSK was singled out in the court ruling: he is now CEO of Sanofi, the third-biggest drug company in Europe. Jean Pierre Garnier was CEO of GSK from 2000 until 2008, only four years ago: he is now chairman of Actelion, a Swiss pharmaceutical company. . . . The court also specifically mentioned Lafmin Morgan, who worked at GSK in marketing and sales for twenty years: Morgan was still working for GSK in 2010, just two years ago.

In 2013, the year following that record penalty. GlaxoSmithKline earned £21.3 billion (about $30 billion). What price deterrence?

# # #

The AllTrials Campaign was born from Goldacre's *Bad Pharma*, and pioneered by Goldacre in conjunction with Síle Lane, campaigns director of the organization Sense about Science. It began with a petition:

> Thousands of clinical trials have not reported their results; some have not even been registered.
>
> Information on what was done and what was found in these trials could be lost forever to doctors and researchers, leading to bad treatment decisions, missed opportunities for good medicine, and trials being repeated.
>
> *All trials past and present should be registered, and the full methods and the results reported.* [my emphasis]
>
> We call on governments, regulators and research bodies to implement measures to achieve this.

At first it seemed this would be, like so many other petitions, a flash in the pan—or the attention span—before the next petition came along. But its signatories currently (January 2018) number 91,024 individuals and 735 organizations. Those organizations include the Cochrane Collaboration, the *BMJ* and also, perhaps startlingly, industry giants like Glaxo-

SmithKline. You can find the website, should you wish to learn more about the campaign and sign the petition yourself, at http://www.alltrials.net.

In 2017, Goldacre and colleagues launched a useful new adjunct to the site, the AllTrials Transparency Index (http://policyaudit.alltrials.net), which both enables physicians and consumers to evaluate the honesty of the various drug companies and prods those companies toward greater transparency in their activities: "We have ranked all the world's major drug companies, and a selection of smaller firms, by the commitments they make to clinical trials transparency. Next, we will be ranking the policies of non-industry trial funders."

Another useful site for evaluating the advice of medical professionals is the Dollars for Docs database.[90] First assembled in October 2010, this offers a publicly searchable list of payments made by drug companies to physicians. (I looked up my own primary-care doctor and discovered he's clearly buddies with one particular drug rep, who takes him for a pub lunch every couple of months, but that's about the extent of his sins.) This is part of an ongoing project, "Dollars for Doctors: How Industry Money Reaches Physicians,"* done by independent investigative-journalism outfit *ProPublica* on the theme of the subversion of scientific medicine by the pharmaceuticals industry.

# # #

The pharmaceuticals that are grabbing the most headlines at the time of writing are the opioids, because there's a nationwide epidemic of opioid dependency. Celebrity opioid addicts have included Rush Limbaugh.**

Morphine and heroin, whose addictive qualities have been known for centuries, are examples of opioids, but make no mistake: prescription opioids, in properly controlled circumstances, can be medically very useful drugs. Their primary use is for pain relief, but they can be handy as well for other ailments, such as cough and diarrhea. They can also create a transient sensation of euphoria, which may in itself be useful—as for making it easier for people to withdraw from other drugs—but which too can lead to

---

* https://www.propublica.org/series/dollars-for-docs. At the start of April 2017 President Trump's then-press secretary Sean Spicer famously mischaracterized *ProPublica* as "a left-wing blog." It was a notion of which he was swiftly disabused.

** Somehow those same politicians keen to wage the War on Drugs by cracking down on the use of the far less harmful marijuana through barbaric sentencing show little inclination to tackle the opioid epidemic similarly.

addiction. The current problem is not that people are using illicit opioids such as heroin—of course they are, but that's a separate issue—but that they've become addicted to prescription opioids, and have become reliant upon their use, often to the point of overdose and death.

It's all too easy to blame the epidemic on overprescription—the illicit market for prescription opioids wouldn't exist unless physicians were prescribing the drugs to people who didn't need them or, as in Limbaugh's case, were letting themselves be bamboozled into prescribing for patients who were also receiving prescriptions elsewhere. But clearly aggressive marketing by the pharmaceuticals industry has played its part. It's been interesting to note the industry's reluctance to get involved in trying to ameliorate the crisis. Presumably this situation will prevail until some or other company comes up with Wundadrug™—the extensively trialed and unfortunately rather pricey cure for opioid addiction!

# # #

Once upon a time new pharmaceuticals made a difference. The introduction of antibiotics heralded a revolution in the treatment of infectious diseases. The antipsychotics revolutionized psychiatric practice, for good or ill. Vaccines brought about the effective annihilation of horrific diseases such as smallpox, diphtheria and polio. Other killer diseases, such as malaria, were brought to heel.

Nowadays the introduction of a new pharmaceutical tends to be a matter of little or no import. Typically, all a new pharmaceutical does is make a minor improvement, at least on paper, in the treatment already adequately supplied by other, older drugs that have gone or are close to going out of patent. Or it could treat some ailment that up until now no one knew existed—bipolar disorder, perhaps, or ADHD. As one of my own physicians was once honest enough to say to me, "Ideally we can get you to the point where you can come off these drugs. They're all poisons, you realize. All of them."

He may have been hyperbolizing, but not by much. As evidence that there's something woefully at fault with modern pharmaceutical medicine, consider the fact that here in the US, the pill-poppingest nation in human history, life expectancy is currently falling.

Anticipating the Post-Antibiotic Era

As we saw on page 11, it was thanks to work done by Howard Florey and Ernst Chain during World War II, work that saved literally millions of Allied lives, that the antibiotic era began.

Alas, it looks as if we're well on our way toward the post-antibiotic era, the era in which antibiotics can no longer be relied upon to save lives because the bacteria they're supposed to kill have evolved effective resistance. As a measure of the problem, a 2017 CDC estimate puts the number of Americans killed annually by antibiotics resistance at 23,000. The comparable European figure for 2016 is 25,000. A recent UK report suggests that by 2050 the global annual death toll from antibiotic-resistant bacteria could run to ten million unless the problem is curbed.

In September 2013, the CDC issued a "threat report" entitled *Antibiotic Resistance Threats in the United States, 2013*, the first time such a summary had been attempted.[91] Of the 18 bacterial groups discussed, three were labeled as requiring urgent attention. Of these, the most threatening were the carbapenem-resistant enterobacteriaceae (CRE), a fairly recently emergent group of bacteria that are extremely toxic and against which we have few if any antibiotic tools.

About seventy bacteria are classified as enterobacteriaceae; a couple of the more familiar ones are *Klebsiella pneumoniae* and *Escherichia coli*. The carbapenems referred to in CRE's full name are a family of antibiotics generally regarded as the antibiotics of last resort—the ones used when all other antibiotics have proven ineffective. Clearly it's bad news if even your antibiotics of last resort are no longer useful.

CRE are as yet not widespread in the general population—which is a good thing, as they kill up to 50 percent of people who get bloodstream infections from them. They're mainly found in hospitals, nursing homes and long-term acute care facilities, and are often transmitted between patients on the hands of medical staff, who in many instances pick up the bacteria from sinks—in other words, the frequent washing of hands that's recommended in healthcare facilities may actually contribute to the problem. CRE are also transmitted from one healthcare facility to another, presumably by similar means but also when patients are moved to different premises.

A dramatic example of what CRE can do appeared in the CDC's journal *Morbidity and Mortality Weekly Report* in January 2017. A woman in her mid-seventies who was admitted to hospital in Reno in August 2016 was found to be suffering from an infection of *Klebsiella pneumoniae*, a bacterium often responsible for urinary tract infections. The hospital staff tried every antibiotic available to them, fourteen in all, but without success. They then sent a sample to the CDC's headquarters in Atlanta. The gloomy news was that it wasn't just the hospital that lacked an effective antibiotic in this instance. The *K. pneumonia* strain concerned, possibly picked up by the patient during time she'd spent in India, was resistant to every single one of the 26 antibiotics available to US physicians. There was literally nothing that could be done to save the woman's life.

Tom Frieden, the CDC's director during the Obama administration, commented that "CRE are nightmare bacteria." Among the CDC's recommendations to help contain the growing problem was that physicians should learn to be sensible—i.e., restrained—in their use of antibiotics.

Second and third in the CDC's 2013 list of threats were antibiotic-resistant strains of *Clostridium difficile* that emerged in the middle of this century's first decade, and H041, the recently emerged antibiotic-resistant strain of the gonorrhea bacterium *Neisseria gonorrhoeae*; in July 2017 scientists at the WHO described it as "only a matter of time" before even last-resort antibiotics would be useless against gonorrhoea. By way of comparison, the CDC rated MRSA (methicillin-resistant *Staphylococcus aureus*), the superbug over whose spread there was such panic a decade or so ago, as merely "serious"; according to the CDC's own estimates, MRSA infected 80,461 Americans in 2011, killing 11,285 of them.

Bacteria become resistant to antibiotics to which they are overexposed. There's no real mystery as to why: it's simply an example of the evolutionary process of natural selection in action. In any population of bacteria, a few are likely, by chance, to be somewhat resistant to a particular antibiotic. Once their fellows have been wiped out by a course of the antibiotic, the resistant bacteria are free to breed. Furthermore, they have *room* to breed, to expand into the space left by their departed confreres. The more often you treat a bacterial population with antibiotics, the tougher the bacteria left behind each time, and the more resistant the new populations become.

Currently antibiotics are being overused in two fields: in the medical treatment of humans and in the agricultural treatment of animals, sometimes for veterinary purposes but in the vast majority of cases in order to fatten up food animals.

In medicine, overprescription is the primary problem—hurried doctors find that dashing off a prescription for antibiotics gets patients through surgery nice 'n' quick and, after all, the antibiotics "can't do any harm" and might well have a placebo effect; besides, being given a prescription generally makes patients feel they've gotten their money's worth, so to speak. There are plenty of instances where doctors have been found to have prescribed antibiotics for viral diseases like colds! An allied problem is patient pressure. Undereducated or misinformed patients can sometimes be quite aggressive in their demands for antibiotics in cases where these are a wholly unsuitable treatment. It's of course the duty of the physician to explain this and prescribe something more appropriate, but not all doctors are good doctors.

Even the time of day can make a difference. According to a 2017 report by Pew Charitable Trusts:[92]

> A recent study analyzed how likely clinicians were to prescribe antibiotics for acute respiratory infections—conditions for which these drugs are only rarely recommended—depending on the time of day. The findings showed that as their workday wore on, physicians were significantly more likely to prescribe antibiotics for these infections. The authors posited that "decision fatigue," a decline in decision-making abilities after having to make repeated decisions, could be an explanation.

A 2016 study[93] done under the auspices of Pew and the CDC found that nearly one in three prescriptions for antibiotics issued at US outpatient facilities—including doctors' and dentists' surgeries as well as hospital-related clinics—is superfluous. That comes to about 47 million worthless prescriptions per year. Some estimates put the figure higher.

This is really a behavioral problem. In the US and many other countries, strong efforts are now being made to educate physicians and patients about antibiotic abuse.

Naturally the pharmaceutical companies aren't thrilled about this; on the other hand, by far the most antibiotics they sell each year are destined not for humans but for animals.

No one knows the exact quantities of antibiotics sold to US agriculture, most of them used for purposes of growth promotion, disease prevention (animals are routinely kept in conditions of overcrowded squalor), or both, but a 2011 estimate indicated the total was about 13,600 metric tons—or perhaps 70–80 percent of the country's total antibiotic consumption. And nobody knows exactly why it is that a diet of antibiotics promotes growth in livestock. The bizarre fact is that it doesn't matter which antibiotic the animals are dosed with: the effect, in terms of fattening up, is just the same.

What we do know is that this gross overuse, unregulated by Congress thanks to the activities of an army of well funded agriculture lobbyists, is both boosting the evolution of superbugs—bacteria capable of resisting antibiotics that eventually spread to humans—and putting a steady ration of antibiotics into our water supplies and our foods, such that, without our being aware of it, we're all undergoing a nonstop course of antibiotics. What we also know is that the FDA, while recognizing the problem, is displaying a distressing pusillanimity about doing anything to deal with it.

Although the agricultural use of antibiotics for growth promotion has been banned in some countries (in 1999 the EU announced a ban that was finally instituted in 2006, although reports imply the law's being widely flouted, while by contrast Sweden's 1986 ban is apparently being widely obeyed), others rival the US in their abuse of the drugs. In China, for example, 2012 figures indicate that 38,500 metric tons were being used for the fattening-up of animals. In the UK, where the standards are supposedly strict, they in fact allow mass-dosing of poultry with fluoroquinolone antibiotics. Research made public in January 2018 by the country's Food Standards Agency (FSA) showed that record numbers of chickens on sale in the supermarkets were showing contamination with antibiotic-resistant campylobacter.[94]

Occasionally there are breakthroughs in developing new antibiotics or improving old ones. In late May 2017, a team led by Dale L. Boger at the Scripps Research Institute and supported by the National Institutes of Health announced in *Proceedings of the National Academy of Sciences* (*PNAS*)[95] that they'd managed to re-engineer vancomycin—a stalwart in medical science's antibiotic toolkit for over half a century—to render it as much as, according to initial reports in *Science*, 25,000 times more effective. This is hopeful news indeed, as vancomycin-resistant *Enterococcus* bacteria are a growing threat in hospitals. There still remain the tasks of streamlining the synthesis of the new superdrug in quantity and conducting animal and human tests, but Boger and his team are confident the improved vancomycin will be in use within five years.

In June 2017, the World Health Organization (WHO) released a set of recommendations concerning the use of antibiotics in order to slow the spread of bacterial resistance to them, dividing the drugs into three groups: "access," "watch," and "reserve." The first group, which includes drugs like amoxicillin, should be used whenever they're needed. The "watch" group comprises antibiotics that should be prescribed sparingly, and only if their use is strongly desirable; an example is ciprofloxacin. And the antibiotics in the "reserve" category should be withheld for last-resort cases, when everything else has been tried without success.

The logistical problem with this ideal is that the "reserve" category contains almost all of the newer (and more expensive) drugs. The pharmaceutical companies, already looking askance at the costs of developing new drugs and the likely return on their investment, are going to become even less enthusiastic if the very antibiotics they've recently invested heavily to create are the ones whose use is going to be deliberately restricted. Cited by the *Washington Post*, Suzanne Hill, WHO's Director of Essential Medicines, acknowledged the problem: "We need to work out how we pay the companies to not market it and to keep it in reserve. This is going to be a challenge."

# # #

Martin J. Blaser, Director of the New York University Human Microbiome Program and Professor of Microbiology at the New York University School of Medicine, has focused on a different consequence of our abuse of antibiotics: the role that our overuse of them, not just in agriculture but more importantly in humans, especially infants, plays in depleting our microbiome—the ecosystem of bacteria living in and on us—many of whose species are crucial to our ability to resist disease. (Caesarean deliveries, in that they avoid the transfer of a basic bacterial colony to the baby from the mother's vaginal walls and anus during birth, are also a problem in this respect.)

In his book *Missing Microbes* (2014), Blaser describes his work pursuing this line of research, beginning with his hunch that *Helicobacter pylori*—the very same *H. pylori* that Barry Marshall and Robin Warren identified in the mid-1980s as the cause of stomach ulcers (thereby gaining themselves a 2005 Nobel Prize)—was not the outright villain the medical profession had thereafter assumed it must be. The behavior of the species of our microbiota generally depends on context: a bug that might be beneficial or at worst harmless most of the time can become dangerous, even lethal, if the circumstances shift. There's no question *H. pylori* can in certain circumstances cause peptic ulcers and even, if those go untreated, stomach cancer in later life, but are we, through our use of antibiotics to eliminate our internal *H. pylori* population, ignoring benefits *H. pylori* can bring in the earlier decades of life?

Blaser believes so, and he and various teams have amassed a mountain of convincing evidence that our dearth of *H. pylori* and other vital bacterial populations is contributing to such "modern plagues" as childhood obesity, asthma, allergies, diabetes, and possibly many more ailments. That "possibly" is important. Blaser is quite specific in pointing out that his suspicions that conditions like autism might be due to antibiotic overuse in the first few months of life are no more than that: suspicions. By contrast, the connections between microbiome debilitation and, say, childhood obesity are evidenced by multiple lines of research.

Blaser points out that the unnecessary courses of antibiotics we take or administer to our children are, through disrupting the relationship we've developed over hundreds of thousands of years with the bugs inside us, causing a gross health crisis, one that may be threatening us at species level.

An additional point, made in July 2017 by Martin J. Llewellyn of Brighton and Sussex Medical School and colleagues in their *BMJ* paper "The Antibiotic Course Has Had Its Day," is that the advice that patients should always complete their courses of antibiotics, as otherwise they might breed resistant bacteria inside themselves, is actually unsupported by science. Although the old idea holds good in a few cases—for example, the tuberculosis bacterium *Mycobacterium tuberculosis*—in most instances the opposite is true. Despite the advice of the WHO, and although they stress that more research is needed, Llewellyn and his colleagues conclude:

> Research is needed to determine the most appropriate simple alternative messages, such as stop when you feel better. Until then, public education about antibiotics should highlight the fact that antibiotic resistance is primarily the result of antibiotic overuse and is not prevented by completing a course. The public should be encouraged to recognize that antibiotics are a precious and finite natural resource that should be conserved. This will allow patient-centered decision making about antibiotic treatment, where patients and doctors can balance confidence that a complete and lasting cure will be achieved against a desire to minimize antibiotic exposure unimpeded by the spurious concern that shorter treatment will cause antibiotic resistance.

The agricultural abuse of antibiotics for growth promotion is part of the pattern of reckless profligacy. Some of those antibiotics remain in the meat, poultry, eggs, milk and so on that we (vegans aside) consume daily. If antibiotics are effective in fattening up our food animals, it's hardly a stretch to surmise they could be doing the same to our children.

#

As long ago as the early 1970s the FDA acknowledged the problem of antibiotic overuse in US agriculture, and in 1977 it announced proposals to curb some of the abuses. There then began an example of that tiresome political game whereby "our" representatives—in this instance "sponsored" by the agriculture industry—keep claiming that "more research is necessary" before anything should be done, no matter how much research there's already been and no matter how conclusive its results. In the case of Big Ag's abuse of antibiotics, this game was permitted to continue for four decades. A typical industry reaction is that of the National Pork Producers Council (NPPC), which in 2011 commented:

> Not only is there no scientific study linking antibiotic use in food animals to antibiotic resistance in humans, as the U.S. pork industry has continually pointed out, but there isn't even adequate data to conduct a study.

This was a quite extraordinarily dishonest claim. There are studies galore "linking antibiotic use in food animals to antibiotic resistance in humans," as the NPPC must well know.[96] And, as evidence for the claim that "there isn't even adequate data to conduct a study," the NPPC cited a Government Accounting Office (GAO) report that bemoaned the poor job that inspectors from various agencies have done in collecting data on the agricultural use of antibiotics. Deftly the NPPC failed to mention the fact that, elsewhere in that selfsame report, the GAO spells out that the superbugs bred through growth promotion are indeed a threat to human consumers.[97] That's no surprise to consumers themselves. We're all accustomed to the regular reports of outbreaks of *Salmonella* poisoning, sometimes fatal, caused by antibiotic-resistant *Salmonella* bred in factory farms and ingested by humans.

In 2010, over three decades after the problem had been identified, the FDA declared itself ready to get tough. But then the agriculture industry rallied its troops yet again, and a lot of money changed hands in the House and Senate. Not fancying a major political fight in an election year, the Obama administration—to its great shame—caved. At the end of 2011, the FDA announced it was opting instead for a system of voluntary regulation. Since the agriculture industry is still pretending, in the teeth of the available science, that the use of antibiotics in livestock presents no medical problem at all, this was not the brightest of ideas.

And how did the FDA's announcement of its capitulation go out? It was buried in the Federal Register for December 22 of that year—perfect placement and timing to assure, albeit unsuccessfully, that it would go completely unnoticed.

Years passed, and then in late March 2015 the Obama administration, having decided it was time to get its act together, issued its "National Action Plan for Combating Antibiotic-Resistant Bacteria." Obama's budget for the year 2016 proposed nearly doubling the funds devoted to the problem, to over $1.2 billion. The Action Plan is full of the kind of sensible measures that tend to get ridiculed by cost-cutting politicians, and overall has the feel of one of those agendas drawn up by people who're perfectly aware ideals won't be met. Even so, the blueprint it lays out is well worth perusing. Of particular interest, perhaps, is the section on stimulating research into new antibiotics, other therapeutics that might replace antibiotics, and new vaccines, as well as the possibilities these, plus probiotics, might offer for agricultural growth promotion.

And finally, in January 2017, the FDA issued new regulations—"guidances"—governing the use of antibiotics in agriculture: antibiotics for growth-promotion were outlawed, antibiotic use for cure and prevention of disease had to be administered by a trained vet. However, within a matter of weeks the GAO was complaining yet again, in its "Report to Congressional Requesters" on the subject,[98] spelling out that inspections were inadequate to inform us if farmers genuinely were cutting back on antibiotic use, as the regulations decreed.

In terms of the other threat posed to human health by agriculture's use of antibiotics for growth promotion—our consumption of not the superbugs but the antibiotics themselves—the FDA has a rule designed to protect the consumer: that meat producers must leave a "washout" period between the final dose of antibiotics given to an animal and its slaughter. (An analogous regulation applies to fish farms.) This rule would be extremely useful if properly enforced; in practice, both lack of funding and a certain amount of financial corruption mean checks that producers are actually observing this rule are few and far between, and tests of foods bought in stores indicate it's widely bypassed.

Even so, partly in response to FDA regulation but likely more because of consumer pressure, poultry producers and processors in the US are moving to phase out their use of antibiotics. According to Jonathan Kaplan in a June 2017 report for the National Resources Defense Council,[99] "Companies fighting back against superbugs include: Panera Bread, Chipotle,

Perdue, Tyson, McDonalds, Wendy's, Compass USA, Subway, Taco Bell, and others."

There's one big exception, though: Sanderson Farms. In the summer of 2016, Sanderson Farms mounted a publicity campaign claiming the other companies were abandoning the use of antibiotics in the raising of their chickens purely as a marketing gimmick, that there was no science to back it up. There is of course a huge amount of scientific evidence that agricultural antibiotic use is breeding antibiotic-resistant bacteria and that these are spreading into the human population, to dangerous effect. Could it be that Sanderson Farms are somehow using antibiotics in such a way that this isn't a problem with their chickens? Not according to the available FDA data for the years to 2014: Sanderson Farms products are roughly as contaminated by superbugs as anyone else's.

Reporting in May 2017 on Sanderson Farms's AGM, Venessa Wong of *BuzzFeed* repeated the astonishing statement to investors by the company's CEO, Joe Sanderson, that "There is no sound science that proves that by using antibiotics in animal agriculture that you create disease-resistant bacteria."[100] She also cited a January 24, 2017 statement the company had released, "Supplemental Information Regarding Stockholder Proposal on Antibiotics."[101] This statement tells us:

> Numerous scientific studies have shown that the preventative use of antibiotics in food animals has not harmed human health, and that banning their use might actually be harmful to humans.

The claim is linked to another Sanderson document, "The Science of Antibiotics in Poultry" (January 2017), where we find:

> One study stated: "To our knowledge, no case of a treatment failure in a human patient, caused by transmission of antibiotic-resistant bacteria through the food chain . . . has ever been documented in the United States." The authors say the notion that the subtherapeutic use of antibiotics causes harm from antibiotic resistance in humans is a "scientific urban legend" based on faulty assumptions and statistical risk models, rather than valid empirical data.

The study in question was, however, produced by Louis ("Tony") Cox and Doug Popken, scientists connected with Cox Associates Consulting, a risk-analysis company whose clients have included the Pork Board, the American Petroleum Institute, and Philip Morris, and whose proud boast is that "Over the past five years, Cox Associates modeling techniques have been credited with saving clients over $100 Million." The study appeared in February 2010 in the journal *Risk Analysis*, of which Tony Cox was then an editor; he is currently (March 2017) listed as editor-in-chief.

All sounds perfectly kosher to me . . .

A number of the presentations at an American Society for Microbiology meeting in New Orleans in June 2017 discussed the fact that colistin-resistant bacteria are now widespread in both pigs and humans in many parts of the world. There could hardly be a clearer-cut case that the responsibility lies at the feet of the agricultural industry: until relatively recently the only use of colistin was in farm animals, because in humans it can cause renal damage. Despite this, in recent years, because of the spread of bacteria that are resistant to other, hitherto better antibiotics, colistin has enjoyed a renaissance in medicine as a last-resort antibiotic. Cited in *Nature*, Lance Price of the George Washington University, DC, summarized: "It's a crappy drug and I think this is a sign of our desperation that we are so concerned about the loss of a toxic antibiotic."

Thanks to studies of ancient bacteria extracted from melting permafrost, we know that antibiotic resistance goes back many thousands of years—the antibiotics concerned being, obviously, not manmade preparations but naturally occurring ones, produced by other organisms to combat bacterial invasion. This has led to another line of argument from those aiming to defend the use of antibiotics for growth promotion. You've guessed it: Clearly antibiotic resistance isn't the fault of modern agricultural practices because it's been around since long before agriculture began.

It's an argument directly analogous to—and every bit as stupid as—that of the climate-change deniers that, since temperatures were higher in the Cretaceous than they are today, climate change is an omnipresent condition and there's nothing to worry about in the fact that the world's inexorably getting hotter.

At the time of writing, with an administration and a Congress dedicated to slashing the budgets of scientific and regulatory agencies, we have no way of knowing if our progress toward the post-antibiotic age will be a reluctant crawl or a foolhardy sprint, but the prognosis isn't good. One optimistic sign is that, as we've seen, consumers are increasingly aware of the problem, and are forcing some of the agricultural producers into taking action to reduce antibiotic use in their business. But doubts still remain as to how honest those NO ANTIBIOTICS claims on the labels often are. A properly funded FDA, with its investigators given teeth, might be a good way of allaying some of those doubts.

The Diesel Emissions Fraud

So far as the US public is concerned, the vehicle emissions scandal began in the fall of 2015, but really it went back some while earlier than that, to 2013, when a Michigan engineer called John German was given a bright idea by a friend. German is joint head of the grandly titled International Council on Clean Transportation (ICCT), a nonprofit organization dedicated in part to the reduction of vehicle emissions. The ICCT was well aware that diesel cars in Europe had a dire reputation for high levels of emissions of toxic nitrogen oxides and dioxides (NO_x). The bright idea was to look at US diesel cars and see where they were getting things right, because they were passing the far more stringent emissions standards that pertained in the US.

So German and his colleagues took three diesel cars out on the road to test them: a Volkswagen Jetta, a Volkswagen Passat, and a BMW X5.

Although the Jetta was a model that had passed all the federal emissions tests, the car, the team found, was producing NO_x emissions at levels anywhere between 15 and 35 times the permitted standard. The Passat did almost as badly, at five to 20 times the permitted standard.

It was at this point the experimenters thought there must be something wrong with their equipment. However, when they took the BMW X5 out on the road, their equipment told them the car was producing emissions at an acceptably low level.

Clearly whatever was wrong had nothing to do with the ICCT's methods or equipment. What was at fault had to be the Volkswagens.

German didn't regard it as his job to go public with this information, nor with his suspicion that Volkswagen was using a defeat device* to get round the emissions testing. Instead, in May 2014, he wrote a report on ICCT's discoveries and sent copies to the EPA, to the California Air Resources Board (CARB), and, as a courtesy, to Volkswagen.

The manufacturer initially tried to claim that a software glitch was responsible for the extraordinarily high emissions, and in December 2014 Volkswagen issued a fix they declared would deal with the problem. Yet, when CARB road-tested a bunch of the company's cars in May 2015, the same problem was still there. The next few months were characterized by a string of explanations from Volkswagen for their vehicles' high emissions

---

* A gadget or piece of software that interferes with the results.

before finally, in September 2015, the company admitted it had indeed been using a defeat device to fool the testers.

Why the company expended all this energy trying to cover up the existence of the defeat device rather than just fixing the problem, however expensive that might have been in terms of recalled cars and so on, is anyone's guess. Volkswagen must surely have realized the truth would come out in the end, that the spurious explanations were merely delaying the inevitable—and, as it proved, delaying it by not very long.

Equally bewildering is that the company didn't earlier take what would surely look like the cheapest option in any rational forward financial forecast—fixing the problem during the manufacturing process, long before any emissions tests—but went ahead with the cheat instead. Could Volkswagen really have been so deluded as to believe they could get away with it indefinitely?

So what were the costs to Volkswagen of the two deceptions—the use of the defeat device and the subsequent attempt to cover up?

Within the few days immediately following Volkswagen's public exposure, on September 18, 2015, the company's shares plummeted over 30 percent. It admitted that eleven million cars, worldwide, were being recalled for retrofitting. By the summer of 2017, with Germany threatening to ban diesel cars entirely, the financial incentives Volkswagen was offering for trade-ins were climbing ever higher.

When all the dust has settled, the company will have paid out tens of billions of dollars worldwide in the form of fines and other legal penalties (for example, bills for environmental damage), sums running into the billions by way of compensation for the owners of the company's diesel cars, and an unknown but clearly large amount of money to settle the many civil cases that are being brought. As of September 2017, six Volkswagen executives are facing criminal charges in the US, with a number of other executives around the world confronting a similar ordeal.

To repeat, it would surely have been cheaper just to engineer the lower emissions in the first place.

In the August 29, 2017, issue of his *Skeptoid* podcast,[102] Brian Dunning presents a case that the penalties being imposed upon Volkswagen are poorly thought through. For a start, $4.3 billion in fines goes to the US government—to the EPA and the Department of Justice. While the EPA's share may well be put to environmental use (although, under the Scott Pruitt regime, who knows?), the DoJ's share will certainly go straight into the government's kitty.

But Dunning's greater concern is that the upgrade being fitted to reduce emissions is environmentally counterproductive. Diesel cars are already cleaner in terms of greenhouse gases than their conventional counterparts, For the sake of our shared future, it's far better that we drive diesel vehicles, NO_x fumes or no. Further, hidden away in the small print is that the Volkswagen upgrade reduces mileage by 7 percent. With more fuel being guzzled for every mile you drive, your diesel car's output of greenhouse gases rises, and hence of course its contribution to global warming—a far greater danger than NO_x emissions will ever produce. Dunning reckons, therefore, that the more environmentally responsible course of action is to continue to drive your non-upgraded Volkswagen diesel car . . . while also saving a buck or two at the fuel pump.

# # #

It turned out that Volkswagen wasn't the only car manufacturer playing this game of Russian roulette. Working in conjunction with the German automobile association Allgemeiner Deutscher Automobil-Club (ADAC), the world's second largest motoring organization, the ICCT discovered that a whole range of other diesel cars produced far higher levels of noxious emissions under road conditions than they did during tests, including models from Fiat, BMW, Mercedes, Mazda, Ford and Peugeot. Volvo's S60, the Jeep Renegade, and Renault's Espace Energy, for example, produced ten times as much NO_x as was permitted under EU regulations. In these other instances it seems there was no use of a defeat device or other intention to cheat. Where the problem lay was in the design and (lack of) stringency of the testing. If there was any corruption of the technology by the car manufacturers it seems to have been a corruption by silence—by failure to point out the inadequacies of the emissions-testing system.

In 2015 a new and improved emissions-testing system, the worldwide harmonized light vehicles test procedure (WLTP), was devised under the auspices of the UN for use on all cars, not just diesel ones. Despite fierce resistance from the car manufacturers, this system should be implemented piecemeal from 2017 onward.

7

Nazi Germany

If science cannot do without Jews, then we will have to do without science for a few years.
—Adolf Hitler, 1933

After World War II, when it became possible to evaluate the damage done to the sciences in Germany by the Nazi regime, the surprising conclusion was, at least in physics, chemistry and math, and at least in the shorter term: not much. To be sure, the vast majority of the best scientists in these fields had departed during the 1930s for more tolerant climes abroad, while a number of promising up-and-comers had been killed by the regime,* but the German scientific tradition was a strong one; in essence, there were still plenty of good German physicists, chemists, and mathematicians, and for the most part they'd been permitted by the authorities to carry on their science as usual, without interference, so long as they kept their heads down. There were temporary victories for the lunatic fringe, but by the end of the war these had largely been annulled. The scientific craziness of the Nazis was essentially for public consumption; in private the Reich realized how necessary good science was for its own glorious thousand-year survival. Even as powerful a man as Heinrich Himmler, whose espousal of the pseudosciences was profound, was sidetracked into his own "scientific" organization, the Ahnenerbe.

There was one area, though, where math and the physical sciences suffered grievously under the Nazis, and it was to be some while before they could recover. This was in education—and the lesson of what happened is

---

* A number, too, died fighting for their country.

one that is valid today, when schools elsewhere are under constant pressure to include "alternative viewpoints" in their science curricula. While German physicists and mathematicians might have been working with or exploring relativity and quantum mechanics in their research, when it came to teaching students they had to be far more circumspect; furthermore, many professors and lecturers were dismissed and either replaced by party hacks or not replaced at all. This meant that in the longer term German science suffered considerably: a whole generation of university students, and a further generation of schoolchildren, were woefully miseducated in the physical sciences, being taught a version of science that contorted itself ludicrously in order to omit or disparage the keystones of twentieth-century science.

The biological sciences fared far worse in the shorter term because of their complete corruption by racist ideology. Medicine did not escape, either. While some of the medicine the Nazis promoted was surprisingly progressive—for example, Hitler had a phobia about cancer, so cancer research and prevention were heavily promoted—the notions of medical pseudoscientists and ideologically sound quacks were presented as if they had validity. Further, the horrific experimentation on live subjects carried out at concentration camps left, like the use of slave labor in the technological industries, a stain on German science that was near-indelible. And, again, the longer-term effects of miseducation in the biological sciences did grievous harm to their practice in the nation for fully a generation. As to the effects of brainwashing young people that they should see the world, science included, only through the lens of fascist ideology? Who knows how much damage that did.

It's useful to consider the different behaviors of the German scientists who lived—or in some cases did not live—through the regime. There were some who, for reasons of ideological belief or straightforward opportunism, actively cooperated with it; examples include physicists Philipp Lenard, Johannes Stark, and Ernst Pascual Jordan; zoologists Eugen Fischer and Otmar von Verschuer; anthropologist/zoologist Konrad Lorenz; and physician Josef Mengele. There were some who did not support Nazism *per se* but who nonetheless contributed, enthusiastically or tepidly, to the German war effort either through a misplaced sense of patriotic duty or, again, through opportunism, in that the regime offered extensive funding and scientists need funding if they're to continue their research; examples include Werner Heisenberg, Otto Hahn, and Wernher von Braun. A few remained in Germany but offered covert—or in a few brave instances overt—resistance to the regime and its efforts: physicist Max Planck was one, although his resistance was more in thought than in deed (to be fair, he was in his eighties when the war began). There were those who were persecuted, interned, starved, and tortured by the regime; by far the best known example of the few in this category who survived is the chemist Primo Levi, who somehow lived through the hell of Auschwitz to tell, most movingly, the tale. And finally there were the many, many scientists who fled; examples are far too numerous to list, but we can mention as of especial note Albert Einstein, Lise Meitner, and Leo Szilard. Many of the physicists who fled in due course became participants in the Manhattan Project.

Leaving Mengele aside for the moment—not to mention the numerous anthropologists who strove vainly to find any scientific rationale whatsoever for the Nazis' dream of Aryan racial purity—the person whose science was arguably most significantly corrupted by the Nazi ideology was the mathematical physicist Ernst Pascual Jordan. At the time quantum mechanics was all the rage in the physics world, a theory that had been pioneered and elucidated largely by German scientists. However, the great clarion cry of the Nazi regime was that theorization was airy-fairy "Jewish science" and therefore to be rejected; instead physicists should be concentrating on *Die Deutsche Physik* (see below), where the emphasis was on experimentation and the focus on the tangible. Jordan, a fervent Nazi, set himself the difficult, indeed self-contradictory, task of showing that quantum mechanics was actually a scientific analogue of Nazism, and that the two—the physical theory and the political ideology—therefore supported each other.* Just as the new physics was in due course going to explain all areas of reality and tie them together as merely different aspects of a unified whole—and Jordan was not afraid to include religious experience and clairvoyance on his list along with cosmology and thermodynamics—so Nazism was not merely a political ideology but a unifying explanation of human existence and a prescription for future human social evolution. In short, both were Grand Universal Theories.

---

* Of course, many in the USSR maintained that Darwinism was a capitalist theory; while ludicrous, this was not quite as ludicrous as Jordan's thesis, in that raw capitalism, with its notion that survival of the fittest is a desirable, even ethical code upon which to base human interactions, can simplistically be regarded as the social equivalent of Darwinism. But both stances—just like the rabid capitalist's citation of Darwinism as a justification of capitalism's worst excesses—display an astonishing degree of ignorance as to what a scientific theory actually is.

One is reminded of Richard Dawkins's remark concerning the proliferation of books linking quantum mechanics and cosmology to Eastern mysticism: their thesis seems to be that quantum mechanics is hard to understand and Eastern mysticism is hard to understand, so the two must really be the same thing. Jordan's logic in trying to equate physics and Nazism seems to have followed this principle.

GEOPOLITICS

The idea of geopolitics was initially the brainchild of the UK diplomat and geographer Halford Mackinder. He claimed that much of humankind's political history had been—and by extension would in future be—determined by geography, with particular focus on what he'd later call the World Island, the great Eurasian continent. Geographical location governed resources; relative position with respect to landmasses and oceans governed the accessibility and defensibility of a culture; taken together, these factors could confer a significant advantage on a lucky culture in war, and hence on that culture's level of dominance in world politics. Any culture that controlled the interior—which Mackinder later called the Heartland—of the World Island was likely to control the world. Much later he was to summarize these ideas in a triplet:

> Who rules East Europe commands the Heartland.
> Who rules the Heartland commands the World Island.
> Who rules the World Island commands the world.

Mackinder's boundaries for the Heartland were approximately those the USSR would come to have after World War II.

A corollary was that sea power, regarded as paramount by the militarists of the day, would inevitably wane in importance as transportation and communications surged ahead within the Heartland (which in Mackinder's day meant the coming spread of Eurasian railways). This was not to say naval power would become insignificant; one passage in Mackinder's 1904 paper "The Geographical Pivot of History" chills today because of its seeming prescience:

> The oversetting of the balance of power in favor of the pivot state, resulting in its expansion over the marginal lands of Euro-Asia, would permit the use of vast continental resources for fleet-building, and the empire of the world

would then be in sight. This might happen if Germany were to ally herself with Russia.

Though Mackinder's ideas seemed plausible, for some decades they went largely ignored in the English-speaking world. However, they attracted the attention of the German retired general Karl Haushofer and his group, who in the 1920s and 1930s merged them with ideas of their own to produce the concept of *Geopolitik*. The timing could not have been direr—just while the Nazi Party was rising to ascendancy, along came a quasi-scientific theory that not only spelled out the geographical "entitlement" of the supposed descendants of the mighty Aryan races but also offered a blueprint for how those descendants could regain their "heritage." Hence was born the notion of *Lebensraum*, "living space"; what the bald translation omits is that the concept had a considerable underpinning of dubious science, both bad evolutionary science, which considered that each species (and by extension each human race) had its own natural, pre-ordained territory, and Mackinder's appealing but empirically dodgy hypothesis.*

Mackinder was not entirely blameless in the elevation of his concepts to military-political dogma. In *Democratic Ideals and Reality* (1919), he emphatically proposed that, while good people might hope for democracy to conquer the world through conquering people's hearts, the truth was that the only determinant of political and economic clout was the increase of territory through war. While he himself claimed to recognize this fact only to prescribe ways of avoiding such conflict, in reality he provided a blueprint not so much for the peacemakers as for the Nazi warmongers.

DIE DEUTSCHE PHYSIK

The masterminds of *Die Deutsche Physik*—"German physics," which rejected abstract theorization as a "Jewish" and thereby false representation of reality—were Philipp Lenard and Johannes Stark. Although one can to an extent understand the direction they were coming from, in that the crux of the scientific method is not to theorize in a vacuum, it's obvious that without the work of theoreticians science would be in a sorry and backward state; it's equally obvious that to make a racial identification, as

---

* Mackinder regarded Australasia, Japan, the Americas and all but the northernmost part of Africa as peripheral. In light of the world's current balance of power, this seems a howler, a death blow to his hypothesis.

if the ability to theorize were hereditary, is nonsensical.* Such obviousness was lost on Lenard and Stark, which is all the more surprising in that they were both deserving recipients of the Nobel Physics Prize, Lenard in 1905 and Stark in 1919. They serve as a graphic reminder of the fact that even the most brilliant scientists can be fools outside their own narrow disciplines.

Lenard was a profound hater. He hated his colleagues, he hated the English, he hated just about anyone who got in his way, and above all he came to hate the Jews. Oddly, he at first hailed relativity; it was only later, on account of Einstein's Jewishness, that Lenard became a leader of the motley ratpack who denounced relativity as a sham and attempted to install *Die Deutsche Physik* in its place. The movement's name was enshrined in Lenard's four-volume magnum opus *Deutsche Physik* (1936), in which he maintained that genuine physics must be both racially conformable to the physics community (*blutmässig*) and intuitive (*anschaulich*), by which latter term he seems to have meant commonsensical (unlike the abstractions of relativity and quantum mechanics) and based on direct experimental observation: those mysteries of nature unamenable to such observation would, and should, remain forever beyond our ken. True physics should be unimaginative, which was why the Jewish influence on physics was so toxic. Jews were imaginative, according to Lenard, and thus in the thrall of theoretical speculation. Any Jew who was good at *Die Deutsche Physik* was so only because he (the notion of it being a she was far beyond Lenard's mental horizon) must have some Aryan blood in him—as for example the part-Jewish Heinrich Hertz, under whom Lenard had studied; Hertz's Aryan blood enabled the great experimental feat of discovering radio waves, but his Jewish blood betrayed him later when he ventured into the cesspit of theory. Those theoretical physicists who were unquestionably Aryan, like Heisenberg, were "white Jews."

Stark was in essence a disciple of Lenard, so far as their bureaucratic activities under the Reich were concerned; a dozen years younger, he was more able to put into practice the ideals of the older man. On the subject of "white Jews" he expounded in 1937* in the SS journal *Das Schwarze Korps*:

* The inclination and ability toward abstract thought would seem to rely upon intelligence, certainly, but more significantly upon youthful environment; i.e., education, both formal and in the home. Since in any nation there are observable disparities between the attitudes of different communities toward education in this broad sense, there are bound to be equivalent disparities in the percentages of "theorists" emerging from those communities. The high ratio of Jews among German scientists reflected such a cultural difference, which was in turn encouraged by the prevalent antisemitism.

And if the bearer of this spirit is not a Jew but a German he is all the more to be combated than the racial Jew, who cannot conceal the font of his spirit. For such carriers of infection the voice of the people has coined the description of "White Jew," which is particularly apt because it broadens the conception of the Jew beyond the merely racial. The Jewish spirit is most clearly discernible in physics, where it has brought forth its most "significant" representative in Einstein.

In the spring of 1933, a few months after being sworn in as Chancellor, Hitler announced that all Jews working for the civil service should be purged from their posts. At the time, the structure of the scientific establishment in Germany was such that academic scientists were civil servants, so the immediate damage to Germany's position as a leading scientific nation was enormous. Many of the nation's physicists, fully aware of this, raised their voices. (It was in response to special pleading by Max Planck that Hitler made the remark cited at the head of this chapter.) Others, opportunistically, saw a chance for promotion into the vacated posts, and kept their peace. And yet others, like Lenard and Stark, welcomed the "cleansing" of German science.

Einstein, who fled, was a particular target for the excoriations of this evil pair. Heisenberg, who stayed, came in for his fair share of abuse as well, and over him Lenard and Stark had some actual power: they could block his promotion and they could make sure Himmler's SS targeted him for special investigation. Although there could be little future in attacking Heisenberg's Aryan credentials, which were impeccable, his history and his widely recognized abilities as a theoretician made him the archetypal "white Jew." Another epithet they used against him was "Ossietzky of physics."** Among his stated crimes were that he defended relativity and had worked collegially with Jews. Heisenberg's future would have looked grim had it not been for the chance that his mother knew Himmler's mother, and Mrs. Himmler promised to have a word with her boy. For about a year Himmler dithered, but finally Heisenberg's name was cleared of all

---

* As cited in translation in Walter Gratzer's invaluable *The Undergrowth of Science* (2000).

** Carl von Ossietzky was a journalist and pacifist, co-founder in 1922 of *Nie Wieder Krieg* ("No More War") and editor from 1927 of *Die Weltbühne*, in which weekly he denounced Germany's secret rearmament. He was sentenced to prison in 1931 for treason, but the sentence was commuted and he resumed his editorship. On Hitler's accession to power Ossietzky was sent to the Papenburg concentration camp. There he received the 1935 Nobel Peace Prize but eventually died of tuberculosis and maltreatment.

"charges" and Lenard and Stark were told to shut up. Hilariously, Himmler, in conveying this decision to his sidekick Reinhard Heydrich, expressed the hope that Heisenberg might be able to help in the elucidation of the World Ice Theory (see below).

Also especially singled out for abuse was the University of Göttingen, supposedly a hotbed of the Jewish conspiracy to poison Aryan science. In fact, there were a lot of Jewish scientists there, especially at the Mathematical Institute, where Christian Felix Klein—of Klein Bottle fame—was one of the stars.* Whether Klein's active encouragement of Jews here and at other German universities came through pro-semitic bias, or simply because he sought the finest mathematicians and many of these happened to be Jewish, is something we don't know, but he was certainly attacked posthumously by the Nazis for the former. It seems likely both motives played their part, because from the 1890s onward he put forward the hypothesis that different races conformed to different mathematical aptitudes:

> [T]he degree of exactness of the intuition of space may be different in different individuals, perhaps even in different races. It would seem as if a strong racial space intuition were an attribute . . . of the Teutonic race, while the critical, more purely logical sense is more fully developed in the Latin and Hebrew races.

Klein's hypothesis was intended to support what we'd now call multiculturalism. Unfortunately, he used for this a term meaning "racial infiltration," and it was for his approval of "racial infiltration" that the Nazis attacked him. Further, his own mathematics marked him out as a "white Jew" *par excellence*. Even so, his racial stereotyping was something they could use as yet another cudgel with which to beat "Jewish science," and no one took it up more eagerly than Klein's ex-student Ludwig Bieberbach, by now a professor at the University of Berlin.

The tragedy of Nazism is almost always seen as the tragedy of the victims of the Holocaust, and who could say otherwise in the face of millions of deaths and untold torments? Yet Nazism also destroyed many of

---

* At one stage when Klein was under surgery he was given a blood transfusion by his colleague Richard Courant, a Jew. Thereafter it was quipped that even Klein, a rare Aryan mathematician at Göttingen, had Jewish blood in his veins. Courant had been the founder of the Mathematical Institute, in 1920. Much later, in 1933, he fled Germany, first for the UK and then for the US, where he pursued a mathematical career of distinction for the rest of his life.

its staunchest supporters. One of these was Bieberbach. Looking at his record before he became infected in the early 1930s by the Nazi virus (he was, curiously, if anything a liberal early on), one might have predicted he'd emerge as one of the twentieth century's mathematical heroes. Instead he's remembered, if at all, as a maimer of German math (replacing it with *Deutsche Mathematik*, of course) and indirect enabler of the Holocaust.*

Bieberbach came to believe mathematics should be intuitive and commonsensical—*anschaulich*, in other words, just like Lenard's *Deutsche Physik*. This brought him into direct confrontation with the modernist school of mathematics spearheaded by David Hilbert, who had succeeded Klein at Göttingen and is today regarded as among the most important figures in math history. It helped Bieberbach's campaign that Hilbert was profoundly anti-Nazi: Hilbert had bitterly resisted the dismissal of Jews from Göttingen during the general purge of 1933 and, crime of all crimes, had many years earlier been influential in the University of Göttingen's refusal of a chair to Johannes Stark; Stark blamed this rebuff on a "Jewish and pro-semitic circle" at the university. Hilbert and his school were, in other words, easy targets at a time when political ideology transcended science. As one example of Bieberbach's folly, he excoriated at Göttingen the (Jewish) mathematician Edmund Landau for the crime of teaching the expression of pi as the sum of an infinite series.

Bieberbach continued his rabble-rousing efforts through the 1930s, and had a certain success among students and junior academics. Internationally, however, and among established mathematicians, his attempts to re-define German math were regarded with anything from horror to incredulity to ridicule, and suffered from the fact that the Nazi hierarchy weren't much interested, probably because, while they thought they could understand physics, they knew math was beyond their grasp. Bieberbach made a number of political moves in an effort to impose his views on the mathematical community from a position of power, if he couldn't do so by persuasion or rhetoric, but these too eventually came to nothing.

As a consolation prize Bieberbach was given the editorship and funding for a new mathematical journal, *Deutsche Mathematik*, whose first issue appeared in January 1936. This contained genuine mathematical papers

---

* The full story of Bieberbach's war on German mathematics is outside the scope of this book. What follows is the briefest sketch. For an excellent account see Sanford L. Segal's *Mathematics Under the Nazis* (2003), in particular its near-book-length Chapter 7, "Ludwig Bieberbach and 'Deutsche Mathematik.'"

interspersed among political exhortations and philosophical ramblings about racial stereotyping in tune with the ideas of psychologist Erich Jaensch: there were two types of human mind, racially determined, the J-type, which conformed to the *anschaulich* ideal and was thus laudably Aryan, and the S-type, possessed by inferior folks like Jews, most French mathematicians, and in general anyone disliked on ideological grounds. Despite or because of such twaddle, the journal was very successful—perhaps also because it was deliberately kept much cheaper than other scientific journals, and thus more accessible to student budgets.

After the end of the war, Bieberbach continued to maintain his views on racially determined mentalities while at the same time projecting wide-eyed innocence about the concentration camps—so far as he was concerned, expelling people from their jobs and homes was merely benign racial purification, he'd had no idea what happened to them thereafter, and why should he? It wasn't until the mid-1950s that he even acknowledged the existence of the camps. It seems he was one of the earliest of all Holocaust-deniers, even though he had to have known what was going on all around him.

# # #

The failure to topple Heisenberg more or less marked the end of the *Deutsche Physik* movement. Lenard was growing progressively more obviously dotty, while the abrasive Stark had rubbed too many of the Nazi hierarchy the wrong way. Still, the Heisenberg episode was repeated in its essentials with other physicists all through the period of the Reich. Only in the imaginings of madmen could such absurdity be regarded as a sensible way to direct the course of science, yet such was the corruptive effect of the Nazi worldview that it seemed commonplace. Certainly Heisenberg must, incredibly, have thought this way of conducting affairs at the very least tolerable, because he decided to stay on and practice science in his native land, in due course leading the Reich's efforts to develop an atom bomb.

Himmler's power to direct the scientific establishment, a power that varied during the years of the Reich but generally approached the absolute, was something the man fought hard to gain, despite the fact that he was a scientific illiterate. (So was Hitler, but this didn't stop crazies such as Lenard and Stark hailing the Führer as a scientific genius.) This was all the more disturbing in that Himmler was an aficionado of the pseudosciences, not just the pseudosciences of racial purity and Aryan superiority but also astrology and the occult.

Either to limit his meddling with the real science underway in Germany or so that the Reich might explore arenas of science the academics were inexplicably neglecting, he had been permitted in 1935 to set up his own research institute, the Ahnenerbe. The name means roughly "ancestral heritage," and indeed a large part of the "research" done there was into tracing the glorious Aryan lineage of the Teutonic peoples. Walter Gratzer's *The Undergrowth of Science* (2000) cites this 1944 letter from Himmler:

> In future weather researches, which we expect to carry out after the war by systematic organization of an immense number of observations, I request you to take note of the following: "The roots, or onions, of the meadow saffron are located at depths that vary from year to year. The deeper they are, the more severe the winter will be; the nearer they are to the surface, the milder the winter." This fact was called to my attention by the Führer.

Himmler was ever ready to accept the ramblings of amateurs like himself over the hypotheses of trained scientists, and there was really nothing any sane person could do to stop him; dissent was equated with treason, and so arguing with Himmler was the fast track to a death camp.

Himmler's attitude spread over into technology, specifically weapons technology, where researchers had no choice but to head down one pseudoscientific alley after another at Himmler's whim. One device they were forced to attempt relied for its functioning on the notion that the atmosphere contains an "insulating material" vaguely analogous to the luminiferous aether,* but in this instance it was a substrate of air rather than vacuum. Without the insulating material, so a pseudoscientist had persuaded Himmler, electrical equipment wouldn't work, it'd short circuit. The aim of the hypothetical device that could help win the war for the Reich was, from a distance, to strip away the insulating material from the air around the enemy's electronics. It sounds like something a mad scientist develops in a James Bond movie, and it's every bit as plausible. Another Himmlerian brainstorm triggered research into the building of a 1,000-metric-ton tank, which might indeed have been near-invincible to enemy fire, but which would have had an interesting experience whenever it encountered a structural weakness in the ground below. Then there was the scheme to harvest fuel alcohol from the chimneys of German bakeries . . .

---

* The omnipresent medium thought for some decades to be necessary for electromagnetic waves to travel through space.

One bizarre notion rife among the Nazi leadership was that disease could be caused by "earth rays." Precisely what these were is hard to establish, but they appear to have been related in some way to magnetism: Rudolf Hess hung magnets around his bed so that, while he slept, they'd draw out of his body any harmful earth rays he'd absorbed during the day. One characteristic we know the earth rays did have was that they were detectable by dowsing; dowsers were often hired to check out Nazi public buildings for any lethal emanations that might be lurking.

Had it not been that so many of the Reich's most able researchers were wasting their time on madcap programs, there's every possibility Hitler might indeed have won World War II. Just think of the situation had the same resources been poured into the German quest for the atomic bomb.*

Another piece of pseudoscience embraced by at least some of the leaders of the Reich's scientific establishment was the World Ice Theory, or *Welteislehre*, a fevered invention of the Austrian mining engineer Hanns Hörbiger, who in 1913 published his speculations in a vast tome called *Glazial-Kosmogonie*. Here they can only be summarized.

What we think of as stars are mere chunks of ice. The only star (in our sense of the word) is the sun, which at the beginning of things was some 30 million times larger than it is now. Everything else, the earth included, orbits the sun; the earth is the only other celestial object not to be a chunk of ice or at least entirely ice covered. Space is not a vacuum but filled with the aether, which is in actuality rarefied hydrogen; as celestial objects move through this, friction causes their orbits around the sun to decay so that they spiral inward until finally falling into the sun, creating sunspots. When stars do this, their ice is vaporized and part of it erupts back into space; some of the ice reaches the earth refrozen as tiny crystals, which we see as high-altitude clouds. They melt and reach the ground as rain. Bigger crystals, which we see as meteors, explode on entering the atmosphere, and their fragments fall as hail.

Mars is covered in an ocean of ice or water some 400km (250 miles) deep, and will one day become a moon of the earth—unless it misses us and instead plunges straight into the sun. In fact, even the earth itself will eventually fall into the sun. This is not to imply the end of the solar system;

---

* The grapevine has it that the team dedicated to this quest, led by Heisenberg, was "close" by the end of hostilities. In fact, it had barely got beyond the basics. The German scientists, by then in captivity in England following the end of the European war, were as stunned as anyone else by the news of the Hiroshima detonation.

there is, apparently, an infinite number of planets beyond Pluto, and these are slowly approaching the sun as if on some cosmic conveyor belt. Thus, even though the planets we know will die a fiery death, others will take their places. Long before our world plummets to its doom, however, the moon, which is spiraling in towards us, will have collided with it. This will not, however, leave our devastated planet moonless; as with the planets and the sun, there are replacement moons queuing up. Indeed, this is not our world's first moon. The most recent of its several predecessors fell to earth just a few thousand years ago, causing a catastrophe we know from various myths and legends as the Flood. The capture thereafter of the next moon, the one we have today, caused further upheavals, such as an outbreak of poleshift* and the sinking of Atlantis, which itself was a relic of the previous fallen moon.

Hörbiger claimed the World Ice Theory came to him fully formed in a dream, and reading the above we might not be surprised. What, one wonders, had he been smoking? Even so, the theory became inextricably involved with the rising tide of German Nazism. This cosmology was as different from the "Jewish" one as any loyal Aryan could hope for. That it came from an Austrian amateur scientist was an added advantage—after all, was not Hitler himself an Austrian amateur scientist? Further, a reason for the Aryans' natural superiority to all the rest of mankind was obviously because their ancestors had been toughened up in the chill of their northern habitat during the last glacial age. (Ice again! Surely it couldn't be coincidence!) Indeed, to reject the World Ice Theory would be to reject the very notion of Aryan superiority.

Hitler and especially Himmler loved this.** It seems what attracted Himmler to the theory was the bit about the ice crystals falling into the atmosphere. Accurate weather forecasting is a more than useful ally when a country is waging war. If the enemy, through foolishly ignoring Hörbiger's theory, were unaware that the true source of weather was outer space, then their meteorologists would be at a marked disadvantage against German ones. Himmler set up an Institute of Meteorology whose scope was im-

---

* The idea of poleshift is that the direction of the earth's rotational axis periodically flips, precipitating widespread disaster. There would in truth be widespread disaster—far more, indeed, than any of the poleshift theorists seem capable of imagining—if poleshift were feasible.

** The *Welteislehre* is of course a theory, and thus logically the Nazis should have put it into the "Jewish science" category. They didn't.

mediately expanded to include all kinds of other disciplines in an effort to prove the *Welteislehre*. A surprising number of supposedly reputable scientists went along with this, through fear or otherwise. It was, however, too much for even arch-bigot Philipp Lenard to stomach; his polemics against effort being wasted on such nonsense were sufficient to drive Himmler's researches, if not underground, then at least into the shadows, where they continued. And woe betide any lesser figures of the German scientific establishment than Lenard should they speak out against the theory.

The World Ice Theory was only one of the many crackpot items of pseudoscience that found their way to the Ahnenerbe. Another was, at least for a while, the hollow-earth theory of Cyrus Reed Teed, a US religious leader and crank; where Teed's theory differed from others was that he believed we live *inside* a hollow earth. Some special "explanations" were required. For example, if you look directly upward at midnight, why do you not see the lights of the cities on the other side of the world? Teed proposed that light could travel only so far around the concave surface of the earth before, as it were, taking a nosedive into the ground. Such a limitation did not apply to light of "arcane" frequencies—such as infrared. And so the Nazis conducted an experiment whereby they hoped to spy on the maneuvers of the British fleet. They pointed their instruments up at an angle of about 45°, and, to their surprise saw . . . clouds!

Teed's hypothesis did, however, have a more substantial effect on the development of Nazi technology. In 1933, the city council of Magdeburg decided it would test the hypothesis by asking the team of rocketry scientists then working in Germany—including von Braun—to fire a few rockets upwards to see if they'd land in the antipodes. Unfortunately the Magdeburg councillors underfunded the project, and so the experiments had to be curtailed—by which time the best performance of the *Raketenflugplatz*'s launches had been a horizontal flight of about 300m (1,000ft). Nevertheless, these experiments paved the way for the V1 and V2.

The Aryan Dream

During the Third Reich, Darwinian evolution was accepted as a fact, even by Himmler. Himmler did draw the line, though, at the notion of the Aryans having evolved, even though the rest of humankind had obviously done so. By contrast, the Aryans had started off as little seeds embedded in Hörbiger's cosmic ice crystals, but had then for some reason been released onto the earth, complete with a godlike understanding of things like electricity—thunder and lightning were belated manifestations of a fierce war machine they'd created during prehistory to conquer Atlantis, until then populated by Asian immigrants.

Himmler also espoused the theory that the Nordic races had long ago almost completely exterminated the peoples of southern regions, but that the survivors had managed to breed the numbers back up again and were present in the world today as Hottentots and Jews. His basis for this conclusion was that many African women have large rear ends—as displayed in figurines from that continent—and he'd noticed the same about Jewish women. One of the anthropologists from the Ahnenerbe, Bruno Beger, was instructed to look into the matter. The farce turned to nightmare when Beger pursued his investigations in Auschwitz.

The term "Aryan" has valid technical meanings, although these are little used today precisely because of the Nazi abuse of the word. Technically, "Aryan" refers in linguistics to the ancestor of the Indo-Iranian language family, and to that family itself; in archaeology/anthropology it refers to a pre-Vedic people of whom little is known save that they represented an early stage in the mutual assimilation of the Indian and Iranian cultures. By the end of the nineteenth century, the word was often used as a distinction from the Semitic cultures; popularly, it became a synonym for "gentile." The appeal of the word to racists was that it derived from a Sanskrit word meaning "noble" or "elevated," hence the emergent mythology of a noble race that had expanded from India across Iran and Europe and was the highest expression of humankind.

The theoretical roots of Nazi racism can probably be traced to the Social Darwinist ideas promoted notably by Ernst Haeckel. Haeckel was Darwin's and evolution's most vocal and influential supporter in Germany, although Darwin politely distanced himself from the man. Another to be notably chary about Social Darwinism—and about Darwinism in general*—was Haeckel's countryman, the great cell biologist Rudolf Virchow, who saw only too clearly the murky swamps into which Social Darwinism could lead.

Social Darwinism adapted the notions of evolution by natural selection and applied them to human societies. The upside of this was that it offered a very optimistic, progressive philosophy concerning the future potential of the human species: change was not something to be resisted but to be embraced. The downside was, as Virchow predicted, that the temptation

---

* Virchow believed disease was a product of cell malfunction. The Darwinian idea that cell mutation could on occasion be advantageous was thus anathema to him.

was there to start "improving" a particular human culture by weeding out its "unfit" members and by resisting the influx of "inferior blood." This was the slippery path of eugenics, down which people such as Francis Galton had already slid, and down which many others aside from the Nazis would also slide, especially in the US but also in parts of northern Europe (see pp. 183–188). In Nazi Germany, the notions were especially toxic because they were married with the myth of the noble Aryan race. The glorious future of Germany depended upon distilling its people down to their essential Aryanism.

This craziness was further spurred by the proselytizing of the UK germanophile Houston Stewart Chamberlain, who promoted the notion of *Rassenhygiene*, racial hygiene—a term not too dissimilar from the more modern "ethnic cleansing." The mentally and physically disabled were obvious "impurities," as were homosexuals and communists, and it was perfectly obvious gypsies, blacks and Jews weren't of Aryan stock. Really, the list of minorities that were stains on the Aryan bloodline could be extended as far as the whim took you, and in any direction. Notably, the consequence of such thinking is that the victims are no longer individual people, just items in a category, and this attitude was encouraged by the Nazi regime, which referred to the minorities as subhuman—it wasn't people who were being slaughtered, just animals.

Physicians such as Alfred Ploetz—founder in 1904 of the Society for Racial Hygiene (he coined the term)—made their own loathsome contribution, observing that medical science was capable of prolonging the lives, and thereby increasing the breeding opportunities, of substandard humans. A further justification for the extermination of the sick and impaired came from the ideas of the surgeon Erwin Liek, author of the bestselling *The Physician and his Mission* (1926). Nazi medicine in general sought a holistic approach—prevention and therapy should focus on the body and lifestyle as a whole, since all was interconnected, rather than on the malfunctions of individual organs or systems. This approach can of course be useful, and the Reich's accent on lifestyle means of prevention had a positive influence on overall public health. When the principle's applied to excess, however, it clearly has dangers: herbalism and homeopathy became the norm. The ideas of Liek, as expressed through his book, took this further, reintroducing a variation on the old idea of the bodily humors. In addition, Liek held that sickness was a symptom of immorality. Pain was a good thing—an essential part of the body's self-healing; pain relief was thus counterproductive.

Between January 1940 and September 1942, in part as a consequence of Liek's ideas, the Nazis gassed 70,723 mental patients on the grounds that their lives "were not worth living."

The anthropologist Eugen Fischer was another significant voice. His special field of study was interbreeding between members of different races, a reasonable enough topic for investigation except when twisted through a racist ideology; Fischer played a part in the mass sterilization of those children born of French soldiers and German women during the 1919 French occupation of the Rhineland, and later, with Nazism in the ascendant, began promoting the analogy that Germany's ethnic minorities, notably Jews, were cancerous tumors feeding on the otherwise healthy German body.

The botanist Ernst Lehmann founded the journal *Der Biologe* (and the Association of German Biologists) before Hitler's ascent to power, which provided him with the perfect opportunity to promote his fascistic views under the guise of biological science. Most of the journal's articles promoted racist pseudoscience; the pseudoscientific theme was perpetuated through items disputing Darwinian evolution, or even developing it; the biologist Gerhard Heberer, for example, claimed that Nordic Man was a recent, more highly developed species destined to replace *Homo sapiens*. (It's not hard to see how this hypothesis could be adopted by those eager for a final solution to the problem of "subhumans.") The Austrian ethologist Konrad Lorenz was on the editorial board of *Der Biologe*; he eagerly promoted the notion of racial purification, and recommended euthanasia for the unfit.*

The quack doctor Joachim Mrugowsky, unable to make any impact upon academia before the rise of the Nazi regime but by 1942 appointed Commissar for Epidemics in the Ostland, had his own rationale for demanding racial purification: the lesser races, according to Mrugowsky, were more vulnerable to epidemic diseases and, once infected themselves, spread the infection among the good folk of the superior races. To show this was the case, he deliberately infected large numbers of *Untermenschen* for experimental purposes, while also enthusiastically thinning their number from

---

* After the war, Lorenz so efficiently remade himself that he's now primarily remembered as the animal behaviorist who received a 1973 Nobel Prize and as the author of such charming books as *King Solomon's Ring* (1949) and *Man Meets Dog* (1950). *On Aggression* (1963), risibly in view of Lorenz's past, promoted the idea that aggression, important for survival among animals, could in the human animal be constructively channeled into other things.

the population of the area under his control. He was sentenced to death at Nuremberg as a war criminal.

Meanwhile physical anthropologists such as Bruno Beger and August Hirt were allowed to exploit the concentration camps for "specimens"—skulls, bones and so on. The concentration camp at Natzweiler, near Strasbourg, became almost a specialist site for the gassing of "donors." Their remains, along with those from captured Russian soldiers, were disseminated among various German and Austrian institutions so researchers could explore the differences, often using such pseudosciences as phrenology, between Jewish or Russian anatomy and that of full humans. Of course they found the differences they were looking for.

The Medical Nightmare

Julius Streicher, founder and editor of the virulently antisemitic newspaper *Der Stürmer* and later sentenced to death at Nuremberg as a war criminal, convinced himself cancer was a bacterial infection, and demanded German medical scientists investigate this theory. This was typical of the way that crank theories rose to the top of the broth during the dark age of the Reich. However, in thinking of the Nazi corruption of medical science, our thoughts inevitably turn first to the loathsome activities of one man: Josef Mengele, the Angel of Death.

Much of what we know about Mengele's career at Auschwitz comes from or is confirmed by the memoir *Beyond Humanity: A Forensic Doctor in Auschwitz* (1946), published shortly after the end of the war by Miklos Nyiszli, a Hungarian Jew and physician who, as a prisoner at the death camp, had the misfortune—or perhaps fortune, because it saved him from the gas chamber—to be pressed into service during the summer of 1944 as Mengele's assistant and who consequently witnessed some of the worst savageries of this psychopath. Prisoners at Auschwitz first encountered Mengele on arrival, when he—and other physicians—selected which people were fit enough to be retained for hard labor and which should go straight to the gas chamber. Later, women might meet him again under similar circumstances, during roll calls, when some would be selected by Mengele for "special work detail." Any ideas they might have that this was a fortunate option, in that they were being spared the miseries of hard labor under the whips of the guards, were soon dispelled, because Mengele was choosing them as subjects for his experimentation, such as shock treatment and sterilization. Few survived. If the experiments themselves did not kill, then infection of the untreated wounds did.

Among the more gruesome of Mengele's experiments was to inject dye into children's eyes to see if this would change their eye-color, and the sewing of two small children back-to-back, linking their blood supplies via the veins at the wrist, in an attempt to create Siamese twins artificially. Twins were a particular preoccupation of Mengele; through experiments like this one, as well as through treating genuine biological twins differently and observing the results, he hoped to show that perceived defects and abnormalities were of racial origin—"in the blood"—rather than specific to individuals or families. In another experiment, Mengele casually murdered four pairs of twins by lethal injection purely so their eyes could be sent for study to a scientist at the Kaiser Wilhelm Institute of Anthropology who was writing a paper on the hereditary eye abnormality heterochromatism.

People suffering other congenital abnormalities, most especially dwarfs, were singled out for Mengele's special attention. Most died of his experiments or the aftermath thereof; he seems to have made a particular attempt, however, to keep alive the members of the Lilliput Troupe, seven dwarf Romanian Jews of the Ovitz family who had toured Eastern Europe as an act before being deported to Auschwitz and Mengele's tender mercies.

Mengele was far from the only physician at Auschwitz to use the opportunity of the Reich's complete abolition of normal human mores in order to experiment upon live human guinea pigs, although he seems to have worked with especial cruelty and lack of justification. "Justification" may seem an odd word to use in this context, but many of the experiments performed by the other medical scientists did at least have some purpose in terms of the war effort.

The camp was obviously an ideal place, too, to study the progress of dysentery—it was, predictably, rife among the inmates. Other diseases were brought within the physicians' ambit by infecting luckless victims appropriately and watching their decline, suffering and death. Similarly, poisons were tested. Nyiszli commented:

Dr. Wolff was searching for causes of dysentery. Actually, its causes are not difficult to determine; even the layman knows them. Dysentery is caused by applying the following formula: Take any individual—man, woman, or innocent child—snatch them away from their home, stack them with a hundred others in a sealed box car, in which a bucket of water has first been thoughtfully placed, then pack them off, after they have spent six preliminary weeks in a ghetto, to Auschwitz. There, pile them by the thousands into barracks unfit to serve as stables. For food, give them a ration of moldy bread made from wild chestnuts, a sort of margarine of which the basic ingredient is lignite, thirty grams of sausage made from the flesh of mangy horses, the whole not to exceed 700 calories. To wash this ration down, a half-liter of soup made from nettles and weeds, containing nothing fatty, no flour or salt. In four weeks, dysentery will invariably appear. Then, three or four weeks later, the patient will be "cured," for he will die in spite of any belated treatment he may receive from the camp doctors.

Perhaps even worse than Mengele was the SS lieutenant and physician Siegmund Rascher, a member of the Institute for Applied Research in the Natural Sciences—a grandly titled adjunct to Himmler's Ahnenerbe. For the Luftwaffe, Rascher subjected a series of live prisoners at Dachau to extreme low pressures, reproducing the effects of high altitudes. The mortality rate was about 80 percent. Not content with this, he and two physicians called Holzlöhner and Finke then began, again with live prisoners, to investigate ways of resuscitating airmen who'd crash-landed at sea and spent a while in icy waters before rescue. Of course, one part of the research had to focus on exactly how long the human body could stay submerged in freezing water before death; some of the Russian prisoners lasted for hours. Out of 200 or so subjects, about 40 percent were tested to death. In the experiments whose purpose was revival, the subjects were dragged comatose from the water and swaddled in bedclothes, immersed in a hot bath, or—the preferred experimental technique—put to bed with one or more prostitutes, who would use their oldest-professional skills in an attempt to revive the sufferer. Sometimes they were so successful in this that the revivee was capable of sex with them, an experimental result that Rascher observed eagerly. The results were summarized by the three physicians in the paper "Concerning Experiments in Freezing the Human Organism" (1942):

> If the experimental subject was placed in the water under narcosis, one observed a certain arousing effect. The subject began to groan and made some defensive movements. In a few cases a state of excitation developed. This was especially severe in the cooling of head and neck. But never was a complete cessation of the narcosis observed. The defensive movements ceased after about five minutes. There followed a progressive rigor, which developed especially strongly in the arm musculature . . . The rigor increased with the continuation of the cooling, now and then interrupted by . . . twitchings. With still more marked sinking of the body temperature it suddenly ceased. These cases ended fatally, without any successful results from resuscitation efforts.

Holzlöhner and Finke thereafter discontinued their involvement, on the grounds that all to be learned had been learned, but Rascher carried on gleefully. He proposed moving, though, from Dachau to Auschwitz: at the larger camp there'd be less risk, he pointed out, of the screaming attracting adverse attention.

On October 25, 1946, the Medical Case, the first of the Nuremberg Trials, began. There were 23 defendants, of whom 20 were physicians. Of these, 16 were found guilty of experimenting on human beings without their consent, often involving severe torment, and of the cold-blooded murder of some of their victims. Seven were sentenced to death, including Mrugowsky, the Ahnenerbe chief Wolfram Sievers, and Buchenwald's chief medical officer, Waldemar Hoven. Five of those who'd played a central part in the Nazi medical experimentation were not at the trial. Rascher had been executed in 1945 on Himmler's orders, for having lied to Himmler about, of all things, his wife's fecundity. Ernst Grawitz had committed suicide before the trial. Carl Clauberg was tried in the USSR in 1948, and sentenced to 23 years; in 1955 he was released to West Germany as part of a prisoner exchange, but soon rearrested there in response to public outcry, dying of a heart attack in 1957 before his case came to trial. Horst Schumann disappeared without trace. And Josef Mengele escaped to South America, where he lived until his death in Brazil in a drowning accident in 1979.

As a gruesome footnote to a gruesome tale, we can add that many of the 200 or so physicians who'd experimented in the Nazi death camps were recruited at the end of the war, under the authority of General Dwight Eisenhower, to work for the US military, bringing the fruits of their researches with them.

8

STALINIST RUSSIA

Leave them in peace. We can always shoot them all later.

—Stalin, responding to Beria's concerns about the ideological
soundness of the physics used in the Soviet bomb project

It could all have gone so well. *In Science and Politics* (1973) Jean-Jacques Salomon summed up the advantages and disadvantages of Soviet-style science in a single sentence:

There can be no doubt that scientific activities enjoyed a status and support in the USSR which had no parallel in any other country before the second world war; the scientific system was inseparable from the political system, of which it was both the means and the end.

Today, the level of governmental support for (non-military) science is more or less a measure of how developed a nation is, and it's accepted almost without question that any nation that fails to support science and science education is economically doomed, sooner or later, unless other factors are playing an extremely important part. This was the thrust of a famous US paper commissioned from electrical engineer Vannevar Bush by President Franklin Roosevelt and presented in 1945 by Bush to Roosevelt's successor, Harry Truman. Even so, the only other country aside from the USSR fully to take the idea to heart before World War II was France, where in the 1930s Nobel laureates Irène Joliot-Curie and then Jean Baptiste Perrin were appointed to the government as under-secretaries to promote and advise on science.

Wise governments support science while interfering with its conduct as little as possible—except arguably during wartime. Without US government direction, the Manhattan Project could never have come into existence; but it's important to remember the Manhattan Project is one of the few exceptions to the general rule. It's obviously legitimate for politicians to ensure taxpayer money is being spent wisely, in which exercise it's to be hoped they call upon scientific advisers to help distinguish genuine research from pork. However, when politicians attempt to take control of the scientific endeavor, or to tell scientists what their science *should be*, disaster is inevitable. This was emphatically exemplified in the USSR during the years when Josef Stalin, through the willing medium of Trofim D. Lysenko, forced the Soviet biological sciences to adopt rank pseudoscience.

HEREDITY CAST DOWN

In the USSR during the 1920s there was considerable turmoil in the field of genetics. On the one hand, there were geneticists who subscribed to ideas stemming from the Chevalier de Lamarck that evolution worked through individual living creatures adapting to environmental stimuli during their lifetimes, and passing on these acquired characteristics to their offspring. On the other were those who, aware of the work of Gregor Mendel and his successors, realized the instrument of heredity was the gene, and that it was random mutations in genes that drove evolution. The two schools of thought were not entirely antagonistic: compromise seemed possible, just as, in the early days after Darwin put forward the theory that natural selection was the mechanism of evolution, there still seemed, even to him, room for Lamarckian ideas. The debate in the USSR did not differ in any important respect from that going on anywhere else in Europe, or indeed in the US.

But then a peasant amateur agronomist called Trofim D. Lysenko appeared, and it was the USSR's considerable misfortune that he did so at a time when earlier foolish ideas of V.I. Lenin, that science should be in accordance with such Marxist and Engelian concepts as dialectical materialism, were being hardened into dogma by Stalin. A corollary of Marxism is that human nature can be molded—such motivations as greed and ambition can be sublimated to serve society as a whole* rather than just the individual. The essential madness of Lysenko was in thinking that the basic nature of plants (their genetic coding) could be molded in the same way.

---

* It's the same basic tenet as that of Christianity, and indeed of the famous John F. Kennedy speech: "Ask not what your country can do for you—ask what you can do for your country."

That he should use Marxism as a guide in this and in much else was not so illogical, in a sense, because under the Soviets Marxism was regarded as *itself* a science.

Lysenko's thinking did not spring from a vacuum. At an early stage in his career, posted to the remote Northern Caucasus, he encountered Ivan Vladimirovich Michurin and was much influenced by him. Michurin, from Koslov (later named Michurinsk in his honor), was an aristocrat who tried to get along as best he could under the new, post-Revolutionary regime. In his private orchard—which was in due course turned over to the state—he worked as a fruit-grower and, his fruit-growing ambitions inhibited by the climate (he wanted to grow in the chilly Caucasus fruits that really belonged further south), in due course necessarily became a plant-breeder. For philosophical reasons, Michurin detested the notion of heredity, and this fitted in well with the preconceptions of the Party. Those same preconceptions turned what could have been a tragedy for this "re-born peasant" into a triumph: the Party attempted to help his efforts by sending him some genuine geneticists, who of course discovered promptly that Michurin's hybridization experiments were a shambles; he called them liars, and was lionized as the true, man-of-the-soil peasant scientist resisting elitist theoreticians and their antisocialistic falsehoods. It was a type of "debate" that Michurin's protege, Lysenko, would use frequently during succeeding decades.

It was while working in the Caucasus that Lysenko resurrected the traditional peasant technique of vernalization. By moistening and chilling seeds in the winter, one can accelerate their development after they've been planted in the spring, essentially altering the habit of winter wheat so that it behaves as spring wheat; the technique is of considerable use in high latitudes where the summers are short. Lysenko announced the technique as a great new discovery, and was chagrined when it was smartly pointed out by his peers that it was anything but. His response was that of blusterers anywhere: when a claim is challenged, rather than weigh the merits of the challenge, retaliate by making the claim ever more extravagant. In Lysenko's case, this meant insisting vernalization could transform any strain of wheat such that it eared earlier in the year; this would obviously—obviously to Lysenko, that is—increase the yield over the year as a whole.

Despite the fact that attempts to put his principle into practice had at best mixed results, Lysenko was steadily promoted up the Party's agricultural ladder, and by 1930 he'd reached the Moscow Institute of Genetics. There he continued his vernalization experiments, extending them to cuttings, bulbs and tubers, while claiming unprecedented success. It was in 1935 that he announced his new theory of heredity, which rejected Mendelian genetics entirely and instead claimed environment as the prime controller of how organisms develop. Notably, Lysenko eschewed the customary process of submitting papers on his results to scientific journals, instead preferring the medium of the newspaper interview; ever since, the use of this tactic has been taken as symptomatic of bad science.* This was a distinction lost on Stalin, who was desperate to hear good news—any good news, even if illusory—in the wake of a disastrous series of failed Five-Year Plans. To Stalin, Lysenko and vernalization represented the great hope. In 1940, Lysenko was appointed Director of the Moscow Institute of Genetics, a position of enormous political power that he used to destroy— sometimes literally—all those who challenged his pseudoscientific ideas. He would retain the position until his forced resignation in 1965, although mercifully his power waned after Stalin's death in 1953.

Sometime before 1936, Lysenko met a lawyer called I.I. Prezent who, despite a lack of scientific training, considered himself an expert on Darwinian evolution. It was in conjunction with Prezent that Lysenko formulated his own theory, "vernalization" (in a confusing new use of the term), to replace genetics—or "Mendelism," as Lysenko preferred to call the latter. Exactly what that theory was is rather hard to explain, because Lysenko never put forward any statement of it that makes self-consistent sense. In *False Prophets* (1986), Alexander Kohn offers a sample statement by Lysenko:

> The speeded up development of such plants (sprouted) we explain basically not by the fact that the eyes of the tubers are sprouted before planting, but by the fact that the sprouts are subjected to the influence of certain external conditions, namely the influence of light (a long spring day) and of a temperature of 15–20°C. Under the influence of these external conditions (and that precisely is vernalization), in the potato tuber's eyes, as they start to grow, there occur those quantitative changes which, after the tubers are planted, will lead the plant to more rapid flowering, and hence to more rapid formation of young tubers.

---

* Later Lysenko started and edited the journal *Yarovizatya* ("Vernalization") to use as his mouthpiece, essentially self-publishing.

Kohn is being kind when he comments: "This represents an explanation without substance." It in fact is no explanation at all, just a piece of verbose woffle.

Woffle or not, the study of genetics in the USSR, although it survived in increasingly harassed form until about 1935, was thereafter ruthlessly exterminated—as were many Soviet geneticists who dared to speak out against Lysenko's increasing lock on power or the absurdity of his ideas. The vast majority of those working in the biological sciences in the USSR switched their allegiance, at least on the surface, to Lysenkoism, which may seem a cowardly action in hindsight but must have been perfectly reasonable at the time, bearing in mind the alternatives. (Besides, by this time many establishment scientists were not recognizably scientists at all, being political appointees chosen for their loyalty to the Cause rather than any competence.)

Others were made of sterner stuff. Those holdouts who were lucky were forced from their jobs. Many more were arrested and either were executed or died in prison or concentration camps. It's hard to establish a figure, but probably nearly 100 were "disappeared" in this way.* Of course, these represent just a tiny fraction of the deaths that can be laid at Lysenko's door: it's estimated that, as a consequence of the lethal combination of the collectivization of Soviet agriculture and the imposition of Lysenkoist agricultural practices, something like 10 million people died, either through starvation during famines or through violence in Stalin's extermination campaigns against those farmers who pointed out that the "improvements" weren't improvements at all.

What is truly appalling is that all of this evidence of the complete uselessness of Lysenkoist ideas and techniques must have been perfectly obvious to Stalin and the party bureaucracy, and yet, so blinded were they by the fact that Lysenkoism seemed ideologically sound, they not only permitted him to continue but increased their support. If reality stubbornly refused to conform to Marxist science, then it must be reality that was at fault.

Lysenko did not restrict his pseudoscience to such practical matters as agriculture and breeding, where his destructive influence and his wild theorizing were far more widespread than can be indicated in the brief sketch above. He also, from 1948, began to mount a concerted attack on the theory of evolution by natural selection, perhaps in hopes of toppling Darwin from his pedestal and installing himself there instead. In the natural world, declared Lysenko, there was no such thing as competition for survival between individuals within a species, since such a concept violated Marxist principles of cooperation. Thus the basic idea that new species arise from old ones through chance advantageous mutations within a species giving some individuals advantage over others was patently ridiculous. New forms arose through crossing between one species and a different one . . . and he and his colleagues had done the experiments to prove it, transforming wheat into rye, cabbage into rutabagas, barley into oats, and so on. None of these experiments were ever properly written up, and none proved replicable by other scientists.

Another oddity of the Lysenkoist scheme was that he refused point blank to believe in the existence of hormones. Oddest of all was his insistence that cuckoos were not really a distinct species, laying eggs in the nests of birds of other species; rather, cuckoo chicks emerged from eggs laid by other birds who'd taken to eating caterpillars.

After Stalin's death in 1953, Lysenko continued to enjoy the support of Stalin's successor, Nikita Khruschev, although this support seemed to steadily ebb. Throughout the 1950s, agricultural science was progressing by leaps and bounds in other countries, and it was impossible for the Party bureaucracy to ignore the fact that many of these developments were happening precisely because the foreign scientists were working directly in contradiction to Lysenko's notions. As an example, Lysenko and his adherents declared it impossible to stimulate superovulation in sheep through the use of hormones, yet the Soviet Minister of Agriculture witnessed exactly this being done during a visit to the UK, and, despite Lysenko's protestations, the technique was thereafter introduced (in fact, reintroduced) to the USSR. Even earlier, in 1955 the Soviet Academy of Sciences became sufficiently confident to elect as head of its biology division a real biochemist, V.A. Engelhardt, and he soon set to work bringing genetics back into the fold of Soviet science.

After Khruschev was deposed in 1964, Lysenko's fall from grace accelerated. In 1965, the Ministry of Agriculture and the Soviet Academy of Sciences convened a panel to examine a particular claim of Lysenko's in relation to the breeding of cattle with a high buttermilk yield, and the panel soon concluded that Lysenko and his confreres had been conducting scientific fraud on an almost unimaginably grand scale. Lysenko was forced from his position as Director of the Institute of Genetics of the Soviet Academy of Sciences, a position he'd held since 1940, and of the Lenin Academy of Agricultural Sciences. Over the next couple of years, Lysenkoist ideas were removed from science education, being replaced where appropriate by Mendelian and Darwinian ones. It was perhaps a further quarter-century before Russian genetics caught up with that in the rest of the world, if it yet fully has.

# # #

Lysenko was not the only Russian geneticist to have his dubious scientific hypotheses ludicrously over-promoted under Stalinism. Before him there had been the sad case of Ilya Ivanovich Ivanov. Ivanov earned his reputation under the Tsar as an animal breeder, specifically for his work on the artificial insemination of racehorses. He also studied the use of artificial insemination to produce inter-species hybrids, and succeeded in creating a zeedonk (zebra/donkey), a zubron (European bison/domestic cow) and various other crosses. Even before the Revolution, he was interested in the possibilities of using similar techniques to produce ape/human hybrids; in 1910 he gave a presentation in Graz to the World Congress of Zoologists describing exactly such a scheme.

After the beginning of Stalin's reign in 1924, new opportunities to pursue this research opened up for Ivanov. Exactly who the notion originated with is unclear, but Stalin was entranced by the idea that hybridization between apes and humans held the potential for breeding a new race of uncomplaining super-warriors for the Red Army, unafraid of death and untempered by compassion—a "living war machine," as the Politburo described the ideal. After some false starts, in 1927 Ivanov began trying to impregnate female chimpanzees with donated human sperm at the zoological gardens in Conakry, French Guinea.

He brought back from Conakry to the USSR—in fact, to his new primate center in Sukhumi, Georgia—a collection of apes including two of the three chimps he had attempted to impregnate (the third died en route). Meeting no success at Sukhumi, as in Conakry, with the impregnation of chimp females, his next approach was to attempt to inseminate human females with ape sperm. In 1929 he gained the support of the Society of Materialist Biologists for such an experiment, and five volunteer women were lined up. Before he could begin, however, his only surviving sexually mature ape at Sukhumi died, and he had to start the slow process of importing some replacements. These arrived in the summer of 1930, but by then Stalin was imposing a general crackdown on Soviet biologists in consequence of the repeated failure of various agricultural plans, and Ivanov was not immune. Arrested late in 1930, he was sentenced to five years' exile in Kazakhstan. There he worked for the Kazakh Veterinary–Zoologist Institute until dying of a stroke in March 1932.

# # #

Ivanov was deluded but harmless. Not so Olga Borisovna Lepeshinskaya (not to be confused with the ballerina Olga Lepeshinskaya), who was very much in the Lysenko mold. An unqualified researcher, she was promoted because of her peasant credentials and because the type of "science" she expounded was politically acceptable in the Stalinist context; the fact that she had been a friend of Lenin's didn't hurt. She studied midwifery in St. Petersburg, finishing her course in 1887, then held various positions before in 1941 becoming head of the Department of Live Matter at the Institute of Experimental Biology of the Academy of Medical Sciences, a post she held almost until her death. It was a position of great power, which, like Lysenko, she used ruthlessly against her scientific adversaries.

Lepeshinskaya had an early position under Alexander Gurwitch (see pp. 97–99), whose initial fall from Party favor she engineered in the late 1920s. She persuaded herself that a phenomenon she'd "discovered" offered an alternative explanation for Gurwitch's mitogenetic rays, and so a clash between the two was inevitable.

It was well known that strong ultraviolet radiation killed living cells. Quasi-paradoxically, though—at least according to Lepeshinskaya's claims—weak doses of UV instead stimulated cells. Reasoning further, she hypothesized that, when cells died, they must re-release this energy in the form of UV; and, *mirabile dictu*, this was precisely what her experiments showed. She investigated this wonderful new form of radiation for some while, although most Soviet biologists of the time seem to have given the whole subject a wide berth. A few years later, once she'd risen to power, such disregard was a luxury few enjoyed.

By now she had made a further remarkable breakthrough in the field of microbiology. It was known that cells reproduced by binary fission (splitting into two identical halves), but Lepeshinskaya was lucky enough to observe another means of cell propagation. In this mode, instead of dividing, the cells disintegrated into granules, and these granules could then reconstitute themselves into new cells—all sorts of different cells, not just the original cell-type. Even better, crystals of inorganic substances could, by the application of nucleic acids, be turned into living cells. The fundamental unit of life was thus, she reasoned, not the cell but a sort of disorganized "living matter" that could be manipulated however the skilled researcher (herself) wanted to manipulate it.

It was only a short step from this latter "discovery" to the long-debunked (by Pasteur) notion of spontaneous generation. She ground up microbes and left the dead residue in a flask for a few weeks. By then there were plentiful signs of life in the flask thanks, although Lepeshinskaya did not know this (presumably having consigned Pasteur, along with so many other significant scientists, to the trashcan of history), to airborne microbial spores. Soon the Lepeshinskaya-inspired renaissance of spontaneous-generation ideas permeated much of Soviet biological sciences with, as might be expected, highly placed sycophantic quasi-scientists reporting experiments that not just confirmed her views but also extended them into new areas of lunacy—for example, that you could grind up pearl buttons and grow living tissue from them by injecting them into animals.

In 1950, the Biological Sciences section of the Academy of Sciences held a symposium called "Live Matter and Cell Development." The chairman was the genuinely distinguished biochemist A.I. Oparin,* who must certainly have known that what he was listening to was purest drivel, but who throughout the Stalinist years seems to have been more concerned with survival than anything else. Lepeshinskaya gave the keynote speech, a political tirade much approved by Lysenko. The symposium declared Lepeshinskaya's researches to be indeed of revolutionary importance; she was elected a Member of the Academy of Medical Sciences and awarded the Stalin Prize. Teachers in medical faculties throughout Russia were instructed, under fear of penalty, to refer to her doctrine in every lecture; any favorable reference to the reality accepted in the rest of the world, that cells arose from other cells via division, was outlawed.

Loren Eiseley summed the situation up in *The Immense Journey* (1957):

> Every so often one encounters articles in leading magazines with titles such as "The Spark of Life," "The Secret of Life," "New Hormone Key to Life," or other similar optimistic proclamations. Only yesterday, for example, I discovered in the *New York Times* a headline announcing: "Scientist Predicts Creation of Life in Laboratory." The Moscow-datelined dispatch announced that Academician Olga Lepeshinskaya had predicted that "in the not too distant future, Soviet scientists would create 'life.'" "The time is not far off," warns the formidable Madame Olga, "when we shall be able to obtain the

---

* Oparin's great contribution was his hypothesis concerning the origin of life, put forward in 1924 and summarized in *The Origin of Life on Earth* (1936). His ideas were picked up and developed in the 1950s by the US chemist Harold Urey, and held considerable sway for some decades.

vital substance artificially." She said it with such vigor that I had about the same reaction as I do to announcements about atomic bombs. In fact I half started up to latch the door before an invading tide of Russian protoplasm flowed in upon me.

What finally enabled me to regain my shaken confidence was the recollection that these pronouncements have been going on for well over a century. Just now the Russian scientists show a particular tendency to issue such blasts—committed politically, as they are, to an uncompromising materialism and the boastfulness of very young science.

After Stalin's death, reality crept back into Soviet cytology. Lepeshinskaya and her ideas were never dramatically denounced, as happened with Lysenko; rather, they were allowed to slide into obscurity, as was she.

DIALECTICAL MATERIALISM DEFINES PHYSICS

Two things saved Soviet physics under Stalin from the same sort of devastation that Soviet genetics suffered during that era: Stalin's enormous respect for the physicist Pyotr Kapitsa, and the Soviet atom-bomb project.

Almost from the start, Soviet physics was in a quandary over the new reality portrayed by Einstein's relativity theory, which seemed to fly in the face of dialectical materialism—i.e., "common sense."* Dialectical materialism prescribed, to simplify, that physics should be, well, physical. The universe, its structure and its laws should all be solidly within the realm of the human senses. A concept such as the luminiferous aether, the supposed substrate of the universe, sat comfortably within the limits of dialectical materialism because, even though the aether was proving infernally hard to detect, it was at least theoretically detectable. Yet a basic tenet of the new physics represented by relativity was that the aether didn't exist. Just to confuse matters for the ideologically conscientious Soviets, no less an authority than Engels thoroughly approved of the theory of relativity, which he felt was in full accordance with dialectical materialism since it gave geometry a materialist expression.

Fortunately, there was a way out of the dilemma. Lenin had condemned the notion of Ernst Mach that our understanding is limited by our percep-

---

* In general, Soviet physicists quietly ignored all the political hoohah and with minimal interference carried on their work, which was fully acceptant of relativity and, soon after, quantum mechanics; throughout the Stalinist era, Soviet physics ranked among the world's best.

tions (by what we are able to observe), so the fundamental *causes* of what we observe will always remain a mystery to us. Relativity could be portrayed as a description of *how* things behaved rather than an explanation of the *reason* they behaved that way, and therefore could be denounced in staunch Leninist terms as "Machist"—which a number of Soviet physicists proceeded to do, some of them also echoing the antisemitic denigrations of relativity by the Nazi physicist Philipp Lenard. These denunciations were countered by other physicists who claimed, like Engels, that relativity was in full accordance with dialectical materialism. That this argument in these terms should have been taking place at all is symptomatic of the corruption of science in all totalitarian regimes, a corruption brought about by the irrational compulsion to distort reality until it can be made to fit a preconceived mold.

Of course, matters got much stormier when quantum mechanics exploded onto the scene. A concept such as Heisenberg's uncertainty principle could be seen as a calculated slap in the face to dialectical materialism. Even worse was when Lysenko decided in 1948 that, having conquered Soviet genetics, he should make his mark on Soviet physics, too. His influence, although the intellectual basis of his dogma was puerile, probably did more damage than all that had gone before. According to Lysenko, Soviet physics represented the true path while all foreign physics, with its funny ideas about relativity and quanta, was an insidious threat to the Glorious Revolution, a deliberate attempt by the corrupt, decadent west to undermine the USSR. Soviet physicists were therefore forbidden to maintain contacts with their counterparts in the West, and accordingly—especially since those who disobeyed Lysenko had an unfortunate habit of disappearing—Soviet physics withdrew into its shell. It was fortunate for the nation that Soviet physics was at the time sufficiently strong that, despite the isolation, it remained among the world leaders.

In a celebrated instance, the *Soviet Encyclopedia* published an article that was stoutly anti-relativity, proclaiming that the aether existed and that Newtonian physics was perfectly adequate to explain the universe. The physicists George Gamow, Lev Landau, and M.P. Bronstein wrote facetiously to the encyclopedia's editor saying they accepted the dictum about the aether's existence and sought similar guidance concerning such substances as caloric and phlogiston.* Bronstein and Landau were promptly kicked out of Moscow University, while Gamow, already too notable for

---

* Both long before consigned to science's dustbin of discarded ideas.

such treatment, suffered later when, while lecturing, he mentioned Heisenberg's uncertainty principle; a local political officer stopped the lecture midflow and dismissed the audience, Gamow himself receiving a disciplinary warning. In 1934, Gamow left for the US, where he became Professor of Physics at George Washington University; he spent the rest of his life working in the US. Landau and Bronstein were not so lucky, the outspoken Landau in particular being periodically hounded by the authorities until Stalin's death in 1953; more than once Pyotr Kapitsa, who had Stalin's ear, had to come to Landau's rescue.

Kapitsa, born in Kronstadt, studied first in Petrograd and then in England at Cambridge under Ernest Rutherford; he served as Assistant Director of Magnetic Research at the Cavendish Laboratory, Cambridge, from 1924 to 1932. In 1934, while on holiday in his native Russia, he was abruptly seized by the authorities under Stalin's orders, and installed as Director of the Institute of Physical Problems, Moscow. There was of course much international outrage, to which Stalin was deaf. Notable among those who declined to protest was, oddly, Rutherford, who declared the USSR needed Kapitsa more than the west did, and arranged for some of Kapitsa's equipment to be shipped to him. Stalin seems to have revered Kapitsa as a kind of scientific god, because time and again Kapitsa was able to intercede on behalf of one scientist or another who was suffering threats for straying publicly from the path prescribed by dialectical materialism. In return, Kapitsa was a good servant of the USSR, not only as the living flagship of Soviet physics but also as a valuable technological innovator.

It was his favored status with Stalin that almost certainly saved his life when, in the wake of the US destruction of Hiroshima and Nagasaki, the project was initiated to produce a Soviet atomic bomb. Kapitsa seems to have been reluctant from the outset, but he became even more so when the man put in overall charge of the project proved to be none other than Lavrenti Beria, the dreaded boss of the NKVD (which became the KGB in 1954). The choice could hardly have been more idiotic, and Kapitsa told Stalin so, requesting Beria be replaced. Maddened by the insubordination, Beria demanded Kapitsa should suffer the same fate as others who'd defied him. Instead, Stalin put Kapitsa under house arrest, where he remained until Stalin's death and Beria's execution in 1953.

---

* Also, much later, in 1954, a leading figure in the creation of the world's first industrial nuclear power plant.

The bomb project itself could have been nullified by the Lysenko infection. One of those particularly influenced by Lysenko's ideas concerning bourgeois western physics was Sergei Kaftanov, Stalin's Minister for Advanced Education. In March 1949, he planned to summon a committee with the intention of outlawing relativity and quantum mechanics from Soviet physics forever. For once, Beria had the good sense to take a second opinion on this, from the project's director and head of the Soviet Atomic Energy Institute, Igor Kurchatov,* who told him bluntly that, without relativity and quantum mechanics, Beria's pet baby wouldn't explode. After consultation with Stalin (who responded with the remark cited at the head of this chapter), Beria canceled Kaftanov's committee and the atomic scientists were allowed to proceed. Just a few months later, in August 1949, the USSR successfully detonated its first atomic bomb.

The Lunatics Conquer the Asylum

In Russia the use of false psychiatric diagnoses of political dissidents, and their consequent incarceration in mental institutions as a means of silencing them (out of sight, out of mind, as it were) that avoided the potential political embarrassment of imprisoning, executing or "disappearing" them, dated from long before the Stalin regime and indeed the Revolution, and probably hit its heights after Stalin's death, during the reign of Nikita Khruschev; this may not reflect so badly on Khruschev as at first appears, but simply be a manifestation of the fact that Stalin was less concerned about "political embarrassment" than his successors. It is likely this false hospitalization continues even today.

The earliest known Russian dissident to suffer thus was Pyotr Chaadayev, classified as deranged by Nicholas I, who thereby simultaneously nullified Chaadayev's previously expressed dissidence and had the man stilled from expressing any more of it. The practice seems to have been repeated on occasion both before and after the Revolution: initially the Bolsheviks appear to have been no worse in this respect than the Tsars before them.

The strategy was certainly used against Maria Spiridonova, leader of the rival Left–Socialist Revolutionary Party and thus seen as a threat; she had been savagely tortured and imprisoned in Siberia in 1905, for the assassination of a brutal police chief, during the unsuccessful 1905 Russian revolution, then released in 1917. After her release, she became more popular than ever. Following the collapse of the alliance between the Bolsheviks and the Left–Socialists in 1918, she was imprisoned and tortured twice more in rapid succession—*plus ça change*—all of which merely increased the number and fervor of her followers. A Moscow Revolutionary Tribunal of 1919 saw a way out of the impasse, declaring that she should be put in a sanatorium for a year to undergo mental and physical therapy. In the event, she avoided this.*

In 1922, a similar technique was used against the dissident Angelica Balabanov, who had served briefly as Secretary of the Comintern (the Communist International). A Ukrainian by birth and a resident of Italy before coming to Russia at the time of the Revolution, she was able to escape back to Italy, where in due course she helped organize resistance to Benito Mussolini.

These were isolated incidents, however. After the ascent of Stalin, the practice became more widespread, with one psychiatric hospital at Kazan being run by the NKVD and seemingly devoted entirely to patients—often genuinely mentally ill—whose mental illness "took a political turn." Mentally sound people incarcerated at Kazan purely on grounds of political dissidence had often arrived there not through the repression of the state, but through the subversive mercy of psychiatrists, who knew what the alternative to a diagnosis of derangement was likely to be. Occasionally, even the police quietly colluded in this. The poet Naum Korzhavin, although eventually sent to Siberia, described how the assessing psychiatrists at the Serbsky Institute for Forensic Psychiatry did their utmost to find any reason for declaring him mentally ill; he should have cooperated with them, as he later discovered.

Not all incarcerations were so benignly intended, though. Far from it. And the practice seems to have accelerated toward the end of Stalin's long tyranny. One victim of it was the loyal but honest communist Sergei Pisarev, who spent a while first at the Serbsky Institute and then in the Leningrad Prison Psychiatric Hospital during 1953–55. On his release, emboldened by the fact that Stalin was gone and the less brutal Khruschev installed in his place, Pisarev went public about the scandal of habitual misdiagnosis as a means of suppressing dissenters. The Central Committee ordered a commission of inquiry, but nothing was ever done in response to the commission's report. Among subsequent victims were the writer Valeriy Tarsis, whose novel *Ward 7* (translated in 1965) fictionalized his experiences,

---

* But hers was no happy-ending story. In 1937 she was sentenced to 25 years in prison. In 1941 she was one of a group of some 150 prisoners massacred by the NKVD in the Medvedevsky Forest.

and the mathematician Alexandr Volpin, who was forcibly hospitalized no fewer than five times between 1949 and 1968. Pisarev tried again in 1970 to get something done about his nation's shame, this time taking his protest to the Academy of Medical Sciences; again nothing was done.

In 1965, the west became aware of the Soviet abuse of psychiatry, first through the UK publication of Tarsis's novel and then through the case of the interpreter Evgeny Belov, which was taken up by Amnesty International.* In the USSR itself in 1968, amid a more general rising of sentiment concerning the regime's human-rights violations, the abuse of psychiatry was particularized, and the Action Group for the Defense of Human Rights in the USSR was established. Even this dragging of the malpractice into the glare of publicity seems to have done nothing to decelerate the abuse: of the Action Group's 15 founding members, four were in due course themselves forcibly hospitalized.

Fortunate dissidents were sent to ordinary psychiatric hospitals, run by the Ministry of Health, where their care and access by friends and family were much as for genuine mental patients.

By contrast, the Ministry of Internal Affairs ran the dreaded "special" psychiatric hospitals, which were in many ways indistinguishable from torture camps; people who disappeared into these were effectively cut off entirely from the outside world. Their companions in incarceration included many of the USSR's most violently and dangerously insane: rapists, murderers, arsonists. The orderlies were criminals working to fulfill their prison sentences; no one would reprove them for treating those under their care brutally, and they frequently did. The hospital directorship was a civil appointment, usually of a reliable Party hack. The medical staff might be genuine enough, but they had little chance of ameliorating conditions for their "patients" because of the prison-like structure of the establishments; they had no means of control over the other personnel, no way of stopping the torments. Copious drugs were administered primarily for the sake of their unpleasant side effects—that is, as just another tool of torture. Particularly notorious was the drug sulfazin, which had been abandoned as therapy in the 1930s when it was found to have no therapeutic value that could possibly justify the intense and enduring pain of its administration.

---

* Belov's case came to light purely by chance. In 1964 a group of UK students had visited the USSR and become friendly with their interpreter. They made another visit in 1965, and not unnaturally asked after their old pal Evgeny. On reaching home they told the truth to the world, campaigning for his release.

The only escape from these hells, assuming one survived them—hopefully with mind intact—was to recant, firstly by "admitting" one was genuinely ill and then by denouncing one's earlier dissidence and claiming it to have been the product of a deranged mind.

While indisputably there were worse medical horrors perpetrated in the Nazi concentration camps, the nightmare of the Soviet abuse of psychiatry was that it perverted medical science itself to make it a weapon of repression and torture.

9

AMERICA IN THE 21ST CENTURY

That's not the way the world really works anymore. We're an empire now, and when we act, we create our own reality.

—Unnamed senior Bush administration official, cited by Ron Suskind in "Without a Doubt," *New York Times*, October 17, 2004

This is infinity here. It could be infinity. We don't really know. But it could be. It has to be something—but it could be infinity, right?

—Donald J. Trump educates Buzz Aldrin and others on the nature of outer space, June 30, 2017

In short, it is prudent to regard the committed and politically ambitious parts of the anti-science phenomenon as a reminder of the Beast that slumbers below. When it awakens, as it has again and again over the past few centuries, and as it undoubtedly will again some day, it will make its true power known.

—Gerald Holton, *Science and Anti-Science* (1993)

Holton's words above have an eerie ring for US readers today. Since he wrote them, there have been periods during which that "anti-science phenomenon" has been embedded deep within the highest offices of government, scientific findings that conflict with the ideological or commercial preconceptions of the administration of the day being either rejected or suppressed—or even falsified.

What follows can be at best a sketch of the problem; there has, quite simply, been far too much corruption of science by recent US administrations for a single chapter to cover it all, or even touch upon it all—and, to state the obvious, the most recent installment of the saga is still under way. Hopefully the examples given here will serve to indicate the general, devastating pattern—snapshots, as it were, of an immense, ongoing catastrophe. Readers seeking a fuller account are referred to such books as Seth Shulman's *Undermining Science* (2006) and Shawn Otto's *The War on Science* (2016).

The first edition of this book was written during the heyday of the George W. Bush administration. It quite simply did not occur to me at the time that the situation *vis-à-vis* public science could actually get worse. And for eight years after the departure of George W. Bush from the White House, during which US federal governmental attitudes toward public science were far from perfect but at least comparable with those pertaining in other declared democracies (and superior, much of the time, to those in such countries as Canada, Australia and, very obviously, Russia), it seemed I was right. The advent of the Trump administration in early 2017 dashed the scales—and the rose-tinted spectacles— from my eyes. Here was an assault on science, on the scientific method, and indeed on our perceptions of reality itself, beyond anything I could ever have conceived in the planet's supposedly most advanced nation.

The main battlefield under both the Bush II and the Trump administrations has been climate change, victim of a virus of science denial that has come to infect almost the entirety of the modern Republican establishment. It's very hard to see the logical difference between the rejection in today's GOP of scientific realities and that of the Stalinist regime in the USSR. Just as with Stalin's insistence, to the cost of millions of lives, that, whenever it came to a contest between reality and Lysenko's ideologically sound drivel, then reality must be at fault, so the modern GOP rejects science in favor of ideology in cases where reality is inconvenient to that ideology. This ideology of theirs is not conservative—rational conservatives are as appalled as anyone else by this behavior—but rather some mutant child of toddler narcissism and intellectual sloth.

But let's backtrack a little to look first at theBush II administration's corruption of the public understanding of sex.

Nasty! Dirty! Horrid!

Sex education was one of many science-related areas bedeviled by the Bush administration's practice of appointing people to scientific posts for which they were manifestly incompetent. In his first term, Bush appointed Claude Allen as his deputy secretary of Health and Human Services; Allen's primary qualifications appeared to be that he opposed abortion and promoted sexual abstinence and Christian homeschooling. Kay Coles James was put in charge of the Office of Personnel; she had previously been a vice-president of the Family Research Council, a Christian right "traditional values" organization. Esther Kaplan, in *With God on Their Side* (2004), notes: "For a window into her worldview, one can look to her 1995 book *Transforming America from the Inside Out*, in which James likened gay people to alcoholics, adulterers and drug addicts." Another alumnus of the Family Research Council, Senator Thomas Coburn, was appointed joint chair of the presidential advisory council on HIV/AIDS. Robert Schlesinger, writing in *Salon* on September 13, 2004, quoted Coburn as follows:

> The gay community has infiltrated the very centers of power in every area across this country, and they wield extreme power . . . That agenda is the greatest threat to our freedom that we face today. Why do you think we see the rationalization for abortion and multiple sexual partners? That's a gay agenda.

Leaving aside the mystery of why gays should be so keen to promote abortion, Coburn, with his frequent anti-gay tirades, would seem an odd choice to offer objective advice on AIDS. (His views on abortion are odd, too: he has advocated the death penalty for doctors who've performed abortions despite himself having, as an obstetrician, performed abortions.)

But perhaps the most controversial of this wave of appointments was that of W. David Hager in 2002 to the Advisory Committee for Reproductive Health Drugs of the FDA; according to *Time* magazine for October 5, 2002, "Hager was chosen for the post by FDA senior associate commissioner Linda Arey Skladany, a former drug-industry lobbyist with long-standing ties to the Bush family." He is an obstetrician and gynecologist with impressive conservative-Christian credentials—Focus on the Family, Christian Medical and Dental Society, etc.—is a tireless advocate of "traditional family values," and is author of books like *As Jesus Cared for Women: Restoring Women Then and Now* (1998).

Some of his medical views are idiosyncratic. In one of his books he recommends Bible study as a cure for menstrual cramps, and his patients have claimed (he has denied this) that he refuses contraceptive prescriptions to the unwed. According to his own account, he was approached by the White House in 2001 and invited to be a possible candidate as Surgeon General. Later this was amended, and he was appointed to two advisory boards. Soon afterward, he was asked to resign from those and instead join the FDA's reproductive drugs panel as its chairman; he quoted the officials as saying, "[T]here are some issues coming up we feel are very critical, and we want you to be on that advisory board."*

When he was nominated as chairman of the FDA's reproductive drugs panel in late 2002, there was immediate outcry. In order to dodge the storm, the FDA used a frequent government trick: timing. It announced his appointment as a panel member (no longer as chairman) on Christmas Eve, when not only was Congress out of session, and therefore unable to debate the issue, but the news was obscured by all the Christmas brouhaha.

As for the opposition to his appointment, Hager has given this account:

> [T]here is a war going on in this country, and I'm not speaking about the war in Iraq. It's a war being waged against Christians, particularly evangelical Christians. It wasn't my scientific record that came under scrutiny, it was my faith . . . By making myself available, God has used me to stand in the breach.

There followed, as night follows day, the tortuous Plan B saga.

In December 2003, two scientific advisory committees at the FDA voted unanimously that the emergency retroactive contraceptive pill Plan B, approved as a prescription drug in 1999, was safe for over-the-counter sale, and by a majority of 23 to 4 that it should be made available for such sale. Its safety was attested to by about 70 scientific organizations, including the American Medical Association and the American Academy of Pediatrics, as well as the governmental science watchdogs in the 33 countries in which it was already available over the counter.

---

* Just a few months earlier Hager had played an important role in the submission by the anti-abortion group Concerned Women for America of a "citizens' petition" which demanded that marketing and distribution of the "abortion pill" RU–486 be halted.

Nevertheless, on May 6, 2004, the Acting Director for the FDA's Center for Drug Evaluation and Research, Rear-Admiral Steven Galson, decreed a continued ban on its open sale. His stated grounds were that Plan B's manufacturers, Barr Pharmaceuticals, had failed to supply sufficient documentation showing the drug was suitable for use by teenagers under the age of 16—this despite the fact that one of the two FDA scientific committees had extensively discussed exactly this point and decided there was no cause for concern; further, the committees had stressed that, in the case of very young girls, there are significant dangers involved in pregnancy, so its prevention was especially important. Of Galson's veto Dr. James Trussell, an FDA committee member and Director of Princeton University's Office of Population Research, said: "The objection . . . is nothing more than a made-up reason intended to sound plausible. From a scientific standpoint, it is complete and utter nonsense." He added: "Unfortunately, for the first time in history, the FDA is not acting as an independent agency but rather as a tool of the White House." Susan Wood, head of the Office of Women's Health, later resigned in protest, complaining that ideology had been allowed to trump science. John Jenkins, Director of the FDA's Office of New Drugs, commented: "The agency has not [previously] distinguished the safety and efficacy of Plan B and other forms of hormonal contraception among different ages of women of childbearing potential, and I am not aware of any compelling scientific reason for such a distinction in this case."

Jenkins was obviously searching the wrong area of human activity for his "compelling reason." To rightwing Christian fundamentalists and the anti-abortion lobby, the idea of a morning-after pill was anathema, to the former because it would "obviously" unleash a torrent of promiscuity upon the land, to the latter because retroactive contraception is tantamount to abortion—the belief being that a fertilized ovum is a human being. (Despite some misleading news reports, Plan B is not in fact an abortifacient.)

According to Hager in a sermon he delivered in Wilmore, Kentucky, in October 2004, the person largely responsible for the scientific judgment being negated was none other than himself. He claimed that, soon after the 23–4 vote, he was asked, as one of the four dissenters, to write a "minority report" outlining the reasons for rejecting the majority decision. In his own words:

Now the opinion I wrote was not from an evangelical Christian perspective . . . I argued it from a scientific perspective, and God took that information, and he used it through this minority report to influence the decision. Once again, what Satan meant for evil, God turned into good.

An important question is who asked him for the minority report? The FDA does not commonly deal in such things, generally assuming that, if an overwhelming majority on a scientific advisory panel says something, it knows what it's talking about.

Initially Hager told reporters and others that the request had come from within the FDA. Evidently realizing this could lead to a major political scandal, he soon backtracked, saying instead it had come from a mysterious "someone" outside the agency, and denying he'd ever stated otherwise. (Unfortunately for this latter claim, at least one journalist had kept Hager's e-mail.) The FDA likewise denied anyone within it had issued the request, claiming Hager's memo had been just a "private citizen letter."

Details that emerged in the course of Hager's messy divorce, regarding his personal sexual behavior, finally tipped the scales of public opinion. As was forcefully pointed out by many, such behavior was not a pedigree to recommend anyone for a position of power on women's health issues—nor was Hager's frequently expressed view that women should regard men as the disciples did Jesus.

Finally, in August 2006, after nearly three years, the FDA caved in to public outrage and impartial science, permitting the over-the-counter sale of Plan B.

Once bitten, twice not shy, however, for the Bush administration. In November 2006 it appointed non-board-certified obstetrician/gynecologist Eric Keroack as Deputy Assistant Secretary for Population Affairs at the Department of Health and Human Services—i.e., as head of, among other things, the DHHS's family planning section, including the Title X program "designed to provide access to contraceptive supplies and information to all who want and need them, with priority given to low-income persons." His previous appointments included serving as a medical adviser to the Abstinence Clearinghouse and as medical director of the Massachusetts pregnancy-counseling network A Woman's Concern. Advancing much pseudoscience,* this network opposes the provision of contracep-

---

* For example, the myth that abortion increases the risk of breast cancer and the myth that sex education causes teen promiscuity. One of its pieces of advice to young people is that, on the (established) grounds that condoms reduce the risk of HIV transmission by 85 percent, "you have a 15 percent chance of contracting [HIV] while using a condom" without qualifications such as that this depends on whether or not your partner has HIV.

tion *even to married women*—on the grounds that contraception demeans women (huh?). Keroack also produced the pseudoscientific theory, unbacked by evidence, that women who over time have sex with a succession of different partners end up suffering an alteration in brain chemistry, through suppression of the hormone oxytocin, that makes it difficult for them thereafter to form long-term relationships. Oxytocin does indeed appear to have some effect on one's level of sociability, but the relationship is by no means a direct one: sometimes higher oxytocin levels make people grumpier, sometimes more amiable. There's no known relationship between oxytocin levels and marital/partnership happiness or number of sexual partners. After all, how could the oxytocin tell the marital status of a partner *vis-à-vis* the woman, or indeed one partner from another?

Ideologically opposed to contraception and a touter of pseudoscience and inaccurate medical information, Keroack was an odd pick as controller of federal funding in a scientific field whose focus is family planning—unless the hidden aim was to run Title X and its affiliated programs into the ground.

#

In October 2002, sharp-eyed observers discovered that the websites of government health departments had been silently edited to be more in accord with the administration's ideology. For example, a statement to the effect that education on condom use did not increase youthful sexual activity and, far from encouraging an earlier onset of it, actually delayed it, vanished from the CDC site's page concerned with the spread of AIDS and other sexually transmitted diseases. Furthermore, new information had been added exaggerating the risks of transferring STDs despite condom use:

> The surest way to avoid transmission of sexually transmitted diseases is to abstain from sexual intercourse, or to be in a long-term mutually monogamous relationship with a partner who has been tested and you know is uninfected. For persons whose sexual behaviors place them at risk for STDs, correct and consistent use of the male latex condom can reduce the risk of STD transmission. However, no protective method is 100 percent effective, and condom use cannot guarantee absolute protection against any STD.

There's no straightforward falsehood here, but the implication is distinctly different from that of a note found tucked away elsewhere on the site in a discussion of HIV: "The studies found that even with repeated sexual contact, 98–100 percent of those people who used latex condoms correctly and consistently did not become infected."

Likewise, the National Cancer Institute's site had once stressed that there was no association between abortion and an increased risk of breast cancer in later life; the newly edited version claimed instead (falsely) that the science on the subject was "inconclusive." The results of an enormous Danish study, published in the *New England Journal of Medicine* in 1997, which found no relation at all between abortion and breast cancer, had been referred to approvingly on the earlier site; now the study had been "disappeared."

# # #

On his first Monday morning in office, January 23, 2017, the brand-new President Donald Trump issued an executive order confirming the Mexico City policy in a new and especially virulent variation.

The Mexico City policy dates in its original form from 1984, and got its name from the fact that it was signed by Ronald Reagan at the International Conference of Population in Mexico City.* The original version of the policy decreed that non-governmental organizations involved in family planning would be denied US funding if they either provided or actively promoted abortion in other countries except in specific circumstances such as incest, rape, and the mother's life being endangered by the pregnancy. This was not a matter of using US funding for such purposes— it's already against the law to provide federal funds for abortions. The prohibition was against the NGOs using their own funds for such purposes.

The Trump version of the policy expanded its scope to include not just family planning NGOs but any foreign health organization. Those organizations must promise not to offer women consultation about abortion, or lobby in favor of relaxing anti-abortion laws. As Sarah Boseley summarized, "Anyone working to fight HIV, get vaccines or vitamins to children, or prevent Zika or malaria is facing a stark and unprecedented choice: sign, or lose all funding from the biggest aid donor in the world."[103] Among the organizations threatened are not just family planning organizations all over the world, but also some of the major international players, such as

---

* Bill Clinton rescinded it on taking office in January 1993, George W. Bush reinstated it in January 2001, and Barack Obama rescinded it in January 2009.

Marie Stopes International and International Planned Parenthood. Even Save the Children is likely to be affected.

Just to make matters worse, Trump announced (April 2017) his determination to cut off the US contribution to the UN Population Fund, an important humanitarian organization that, *inter alia*, makes sure women in the most deprived possible situations, such as refugee camps, have access to contraception. Should Trump's planned cuts to the US foreign aid budget go through, all US funding for overseas family planning organizations will dry up.

What is the effect of a gag order such as the Mexico City policy? As has been proven time and again, it increases the number of abortions. Lacking proper information on family planning, more women become pregnant and more of them seek terminations. A big difference is that they're far more likely to seek the aid, not of a qualified doctor, but of a back-street abortionist. And that means that more women die unnecessarily.

The affected NGOs have not been idle since then, mounting successful campaigns to replace US funding, at least in part, by contributions from other wealthy nations. (The Bill and Melinda Gates Foundation has also been a major donor.) While this might seem fiscally prudent for the US—not that this is the main purpose of the Mexico City policy—in reality it means that, for a relatively small saving, the US is losing a considerable amount of foreign influence and goodwill.

Since Trump himself has a patchy record on abortion—sometimes he's been in favor, sometimes against—it's worth looking at those around him who might have induced him to bring into law such a draconian, and powerfully misguided, measure. While a number of those who had his ear in the White House at the time, such as advisor Steve Bannon and spokeswoman Kellyanne Conway, were conservative, anti-choice Catholics, the finger much more likely points to Vice-President Mike Pence, whose record on sex education, family planning, and abortion (not to mention gay civil liberties) as governor of Indiana was appalling: even in cases of profound fetal abnormality, abortion was to be banned, according to a law he introduced in March 2016, while women seeking abortions under any circumstances would be forced to undergo an intrusive ultrasound examination—technological rape, as it has been described.

It was also Pence who had to be pulled in to cast the deciding vote on a measure by Congress that severely affects the operations of Planned Parenthood and various other groups involved in promoting family planning, particularly among the poor.

One of the last actions of the Obama administration, in mid-December 2016, was to pass a rule prohibiting states from withholding federal funds earmarked for family planning from Planned Parenthood and other health centers that provide abortions alongside their family planning facilities. This new rule was to have taken effect on January 18, 2017, just two days before Trump's inauguration. However, there's an existing law called the 1996 Congressional Review Act (CRA) that permits Congress, with the permission of the sitting president, to overturn federal regulations issued by government agencies. On March 30, 2017, therefore, Congress passed a measure to rescind Obama's order and empower states to withhold the funds from the relevant healthcare organizations, and on April 13 Trump signed this into law.

It's interesting to note, however, that he felt it necessary to do so behind closed doors, presumably in the hope that as few as possible members of the public would learn about this regressive move. The witnesses to the signing were limited to his advisor Kellyanne Conway, who has a history of ultraconservative anti-choice activism, Representative Diane Black (R–TN), Marjorie Dannenfelser, president of the anti-abortion Susan B. Anthony List, and Penny Young Nance, president of the rightwing Christian Concerned Women for America, plus Seema Verma, the Trump-appointed administrator of the Centers for Medicare and Medicaid Services. While these individuals were naturally triumphalist about the "achievement" afterward, they said nothing about the fate of women stuck in states saddled with neanderthal administrations, who would soon discover that accessing contraception had just gotten a whole lot more difficult. As always, a restrictive ideology makes for bad healthcare and rotten outcomes for real human beings, often the most vulnerable.

Of course, this is merely a recent manifestation of the Christian right's long-running war on Planned Parenthood, a war that, despite extensive backing by GOP politicians in Congress, has achieved almost nothing in terms of its stated aim: to reduce abortion levels. Although Planned Parenthood is the nation's single largest provider of abortions, it is also, through the family planning advice and resources it offers, the nation's single largest *preventer* of abortions. Crippling Planned Parenthood's efforts, therefore, has the effect of driving *up* the numbers of abortions. The big difference—and here's the familiar tune—is that defunding puts more women at the tender mercies of back-street abortionists, with consequent higher fatality rates.

Impeding Planned Parenthood also drives up the rates of sexually transmitted diseases, including but not limited to HIV/AIDS. To judge by the fairly frequent revelations about the extracurricular activities of our political representatives, you'd have thought they might be acutely concerned about this, but apparently not. Perhaps their STDs are amply covered under the free healthcare they enjoy at the taxpayer's expense.

Over the years, Planned Parenthood has been the victim of a number of rightwing video stings, all seeking to expose illegal practices by the organization's staffers—such as performing illegal late-term abortions (falsely and misleadingly termed "partial-birth abortions" in the emotive lingo of the anti-choice movement), condoning underage sex or prostitution, or illicitly selling fetuses for stem-cell research. What such efforts have come up with is almost precisely zilch; a few staff have been fired or retrained, but no criminal charges have ever resulted.

A particular source of outrage among rightwingers in Congress was a series of videos made by the anti-abortion outfit Center for Medical Progress in 2015, and various bills were launched to defund Planned Parenthood. None became law, and it was soon shown that the videos had been "edited." Although charges were later dropped, Center for Medical Progress's principals faced criminal prosecution in both Texas (2016) and California (2017). The videos are believed to have been the inspiration behind Robert Dear's murderous attack in November 2016 on the Planned Parenthood clinic in Colorado Springs, during which he shot six civilians, two fatally, and six cops, one fatally.

INot just abortion but basic family planning has drawn the ire of the Trump administration—or the Trump/Pence administration, as one is tempted to call it in this context, since it's plain Trump himself has no particular animus toward family planning but cares only for the adulation of his core supporters . . . many of whom may very well be incensed when they discover what the administration has done for contraception, in the same way that, despite a blitz of propaganda to the contrary from the likes of Fox News and the Sinclair Broadcast Group, many of them realized that the various GOP attempts to repeal Obamacare were not in their interests.

On October 6, 2017, the administration issued a measure designed to push back against the Obamacare regulation that required employer-provided health insurance packages to offer free contraceptive coverage. The new rule hugely broadens the categories of organizations that can refuse to supply such services in their insurance packages, from religious groups to, essentially, just about any business keen to save a buck or two at the expense of its employees' welfare.

The good news is that the rule change is almost certainly unconstitutional. Other preventative-medicine elements of such insurance policies that affect both men and women are unaffected, so cracking down on contraception, which primarily affects women, could be seen as gender discrimination. Within hours of the rule change's announcement, Maura Healey, Attorney General of Massachusetts, had a suit in court along these lines.

The bad news, should the legal challenge reach the Supreme Court, is Neil Gorsuch.

MATTERS OF WAR

The falsehoods promulgated to the public, to Congress, and even to the UN by the Bush administration in the leadup to the US invasion of Iraq in 2003 have been the subject of several books. Some of the false claims were of scientific or technological interest, perhaps most notoriously the case of the 100,000 high-strength aluminum tubes which the Hussein regime had apparently attempted to acquire on the international market. Bush told the UN in September 2002 and Colin Powell repeated to the UN in February 2003 that these could only be intended for use in gas centrifuges for uranium enrichment as a preliminary for the production of nuclear weapons. It seems the tight tolerances required for the tubes' dimensions and finish persuaded at least some in the CIA that they could have no other purpose.

Technical experts from the Department of Energy almost immediately dismissed the claim: the tubes were of the wrong dimensions for use in gas centrifuges. They were, however, identical to tubes Iraq had bought in the past for use in medium-range rockets. The State Department agreed. These scientific assessors also pointed out that, had the tubes been intended for the purpose stated by the administration, there would be evidence too of the Hussein regime attempting to buy countless other essential and quite specific components for uranium enrichment.

Bush himself repeated the aluminum-tubes falsehood in his 2003 State of the Union Address, despite the fact that *just the previous day* the International Atomic Energy Agency (IAEA) had told the UN Security Council, which includes the US, that there was no evidence whatsoever of Iraq pursuing a nuclear-weapons program or of forbidden nuclear activities taking place at the country's relevant sites, and that the IAEA's analysis of the aluminum tubes had shown they would be useless for centrifuges.

There were other science-related dishonesties involved in the propaganda buildup to the invasion, most notoriously the continuing claim, long after it had been shown to be bogus, that Iraq had sought to buy yellowcake uranium from Niger. Also of interest were the insinuations about the as-yet-unsolved case of the anthrax parcels sent to various media and Democratic Party figures during the Fall of 2001. Could Saddam Hussein have been behind these attacks? Various members of the administration acknowledged this possibility, even though they knew that all the evidence pointed to a domestic perpetrator. The false insinuations about the anthrax packages jigsawed in nicely with the claims concerning Hussein's stupendous armory of biological and chemical weapons, all of which ignored the fact that those weapons had been amassed prior to the first Gulf War, so most if not all would have degraded to near-uselessness by now.

The administration's most Orwellian act of all, with the mainstream media as accomplice, was effectively to obliterate from US public awareness that the UN weapons inspectors led by Hans Blix, who were actually on the ground in Iraq, were clamoring that there were no signs of any active program there to create weapons of mass destruction. After the attack, Bush himself rewrote history on this issue, claiming that one motive for the invasion was that the Iraqi regime had barred the UN inspectors from the country.

THE AIR THAT WE BREATHE, THE WATER WE DRINK

Under Bush [as State Governor], Texas had the highest volume of air pollution, with the highest ozone levels of any state—while ranking forty-sixth in spending on environmental problems. Moreover, after 1994 Texas was the nation's leading source of greenhouse gases, accounting for 14 percent of the annual US total while boasting only 7 percent of the US population. Under Bush, Texas's oil refineries became the nation's dirtiest, with the highest level of pollution per barrel of oil processed. And because all such industrial effluvia are concentrated in or near the state's poorest neighborhoods, Bush's Texas also led the nation in the number of Title VI civil rights complaints against a state environmental agency—in this case, the Texas Natural Resource Conservation Commission (TNRCC), which Governor Bush staffed brazenly with staunch anti-environmentalists like Ralph Marquez, a veteran of Monsanto Chemical, and Barry McBee, of the pro-business law firm Thompson & Knight.

—Mark Crispin Miller, *The Bush Dyslexicon* (2001)

The name of the Bush administration's Clear Skies Initiative of 2003 deployed the technique of Newspeak. The aim of the Clear Skies Initiative was to replace the Clean Air Act—introduced in 1963 and bolstered several times, including in 1990 by the George H.W. Bush Administration—and to loosen controls on industrial polluters, thereby inevitably increasing air pollution and the consequent sickness and death rates.

It's worth summarizing the achievements of the Clean Air Act. Even before the Act's 1990 strengthening, the EPA estimated it had prevented 205,000 premature deaths within the continental US. Further, according to the EPA, "millions" of US citizens had been spared illnesses ranging up to "heart disease, chronic bronchitis, asthma attacks, and other serious respiratory problems. In addition, the lack of the Clean Air Act controls on the use of leaded gasoline [see page 251] would have resulted in major increases in child IQ loss and adult hypertension, heart disease, and stroke."

In 1999, the EPA did a study of the effects of the 1990 bolstering of the Act and estimated that, in the single year 2010, 23,000 premature deaths would be prevented because of it, not to mention 67,000 incidents of chronic and acute bronchitis and a whopping 1.7 million asthma attacks. For that one year, 4.1 million work days would be saved—an enormous contribution to the US economy.

The Bush II Administration disliked the Clean Air Act for ideological reasons: the thrust of almost all of the administration's environmental policies was to reduce the legal obligations of business at the expense, if necessary, of ordinary citizens and the public purse. It's a way, not an especially subtle one, of transferring wealth from the poor and middle class to the rich, one that exacts a heavy toll in human life and welfare.

A particular concern of that administration related to the regulations governing neurotoxic mercury emissions by coal-fired power plants. Accordingly, the White House watched carefully as the EPA compiled an advisory report, scheduled for release in 2002, which examined the effect of environmental factors on children's health. That report concluded 8 percent of US women aged 16–49 (i.e., of potential child-bearing age) had blood mercury levels that significantly increased the likelihood of any children they bore having deficient motor skills and reduced intelligence.

And this was *before* the introduction of the Clear Skies Initiative, which would further increase atmospheric mercury levels. Accordingly, in May 2002, before the report's release, it was taken by the White House Office of Management and Budget (OMB) and the Office of Science and Technology Policy (OSTP) for "review." Nine months later, in February 2003, when it still hadn't emerged from that "review" process, it was leaked to the *Wall Street Journal* (a newspaper generally very much in the administration's camp). It seems likely the report—which was then hurriedly issued by an embarrassed administration—would never have seen the light of day had it not been for that leak.

Clearly action had to be taken to ensure the EPA didn't commit such a *faux pas* again—putting scientific fact above political considerations. The chastened EPA accordingly produced in early 2004 a revised set of proposals on the regulation of power-plant mercury emissions. Almost immediately it was discovered that twelve paragraphs of the report had been copied, more or less verbatim, from a strategy document produced earlier by power-industry lawyers.

Horrified by the effect the Clear Skies Initiative was likely to have on the nation's health, four senators—three Republicans and one Democrat—proposed a countermeasure that would tackle the problems of atmospheric CO_2, nitrous and nitric oxides, sulfur dioxide and mercury. Their proposal was passed to the EPA for analysis in terms of costs and benefits. For months there was silence. Finally, in July 2003, another despairing EPA staffer leaked internal documents to the press, this time the *Washington Post*: the four senators' proposal would be more effective and speedier in reducing the pollutants than the Clear Skies Initiative, thereby saving some 18,000 lives by 2010, not to mention some $50 billion annually in health costs. The costs to industry of implementing the senators' proposal rather than the Clear Skies Initiative would be, the analysis concluded, "negligible." According to EPA staffers, this information was suppressed by Jeffrey Holmstead, Assistant Administrator for Air and Radiation.

Holmstead was yet another example of the administration filling positions with unqualified but politically loyal individuals; before appointment, Holmstead was a lawyer, employed by the firm Latham & Watkins, which represented one of the US's largest plywood manufacturers. In January 2002, by now at the EPA, he held a meeting of EPA staffers and the EPA Air Office's general counsel, William Wehrum (whom we shall soon meet again—see below), who by curious coincidence had previously been a partner in . . . Latham & Watkins. The purpose of the meeting was to discuss a rule governing emissions of formaldehyde by wood-products plants. Representing the industry were Timothy Hunt, a lobbyist for the American Forest & Paper Association, and that organization's lawyer, Claudia O'Brien, who had earlier been a partner in . . . Latham & Watkins.

O'Brien recommended that supposedly low-risk products should be exempt from any new emission controls, since the cost of introducing them would make US manufacturers vulnerable to cheaper foreign competition.

To the astonishment of the staffers, who knew that this proposal would violate the 1990 Clean Air Act provisions, Holmstead backed it.

So far as the Clear Skies Initiative and emissions by power plants were concerned, the administration apparently decided EPA leakers might continue to bedevil White House adulteration of science, and thus introduced a new scheme. The New Source Review, by which power plant emissions were gauged, would be "re-interpreted"—in other words, made less stringent. In 2002, Holmstead reassured a Senate Committee that the new, laxer rules would not be applied retroactively. Surprise, surprise, in November 2003 the top brass of the EPA announced to staffers that the rules would be applied retroactively: cases against about fifty coal-burning power plants guilty of violating the Clean Air Act were to be dropped. This was done at the instigation of a mysterious energy task force, staffed entirely by unnamed representatives of the power industry, convened earlier by Vice-President Richard Cheney.

In February 2005, the EPA's Inspector General admitted that pressure had been put on the body's scientists to alter their reported results on the impact of mercury pollution to bring them into line with the industry-friendly conclusions the administration demanded. That same month, the Government Accountability Office announced that indeed the EPA's results had been falsified so as to significantly underestimate the damage mercury poisoning does to the fetal and infant brain.

The EPA responded to these grave charges by making no changes at all to its final ruling, which was issued on March 15, 2005.

Shortly afterward, yet another piece of scientific fraud came to light. A government-commissioned Harvard study had shown the costs of mercury pollution to be higher than previously thought, and the benefits of tighter control greater. The results of this study had been suppressed entirely by the EPA's political appointees.

Fortunately, the Clear Skies Act, sponsored by senators James Inhofe (R–OK) and George Voinovich (R–OH), never made it to a Senate vote, being deadlocked in committee. The administration then tried to implement some of its measures by fiat through the EPA.

In October 2004, the Sierra Club summarized the Clear Skies Initiative thus: "Allow our dirty old power plants to continue polluting."[104] That was about right.

# # #

In early 2006 the Bush administration advanced a new measure to reduce environmental protection, the emasculation of the Toxics Release Inventory program. This program, initiated during the aftermath of the 1986 disaster in Bhopal, India, requires companies to declare annually how they are disposing of some 650 different toxic chemicals in their wastes—in other words, where the poisons are going and how their environmental impact is being controlled. The new initiative would have reduced the frequency of reporting from annually to biannually, increased by a factor of ten the threshold over which the reporting of released toxins had to be reported, and removed altogether the obligation to report the release of cumulative toxins—such as lead and mercury—up to an annual level of about 275kg (585 pounds).

Unsurprisingly, the Attorneys General of a dozen states were among the countless individuals and organizations who immediately protested—they have lives to save rather than profits to protect. After disasters such as 2005's Hurricane Katrina or 2017's Hurricane Maria, it's vital that responders have ready access to information as to where toxic chemicals might have been released by the upheaval. Even companies and corporations were far from unanimous in welcoming the proposed relaxation, with many saying they would carry on reporting under the old rules anyway. Typical was the reaction of Edwin L. Mongan III, Director of Energy and Environment at DuPont: "It's just a good business practice to track your hazardous materials."

Aside from one small detail, squeezed through by deceptive means, the measure was dropped.

#

The American Medical Association no longer advises American delegations to UN summits on children's issues; Concerned Women for America does instead. Leaders of the National Association of People with AIDS no longer sit on the presidential AIDS advisory council, though religious abstinence advocates do; and members of the right-wing Federalist Society now vet judicial nominees rather than the mainstream American Bar Association.

—Esther Kaplan, *With God on Their Side* (2004)

As noted, the ploy of making political appointments to supposedly scientific posts was endemic in the Bush II administration. Yet another example was James Connaughton, Chair of the Council on Environmental Quality (CEQ), an advisory agency to the executive office of the President.

His responsibility was "to bring together the nation's environmental, social, and economic priorities" and to "prepare the president's annual environmental quality report to Congress"—very important responsibilities indeed.

Connaughton was not an environmental scientist but a lawyer, and a lawyer who had countless times done battle with the EPA on behalf of big business. In 1993, he was coauthor of the article "Defending Charges of Environmental Crime—The Growth Industry of the 90s"; this appeared in *Champion Magazine*, published by the National Association of Criminal Defense Lawyers. He lobbied on environmental issues on behalf of major corporations and corporate associations including the Aluminum Company of America, the Chemical Manufacturers Association, and General Electric. The latter was believed responsible at the time for more toxic sites in the US than any other corporation. The mining company ASARCO, another of Connaughton's clients, had lobbied that the US relax its 1942 limit on the amount of arsenic present in drinking water. These were among many other corporate and trade association clients Connaughton had represented at the state, federal, and international level, frequently acting as a defense lawyer in cases of environmental crimes.

It could of course be argued that Connaughton was qualified to be chair of the CEQ on the grounds of his supreme knowledge of environmental law. It could equally be argued that Al Capone had a supreme knowledge of criminal law, without necessarily regarding him as an ideal Attorney General. Presumably, though, Connaughton passed the relevant political "litmus test"—an approval based on ideological loyalties rather than scientific qualifications. Bush-appointed Science Advisor John H. Marburger III, a genuine scientist (and a Democrat), stated: "[T]he accusation of [the existence of] a litmus test that must be met before someone can serve on an advisory panel is preposterous." In fact, numerous eminent scientists reported that exactly such a test had been applied to them when they were being considered for relevant posts within the administration. The Union of Concerned Scientists investigated these reports and detailed many of them in its various publications (see Bibliography) called *Scientific Integrity in Policymaking*. It seems Marburger* must have been speaking on the basis of incorrect information.

---

* Whom I met after he'd served his time with the administration, and who said very kind things about the first edition of this book.

Further curious appointees in the area of the environment included Gale Norton as Secretary of the Interior and Ann Veneman as Agricultural Secretary. Both were protegees of the infamous James Watt, the anti-environmentalist appointed by Ronald Reagan as his Secretary of the Interior, and worked for Watt's Mountain States Legal Foundation—self-described on occasion as "the litigation arm of wise use" (see page 220).

In early 2007, Bush issued an executive order to ensure White House control, either directly or through political appointees, over the directives issued by the EPA and the Occupational Safety and Health administration, among other federal agencies. One of his baldest moves to subject the findings of science to ideological manipulation, it was chilling in its cynicism and disregard for the truth.

#

With the coming to office of President Barack Obama in 2009 there was a welcome rebirth of the idea that the federal government's leading science officers should be scientists, or at least scientifically literate—in fact, it had been decades since last there was an administration containing so many scientists, many of them distinguished. Among them were the environmentalist John Holdren as Director of the White House Office of Science and Technology Policy, the Nobel-winning physicist Steven Chu as Secretary for Energy, and the geophysicist Marcia McNutt as director of the United States Geological Survey (USGS) and science adviser to the United States Secretary of the Interior; McNutt is currently (as of July 2016) President of the National Academy of Sciences (NAS).

Although the scientific community, witnessing these appointments, not unreasonably expected that the new administration would steer the US back toward the practice of prioritizing science, that never quite happened. Nor did the forging ahead that had been hoped for on issues such as climate change.

A major problem, one that was especially exasperating during the administration's early years, when it had the benefit of Democratic majorities in both the House and the Senate, was Obama's own longstanding inclination to extend a hand across the aisle—to bring his political opponents on board, compromising where need be. This is in many ways a laudable tactic. What it took Obama far too long to recognize was that such a tactic is idiotic when the openly declared intent of those political opponents was to fight each and every notion that he proposed, whatever the cost to the country and its people—to make it sufficiently difficult for him to govern, in fact, that the voters would reject him and he'd be a one-term president.

Here, for example, is Obama talking to Congress about climate change in January 2010 in his very first State of the Union address:

> I know that there are those who disagree with the overwhelming scientific evidence on climate change. But, even if you doubt the evidence, providing incentives for energy efficiency and clean energy [is] the right thing to do for our future—because the nation that leads the clean energy economy will be the nation that leads the global economy. And America must be that nation.

The latter part of this sounds like a stirring call to arms, and indeed that was the way it was hailed by the left-leaning sectors of the media. But the former part was committing the same false-balance error as occurs so often in journalism: the pretense that opinions born of ignorance and stupidity nonetheless possess merit.

In other words, Obama let himself become the man trying to cure the rabid dog with kindness. The obdurate Republicans found themselves voting against even several policies of their own that Obama, believing them sound, presented to Congress. When the GOP regained control of the House, the situation became even worse. Only toward the end of his incumbency did Obama take the bit between his teeth and start using executive orders to try to bring about some of the science-related changes that were necessary for the future well-being of the country—and of the world.

While it would be dishonest to pretend that the Obama administration's treatment of science was remotely as heinous as that meted out by its predecessor and even more so by its successor, it was very far from perfect. Leaving aside its poor record over its first half-dozen years on climate change, we'll here look briefly at its draconian treatment of whistleblowers.

The administration endorsed the 1989 Whistleblower Protection Act. Obama signed both 2012's Presidential Policy Directive 19 and the Intelligence Authorization Act for Fiscal Year 2014; the former was concerned with the protection of whistleblowers in the intelligence agencies and the latter contained provisions to the same end. The safeguards are not great: whistleblowers may report on matters such as waste of resources and individual abuse of power, but investigations of retaliation are conducted behind closed doors, supposedly in order to shield national secrets. The scope for corruption within this latter proviso is obvious. If the establishment chooses to unite behind one of their (wrongdoing) own, the whistleblower is in, to say the least, an unfortunate position.

As seriously, there's no protection at all for the whistleblower who exposes systemic abuse of power and breaches of the law by an intelligence agency, which is what CIA computer specialist Edward Snowden did in 2013 when he leaked a very large amount ot information about the National Security Agency's extensive illegal surveillance activities. Although a very good case can be made that Snowden was acting in the public interest in exposing these offenses against the public, Obama's Department of Justice, and Obama himself, declined to acknowledge this fact; Snowden, who with some justification regarded promises of a fair trial as hollow, was driven into exile, and currently resides in Moscow.

Snowden's whistleblowing was not directly related to science (although you could say it was intimately concerned with the abuse of technology), but the administration's punitive reaction to it—and to the earlier leaking of military atrocities by Bradley/Chelsea Manning and of governmental waste by NSA official Thomas Drake*—reportedly sent a chilling message to would-be whistleblowers in other agencies, including the scientific ones. To be fair, though, there was almost no suppression of science similar to that during the Bush years.

What the administration does appear to have continued from its predecessor, however, is the practice of "censorship by public information officers." At the CDC, for example, where the public policy is one of openness and transparency, with scientists theoretically being free to talk to the press at will, in actuality all queries from the news media had to be routed through the CDC's head office in Atlanta. In August 2015 this sparked an open letter of complaint from the Society of Professional Journalists (SPJ) directly to President Obama that read in part:

> One year ago more than 40 journalism and government accountability organizations expressed deep concern about the constraints on information in the federal government today. These include:

---

* On January 17, 2017, just before his departure from office, Obama commuted Manning's sentence to time served. In 2011 the DoJ dropped its case against Drake, settling for a misdemeanor plea. In March 2012 Matt Miller, formerly the DoJ's Director of Public Affairs, admitted to the *Washington Post* that "Drake did seem to be trying to expose actual government waste. I think the outcome of the case probably shows that it was an ill-considered choice for prosecution."

• prohibiting staff from communicating with journalists unless they maneuver through public affairs offices or through political appointees;

• refusing to allow reporters to speak to staff at all, or delaying interviews past the point they would be useful;

• monitoring interviews; and

• speaking only on the condition that the official not be identified even when he or she has title of spokesperson.

The response Mr. Josh Earnest sent the Society of Professional Journalists on August 11, 2014, failed to address these issues, and despite repeated requests to discuss the issue publicly, the White House has yet to engage in a meaningful conversation.

We request again, just weeks after the 49th anniversary of the U.S. Freedom of Information Act and 239th anniversary of our nation, that you change these practices in your administration and participate in a public dialogue toward improving the flow of information for the American people. The public has a right to be alarmed by these constraints—essentially forms of censorship—that have surged at all levels of government in the past few decades. Surveys of journalists and public information officers (PIOs) demonstrate that the restraints have become pervasive across the country; that some PIOs admit to blocking certain reporters when they don't like what is written; and that most Washington reporters say the public is not getting the information it needs because of constraints. An SPJ survey released in April confirmed that science writers frequently run into these barriers.

This information suppression is fraught with danger. A recent review found that the Centers for Disease Control and Prevention had a culture of unsafe handling of dangerous pathogens and that some staff feared reporting incidents. Last year the Food and Drug Administration announced it had smallpox, among other dangerous materials, in an apparently uninventoried storage for decades in violation of some of the most solemn of international treaties. All the employees working around those situations for years were forbidden to speak to reporters without surveillance by the PIOs, as was all other staff in those agencies.[105]

Examples of stories suppressed in this way included, in 2012 in Jacksonville, Florida, the country's worst outbreak of tuberculosis in twenty years, the result of a homeless schizophrenic contracting TB but going without treatment for some eight months as he mixed with other homeless people, and a catastrophic spill of the chemical 4-methylcyclohexanemethanol (MCHM) into the Elk River, West Virginia, in January 2014.

It's not immediately apparent why the CDC should have wanted to keep either story under wraps. Were the officers wary of spreading panic? But, in the West Virginia case, surely their stark warning that pregnant women should switch to bottled water *pro tem* was far more likely to create a panic than if it had been accompanied by a full explanation of what was going on. Perhaps they were unwilling to admit that MCHM is one of those countless industrial chemicals whose hazards the federal science agencies have yet properly to assess?

According to a January 2018 *Scientific American* roundup of the methods used by government agencies during the Obama years (and earlier) and still being used today to minimize the press's direct access to scientists,[106] another technique is in use at the FDA. This is the close-hold embargo, whereby a limited number of reporters are informed on a particular story but with the warning that they mustn't break it until given the go-ahead by the agency. The hidden—and sometimes not-so-hidden—corollary is that journalists who don't follow the agency's desired spin in their subsequent report are likely to find themselves left off future short-lists. Bizarrely, the FDA's press officers portray this practice not as the manipulation and/or suppression of information that it is—information that is, in effect, the property of the US public—but instead as the FDA doing the news media a favor! And woe betide the journalist who kicks up a fuss about it.

Overall, the Obama administration was scarred by insufficient progress in science-based areas where action was urgently required. Before the 2010 midterms Obama was dealing with a Democratic House and Senate, so really had no excuse for this; even taking account of Blue Dog Democrats, there was plenty of scope for him to push through measures on the environment and other science-related issues in 2009 and 2010. He did succeed in signing into law the Affordable Care Act, a preliminary, hesitant step along the way (let's hope) to bringing US healthcare up to the standards enjoyed by the rest of the world's developed countries. Yet, as we saw (page 314), given the opportunity to take action on the serious consequences of the abuse of antibiotics in agriculture, the Obama administration merely kicked the can down the road until finally taking action in 2015.

Only late in his second term did Obama appear to appreciate that the chance to improve the nation's well-being in these areas was slipping through his fingers, and it was then that he began issuing the series of executive orders that, inevitably, brought him castigation in certain sections of the media. Among these orders were many of scientific interest, of which the following is a small selection:

- #13,505 (March 9, 2009) removed the "faith-based" restriction the Bush II administration had placed on stem-cell research.

- #13,514 (October 5, 2009) asserted that "In order to create a clean energy economy that will increase our Nation's prosperity, promote energy security, protect the interests of taxpayers, and safeguard the health of our environment, the Federal Government must lead by example. It is therefore the policy of the United States that Federal agencies shall increase energy efficiency; measure, report, and reduce their greenhouse gas emissions from direct and indirect activities; conserve and protect water resources through efficiency, reuse, and stormwater management; eliminate waste, recycle, and prevent pollution; leverage agency acquisitions to foster markets for sustainable technologies and environmentally preferable materials, products . . ."

- #13,521 (November 24, 2009) established the Presidential Commission for the Study of Bioethical Issues.

- #13,539 (April 21, 2010) re-chartered the President's Council of Advisors on Science and Technology, which had originally been chartered by President George W. Bush.

- #13,554 (October 5, 2010) established the Gulf Coast Ecosystem Restoration Task Force in the aftermath of the Deepwater Horizon disaster of April 20, 2010, where a BP oil rig spilled some 200 million gallons of crude oil into the Gulf of Mexico, causing colossal environmental damage.*

- #13,605 (April 13, 2012) championed "efforts to promote safe, responsible, and efficient development of unconventional domestic natural gas resources"—i.e., fracking.

- #13,653 (November 1, 2013), titled "Preparing the United States for the Impacts of Climate Change," was a major and long-needed step. It sought to bring together the efforts of dozens of federal government agencies to tackle such tasks as "Modernizing Federal Programs to Support Climate-Resilient Investment," "Managing Lands and Waters for Climate Preparedness and Resilience," and "Providing Information, Data, and Tools for Climate Change Preparedness and Resilience."

- #13,676 (September 18, 2014) was the start of the long-delayed effort to tackle the problem of antibiotic-resistant bacteria.

---

* The suggestion that BP might actually be compelled to pay for some of the cleanup effort was vociferously resisted by some congressmen, notably Joe Barton (R–TX).

• #13,677 (September 23, 2014) focused on "Climate-Resilient International Development."

• #13,699 (June 26, 2015) established the Advisory Board on Toxic Substances and Worker Health.

• #13,717 (February 2, 2016) set up the Federal Earthquake Risk Management Standard.

There were also numerous "presidential memoranda," which can be regarded as sort of downmarket versions of executive orders. Among these were such items as "Establishing a Task Force on Childhood Obesity" (February 9, 2010), setting up a "Task Force on Space Industry WorkForce [*sic*] and Economic Development" (May 3, 2010), "Implementation of Energy Savings Projects and Performance-Based Contracting for Energy Savings" (December 2, 2011), "Flexible Implementation of the Mercury and Air Toxics Standards Rule" (December 21, 2011), "Creating a Preference for Meat and Poultry Produced According to Responsible Antibiotic-Use Policies" (June 2, 2015) and "Climate Change and National Security" (September 21, 2016).

Wikipedia has a comprehensive list[107] of Obama's executive actions, almost all with links to the relevant documentation, and reading through them one's struck by the fact that the vast majority are completely unexceptionable, if not humdrum, concerned with the daily continuance of government: renewal of sanctions against various rogue nations, distancing of the US from political crises around the world, reinforcing efforts against terrorists and narcotics traffickers, and so on. The bulk of the scientific measures are the sort of sensible stuff that any functional Congress, whether Republican or Democrat, would have greenlighted without a murmur. The fact that Obama had to resort to executive actions in order to advance commonsense efforts on climate and other aspects of environmental protection reflects not a draconian attempt by him to expand executive power, as has often been the accusation, but merely piecemeal endeavors by him, in the country's interests, to do some of the jobs that Congress should have been doing rather than grandstand about trivia to please the less well educated elements of the public.

The environmental issue over which it seemed for a long time that Obama was most likely to fail was the proposed extension of the Keystone Pipeline System. The existing system brought oil from the tar sands of Alberta down through the US to refineries/storage facilities in Illinois, Oklahoma, and Texas. The proposed Keystone XL extension would offer both a shortcut and a significant increase in capacity.

Because, if we're serious about leaving the world in a habitable state for our children and grandchildren, we should be seeking to reduce rather than expand our fossil-fuel capacity, the planned Keystone XL extension became an iconic symbol of environmental destruction. Respected figures within the climate-preservation movement, such as Jim Hansen and Bill McKibben, described Keystone XL as a potential game-changer for the planet's future—at least so far as its human population was concerned.

Even so, the Obama administration vacillated for some six years over the Keystone XL case before finally, in November 2015, coming down against it. One can't help feeling that, had the matter been addressed with a bit more expedition, five years' worth of human endeavor might have been directed toward other, equally important environmental concerns.

In January 2017, one of the newly elected President Trump's first actions was to issue a presidential memorandum rescinding the Obama administration's veto of the Keystone XL extension.

APPOINTMENTS AND DIS-APPOINTMENTS

One area in which Obama was near-exemplary was the appointment to positions of scientific importance within the administration of people who actually knew what they were talking about. In this he was in stark contrast to his predecessor, George W. Bush, as we've seen, and, alas, to his successor, Donald Trump. In some instances, it seems as if Trump is going out of his way to select those who are not just unsuitable for such posts, but surreally so.

Take the example of former Texas governor Rick Perry,who became Trump's nominee as Secretary of Energy; Perry, proud bearer of a bachelor's degree in animal science from Texas A&M University, succeeded Ernest Moniz, a nuclear physicist from MIT.

A profound denier of climate science—"Calling CO_2 a pollutant is doing a disservice to the country, and I believe a disservice to the world," he said in a 2014 interview with the *Dallas Morning News*—Perry was already linked to the Department of Energy in the public mind: his 2011–12 run for the GOP nomination for the presidency was largely scuppered when, during one of the debates, he couldn't remember the DoE's name while trying to list those departments he'd like to abolish. During his brief 2015 run, it became evident he was unaware of what the DoE actually *did*—notably that it's responsible for the safety of America's nuclear resources.

In September 2017, Perry's ignorance of his department's specialty became apparent when, rather than investing $3.7 billion of taxpayer money in, say, renewable energy sources* or even next-generation nuclear technology, he threw it, in the form of loan guarantees, at the utility companies trying to complete two reactors, vastly over-budget and behind schedule, at the Alvin W. Vogtle generating plant in Georgia—this after the department had already loaned the project $8.3 billion. Perry's decision drew vociferous protest from one of the most unlikely coalitions in US energy history, containing the likes of the American Petroleum Institute on the one hand and the Solar Energy Industries Association and American Wind Energy Association on the other.

Perry's colleague William Bradford, appointed by Trump to lead the Energy Department's Office of Indian Energy, resigned in August 2017, after racist and antisemitic tweets of his were unearthed. Even though the tweets dated from before his time in office, he was deemed unsuitable to retain the post.

University of Cincinnati toxicologist Michael Dourson, nominated by Trump to head the EPA's efforts on chemical safety and pollution prevention, had previously earned notoriety by self-publishing a string of "books matching science and Biblical text," including the three-book series, *Evidence of Faith*, with a fourth, covering Noah and the Flood, under development. An extract:

> First, it is assumed that the natural law (Nature) and the written law (the Bible) have the same dignity and teach the same things in a way that one of them has nothing more and nothing less than the other because they have the same author—God (St. Augustine, 354–430). . . . Second, it is assumed that the inspired word of the Bible, in its various translations, reflects what actually happened, but using words that were most appropriate at the time of writing.

In the EPA's press release announcing the nomination, two of the seven endorsements came not from medical scientists but from Christian leaders.

There's obviously no objection to federal scientists propagating their

---

* Much has been made of the Obama administration's loss of its relatively minor investment in the solar-panel company Solyndra in 2011. It's worth noting that Solyndra was part of a portfolio of renewables companies that the government invested in. Almost all of them succeeded, and so the taxpayer has had a more than healthy return on the investment in the portfolio as a whole. As any investor will tell you, you're always gonna lose a few: it's the overall return that counts.

faith or writing religious tracts in their own time; Francis Collins, the extremely distinguished scientist who headed the Human Genome Project and is currently director of the National Institutes of Health (NIH) makes no bones about his Christian faith, which has most certainly not hampered his science. But the pretense that the Bible—with its flat-earth cosmology and its two conflicting creation accounts, among much else—is an adequate science textbook is simply false. It is perfectly reasonable to ask if it's wise to appoint someone to a leadership role in the EPA who, whatever his distinction as a toxicologist, has a record of peddling what's in effect pseudoscience.

Also alarming are Dourson's longstanding links to the chemicals industry and former links to the tobacco industry, and his reported belief that the EPA habitually exaggerates environmental risks. In 1995, he founded the consulting company Toxicology Excellence for Risk Assessment (TERA), which assesses the hazards presented by chemicals and the levels at which they're toxic. Among TERA's clients have been Koch Industries, Dow Chemical, the American Petroleum Institute, and the American Chemistry Council. According to an analysis done by the Environmental Defense Fund in September 2017, the curious thing about the assessments done by TERA for its commercial clients is that they consistently show the chemicals concerned to be safe at levels far higher "than government standards or guidelines prevailing at the time."[108] The EDF identified Dow Chemical's pesticide chlorpyrifos as one such; chlorpyrifos is known to be toxic to humans at higher concentrations—it can, for example, harm the future neurological development of children exposed in the womb—and has been suspected of contributing to colony collapse in bees. Sale of chlorpyrifos is banned in the European Union and in some other parts of the world; its residential use was banned in the US in 2001. In 2007 Dow Chemical was found guilty of bribing bureaucrats in India to forestall the banning of chlorpyrifos there.

In November 2016, the EPA reaffirmed a 2015 decision to move toward banning the use of chlorpyrifos outright. At the end of March 2017, Scott Pruitt, the incoming EPA Administrator, reversed that ruling, claiming his decision represented a return to "sound science"—a term that sends shudders down the spine of anyone concerned about scientific integrity (see page 217). In the ensuing months it was revealed that Pruitt's decision followed numerous meetings with representatives of industry, Dow included, where he'd promised a more complaisant—i.e., corrupt—EPA. The American Academy of Pediatrics commented on his decision:

There is a wealth of science demonstrating the detrimental effects of chlorpyrifos exposure to developing fetuses, infants, children and pregnant women.
The risk to infant and children's health and development is unambiguous.[109]

Even its strongest supporter, then, would have to admit that, in terms
of health safety, chlorpyrifos's status is at best iffy. But not Scott Pruitt, and
not Dourson's TERA. And chlorpyrifos is only one of dozens of industrial chemicals that TERA has deemed safer than reckoned by established
science. Among others, there's 1,4–dioxane, thought probably to be a car-
cinogen: the EPA's existing safety standard allows a maximum of 0.35ppb;
TERA's recommended safety level was one thousand times higher.

In July 2017, Trump nominated Samuel H. Clovis, Jr., as the United
States Department of Agriculture's undersecretary of research, education
and economics, the USDA's top science position.

Sam Clovis has a doctorate in public administration. He served as a
professor of economics at Morningside College, Sioux City, Iowa. A former rightwing radio bloviator—he stands accused of the usual racist and
homophobic remarks during that stage of his career—Clovis ran unsuccessfully in 2014 in a Republican primary for the US Senate. In 2015, he
joined Rick Perry's presidential campaign team and, when Perry dropped
out of the race, signed on to the Trump campaign as a policy advisor and
then national co-chair. Like his anticipated boss at the USDA, Sonny Perdue, Clovis vociferously denies climate science—a stance that is obviously
inordinately dangerous for the country's agriculture in any senior USDA
official. Farmers will need to know what to do in our changing climate, and
be encouraged financially and otherwise to take necessary steps, not told
that climate change is, shall we say, a Chinese hoax.

In other words, however fine an economist Clovis might be, it's quite
clear that scientifically he's ignorant even beyond his lack of qualifications,
and arrogant in that ignorance. Fortunately for us all, in November 2017
Clovis withdrew his candidacy.

David Bernhardt, Trump's pick to be Deputy Secretary at the Department of the Interior, was one of George W. Bush's political appointees at
the department. In that earlier role, as a key aide to Interior Secretary Gale
Norton, he was noted for having helped alter the facts and conclusions of
a scientific report produced by the Fish & Wildlife Service on the effects
of oil development in the Arctic National Wildlife Refuge. The fraud was
discovered by the activist group Public Employees for Environmental Responsibility (PEER), and in the resulting fracas over the department hav-

ing in effect lied to senators over the true impact oil development might have on the refuge, Bernhardt repeatedly fibbed again, claiming for example that the original report had been merely "edited for responsiveness."

As PEER's Jeff Ruch wryly remarked in May 2017 when Bernhardt was nominated for the new role, "It appears Mr. Bernhardt shares an unfortunate affinity for alternative facts."

The man put in charge of the several doomed attempts to replace Obamacare and replace it with "Trumpcare"—a collective term for various schemes whereby tens of millions of Americans would have lost their health insurance at considerable cost to the nation but immense benefit, in the form of tax breaks, for the very wealthy—was Secretary of Health and Human Services Tom Price. Price was forced to resign in September 2017 due to his habit of flying by chartered rather than commercial aircraft on the taxpayer dime—he generously offered to repay about one-eighth of the cost. To be fair, Price was not completely unqualified, having had a career as an orthopedist before entering the House of Representatives as a representative for Georgia in 2005; also to be fair, he had been a member of the Association of American Physicians and Surgeons, a far-right political organization that wars against socialized medicine and mandatory vaccination (see page 75). In 2009, he voted against the Family Smoking Prevention and Tobacco Control Act, a measure giving the FDA a degree of control over the tobacco industry, including its marketing of its products to kids.

Before his appointment by Trump, while he was still a congressman, he had come to public notice for some of his investments in medical companies during the period 2015 to early 2017, where he made a handsome profit when shares in medical companies leaped in value for reasons not entirely unconnected with legislation that Price himself had introduced in the House. In March 2017, Robert Faturechi reported in *ProPublica* that one of the cases being investigated by US Attorney Preet Bharara, unexpectedly fired by the Trump administration a few days earlier, had concerned these dealings by Price.[110]

As noted, GOP efforts to repeal Obamacare by the normal means fell flat. At the time of writing, it's too soon to assess the effect of some more underhanded techniques deployed by Price during his stint at Health and Human Services. In order to participate in Obamacare, individuals must enroll via the plan's online exchanges, which in previous years were open for a three-month period from November 1 of each year. For no other conceivable reason than to try to throw a spanner into Obamacare's works,

Price shortened the enrollment period from three months to six weeks, reduced by 90 percent the department's budget for advertising open enrollment, and declared that every Sunday during the open-enrollment period the site would be shut down for twelve hours "for maintenance."

Challenged on this last, a spokesman claimed that technical problems had created equivalent closed periods during the Obama era. Needless to say, this was untrue. There were glitches to begin with, but by the 2015 and 2016 enrollment periods the site's total downtime was a mere 0.1 percent.

After Price's departure the GOP passed, in December 2017, a mighty overhaul of the tax system that included, tucked away amongst countless other provisions, the cancelation of the mandate widely seen as an essential underpinning of Obamacare. Again, we don't yet know what the effects of this will be.

In January 2018, Trump hit on another way of attempting to undermine the hated Obamacare, this time through issuing a guidance to states that they can withold Medicaid from the unemployed. The rationale appears to be that people who're too ill to get jobs should be denied healthcare until they damned well get better—a piece of vindictive lunacy that could have come straight out of *Catch 22*. Besides, as Jonathan Chait pointed out in *New York*, "The number of able-bodied, working-age Americans who receive Medicaid and who could obtain employment but choose not to is extremely small." He added, "Imposing work requirements, which entail some kind of test to determine if a recipient is employed or seeking employment as a condition of enrollment, is costly."[111]

Other administration officials have shared Price's habit of flying by private jet at taxpayer express: Kellyanne Conway, Steven Mnuchin, Scott Pruitt . . . As the editorial board of the *New York Times* angrily noted,

> Not least because of his own self-dealing—the hotel down the block, the failure to fully divorce himself from the Trump Organization, all the rest—the White House became ground zero for grasping lobbyists and ethically challenged, self-promoting staffers. So great was the back stabbing and chaos that, in July, Mr. Trump turned to an upright military figure, Retired Gen. John Kelly, to set things straight as chief of staff.
>
> It has proved an impossible task. The tighter ship Mr. Kelly vowed to run seems to be springing ethical leaks almost daily, as more and more accounts surface of this or that official using public resources for private gain or crossing a line demarcating personal from government business. . . .
>
> Sadly . . . they seem to have already absorbed a more persuasive lesson from the chief executive—get it while you can and to hell with the rules.[112]

To replace Tom Price, Trump nominated Alex Azar, a former Eli Lilly executive who had served during the George W. Bush administration as Deputy Secretary of Health and Human Services; he had also been on the board of pharma-lobbying Biotechnology Innovation Organization. The nomination was controversial, not least because Azar's history at Eli Lilly had been marked by frequent increases of drug prices for no perceptible reason other than increasing profits and with little regard for patients' needs, while at the same time the company had used offshore profit-stashing as a means of avoiding taxes. Asked about this by Senator Ron Wyden (D–OR) at his confirmation hearing in early January 2018, Azar responded with platitudes such as "Drug prices are too high"—and, to be fair, Eli Lilly was far from alone among the major pharmaceutical companies in this respect.[113] Even so, as the left-leaning group Public Citizen observed, "At a time when the US is facing a nationwide crisis of access to affordable medicines, the top official in charge of healthcare should not be a former pharmaceutical company executive with a history of making lifesaving medicines unaffordable."[114]

More substantial a concern, arguably, was that in the past Azar had expressed support for abolishing Medicaid and replacing it with a block-grant system, which would give financial benefits to the healthy and penalize the sick — in other words, the very people most in need of Medicaid. The Center on Budget and Policy Priorities, in a position statement as long before as November 30, 2016, had spelled out why such a scheme was a poor idea:

> A Medicaid block grant would institute deep cuts to federal funding for state Medicaid programs and threaten benefits for tens of millions of low-income families, senior citizens, and people with disabilities. To compensate for these severe funding cuts, states would likely have no choice but to institute draconian cuts to eligibility, benefits, and provider payments. To illustrate the likely magnitude of these cuts, an analysis from the Urban Institute of an earlier block grant proposal from Speaker Ryan found that between 14 and 21 million people would eventually lose their Medicaid coverage . . . and that already low provider payment rates would be reduced by more than 30 percent.[115]

Questioned on this issue at the Senate hearing, Azar again fudged the issue, saying that it was "a concept to look at" but refusing to commit one way or the other. As Senator Elizabeth Warren (D–MA) eventually said in exasperation, "You want to smile and pretend otherwise until you get the job. No one should be fooled."

Also for the Department of Health and Human Services, Trump nominated, as Assistant Secretary, Charmaine Yoest, whose scientific qualifications consist of a bachelor's degree in politics and service as president and CEO of the anti-abortion activist group Americans United for Life. Troublingly, Yoest appears to subscribe to the conspiracy theory that abortion increases a woman's chances, later in life, of contracting breast cancer. This piece of groundless pseudoscience, which has been repeatedly debunked, seems first to have been widely promoted in the aftermath of *Roe v. Wade*, in 1973. A meta-analysis done in 2004 of 53 studies involving 83,000 female respondents found there was no link between abortion and breast-cancer rates. The National Cancer Institute, the American Cancer Society and the American Congress of Obstetricians and Gynecologists are among the prominent scientific institutions that agree. Yet, as late as November 2012, Yoest was telling the *New York Times* this was fake science: "Why can't you report what the research actually shows?" Apparently the National Cancer Institute and the others are "under the control of the abortion lobby."[116]

Going on, the Centers for Disease Control and Prevention (CDC—a single agency despite the name) acquired its director Brenda Fitzgerald, an ex-USAF obstetrician–gynecologist. There's nothing wrong with Fitzgerald's credentials for the job, at least on paper. Troubling, however, is her promotion of the quack anti-aging remedy known as "bioidentical-hormone therapy"; one of the credentials Fitzgerald used to list on her website[117] was in Anti-Aging and Regenerative Medicine, seemingly granted her by the American Academy of Anti-Aging Medicine, an activist organization that is not recognized by any of the country's professional medical bodies. According to the American Congress of Obstetricians and Gynecologists, "Evidence is lacking to support superiority claims of compounded bioidentical hormones over conventional menopausal hormone therapy."[118] As with all "fringe" medical practices, there are also unacceptable risks involved, in this instance the need to use compounding pharmacists, not all of whom may be reliable, to prepare the bioidentical hormones.

Fitzgerald had raised eyebrows earlier, while Georgia's Health Commissioner, over her cooperation with the Coca-Cola Company in the state's campaign against childhood obesity (see pp. 277–278). She was ousted in January 2008 after Politico exposed her investments in the tobacco industry.

We've already met corporate lawyer William Wehrum in the context of the George W. Bush-era EPA, where he was active in efforts to circumvent the Clean Air Act. A number of the "industry-friendly" measures he introduced during those years were later overturned by the courts. Who could be worse, you might think, to select as the EPA's Assistant Administrator (Air and Radiation) than Bill Wehrum? So Donald Trump nominated Wehrum for the EPA post in September 2017. As the Natural Resources Defense Council (NRDC), a conservationist group, observed in its September 20, 2017, newsletter, Wehrum

> doesn't believe that the Clean Air Act should apply to the carbon pollution that fuels climate change and storms like Hurricanes Harvey and Irma. He doesn't believe in tough rules to rein in dirty power plants' toxic emissions, which cause thousands of premature deaths and hundreds of thousands of asthma attacks each year. He's a dangerous choice to run the EPA's clean air office . . .

Questioned in the Senate by Senator Jeff Merkley (D–OR), Wehrum, in the midst of displaying astonishing ignorance of the basic issues relating to the environment, described the problem of ocean acidification as "an allegation."

Of course, this means he should fit in just fine with Trump's EPA Administrator, Scott Pruitt.

Of all the nation's science-oriented agencies, the most iconic must surely be NASA, which not only sends probes (and sometimes humans) to explore the cosmos and brings us all those pretty pictures of the universe from the Hubble Telescope and elsewhere, but also plays a vital role in research concerning our own planet, such as monitoring the atmosphere for information relevant to climate science. Past NASA administrators have typically been physicists, engineers and the like. Surely even Trump wouldn't . . . ?

Oh yes he would.

Despite sitting on the House Science, Space and Technology Committee, Congressman (since 2012) Jim Bridenstine (R–OK), a former Navy flier, has no formal science qualifications. He is, however, a climate-science denier, and it's presumably because of this "qualification" that Trump tipped him at the start of September 2017 as NASA's next Administrator. "Mr. Speaker, global temperatures stopped rising 10 years ago," Bridenstein told the House in 2013, repeating one of the hoariest and most thoroughly debunked of all the denialist canards. "Global temperature changes, when they exist, correlate with Sun output and ocean cycles," he added, making it two for two.

Questioned by the organization Vote Smart in 2012 as to whether he believed human activity contributes to climate change, he replied with what's in effect a series of Fox News talking points:

No. The Earth's climate has always varied substantially as demonstrated by pre-industrial human records and natural evidence. There is no doubt that human activity can change local conditions, but on a global scale natural processes including variations in solar output and ocean currents control climatic conditions. There is no credible scientific evidence that greenhouse gas atmospheric concentrations, including carbon dioxide, affect global climate. I oppose regulating greenhouse gases. Doing so will significantly increase energy prices and keep more people in poverty.[119]

Even some of the Senate's Republicans are raising questions about Bridenstine's suitability for the job, which raises the hope that his nomination may in due course be rejected. On the other hand, some of the Senate's Republicans questioned the suitability of Betsy DeVos to be United States Secretary of Education, yet in the event enough of them knuckled under to partisan considerations that, with a tie-breaking vote from Vice-President Mike Pence—another climate-science denier—she sneaked through, to the ongoing detriment of the nation's schoolchildren.

The Trump administration has, to date, been marked by the astonishing slowness with which it has filled federal positions—not just in the sciences but all across the government. It's not immediately clear why this should be so. However, the administration has been swift to implant political aides in every Cabinet agency to ensure that the employees there maintain loyalty— loyalty not to the American people, which is of course what their oath of employment is all about, but to the White House. These aides—they have the job title of Senior White House Adviser—report to White House Deputy Chief of Staff Rick Dearborn at the Office of Cabinet Affairs. In instances where agencies have yet to receive appointed leaders, these aides, many of whom are former lobbyists, can be the de facto bosses.

At least one of them, Curtis Ellis, installed as a special adviser at the Department of Labor, was formerly a columnist at the conspiracy-theory fake-news outlet *WorldNetDaily*.

Meanwhile Rex Tillerson, appointed as Secretary of State, was famed not just as CEO of ExxonMobil but as one of the main proponents of the idea of "energy poverty" alongside such celebrated deniers of the dangers of climate change as Bjorn Lomborg. The "energy poverty" concept is that the main hazard the inhabitants of developing countries face is not starvation, contaminated water, or disease, but insufficient energy supplies. The corollary advanced is that these energy needs can only be fulfilled by fossil fuels.

That's a fallacy, of course, because in many regions of the developing world the basic resources for solar energy (sunlight) and hydroelectricity (running water) are plentiful and easily accessible, whereas energy produced from coal or oil has to be brought in by lines from distant power plants. In India, for example, many small companies like Omnigrid are bringing renewable energy (alongside cellular telecommunications) to rural areas, and reportedly making a great success of it. On May 10, 2017, Michael Safi reported in the *Guardian*:

> Wholesale solar power prices have reached another record low in India, faster than analysts predicted and further undercutting the price of fossil fuel-generated power in the country. The tumbling price of solar energy also increases the likelihood that India will meet—and by its own predictions, exceed—the renewable energy targets it set at the Paris climate accords in December 2015.

SCOTT PRUITT AND THE DISMANTLING OF THE EPA

By far the most egregious of Trump's "scientific" appointments has been that of Scott Pruitt, the Betsy DeVos of environmental stewardship, to head the EPA—a move disastrous to the national (and international) interest.

Pruitt's primary qualification to be EPA head was that, as Attorney General of Oklahoma from 2010, he'd sued the agency at least fourteen times (counts vary) in attempts to overturn environmental regulations that he claimed to be part of a "political agenda"—a political agenda to which, as with the supposed conspiracy of climate science, nobody can ascribe a motive that doesn't sound completely absurd to anyone who hasn't had a few drinks.* Needless to say, Pruitt's political campaigns have received funding from the fossil-fuels industry.

His appointment as EPA Administrator was a slap in the face not just to those concerned about the safety of our air and water, not just to the scientists alarmed about our increasingly dangerous future as climate change accelerates—as exemplified in the increasing severity of the hurricanes that devastate the US and other countries at enormous human and economic cost—not just to advocates of greater security in the home and workplace, but to anyone with a shred of rational intellect. This was a de-

---

* Also as Oklahoma's AG, Pruitt fought against gay marriage, abortion rights and the Affordable Care Act. His detestation of environmental safety is part of a pattern.

liberate insult by the administration to those of the American people—the majority—who reject the myth that the EPA is "a jobs killer."

To digress for a moment on this issue: Despite the claims of the American Chamber of Commerce, the Heritage Foundation and similar bodies, most analyses suggest that the EPA is roughly jobs-neutral—although jobs may be lost in consequence of regulations, new jobs are created as a further consequence. Here's Josh Bivens of the Economic Policy Institute on February 7, 2012 ("The 'Toxics Rule' and Jobs"):

> On December 16, 2011, the Environmental Protection Agency finalized national standards for mercury, arsenic, and other toxic air pollutants emitted by power plants. Known as the "toxics rule," this ruling is a significant expansion of the Clean Air Act and will, according to nearly all expert opinion, lead to enormous benefits in terms of lower mortality and improved health outcomes for Americans. Judged as it should be—by balancing benefits to health against costs of compliance—the toxics rule is a clear win for Americans. Unfortunately, the debate over regulation more generally has strangely become fixated on jobs.
>
> This is unfortunate because standard economics clearly demonstrates that regulatory changes will generally have only trivial effects on job growth.

A major part of Pruitt's agenda—and that of the administration in general—seems to be to overturn sensible environmental regulations that were put in place by the preceding administration after prolonged consideration of the extant relevant science. We've already noted (page 385) Pruitt's overturning of the proposed ban of the pesticide chlorpyrifos, rejecting for unstated reasons (voices in his head?) the scientific evidence that it's a neurotoxin. He has also overturned an Obama-era regulation requiring fossil-fuel companies to keep track of and report methane emissions on their sites; since methane is an important greenhouse gas, it's essential to the welfare of all of us that the fossil-fuels industry keep this pollution to a minimum, but apparently the companies pleaded that the regulation added to their paperwork and accounting costs, and Pruitt was sympathetic to this momentous consideration.*

** In early July 2017 the US Court of Appeals for the District of Columbia Circuit ruled that the EPA did not have the legal authority to suspend this rule, but must first go through a process of public consultation.

Another useful measure introduced by Obama's EPA and due to take effect in 2018 would have restricted the quantities of toxic metals—lead, mercury, arsenic, etc.—that power plants are allowed to dump into our public waterways. Pruitt delayed the implementation of the measure until 2020, seemingly with a view to diluting it or letting it be gradually forgotten. Power-plant owners also profited from a Pruitt decision to review the rules concerning their dumping of coal ash near public waterways. Their real bonanza, though, came when the EPA agreed to reconsider a proposed regulation governing power plants' emissions levels during shutdown and startup.

There are countless other examples, all of them characterized by the Trump administration's willful disregard of environmental concerns—indeed, by a seemingly infantile determination to destroy the environment in the most malicious conceivable manner, as if this were somehow a virtuous thing—at an inevitable cost to the health and welfare of American citizens, and by no means only in the long term. Who in their right minds could possibly believe that dumping more mercury and lead into our drinking water is a good thing? Who would willingly risk increasing the levels of neurotoxins to which children are exposed? Who would want to increase the rate and severity of climate change by pumping additional methane into the atmosphere?

The customary excuse offered by Pruitt and his allies within the administration is that all of these destructive measures are being passed in the interests of the economy—in the interests of providing more jobs. Yet this doesn't wash. As noted, EPA regulations are, overall, more or less jobs neutral, and some of them are significant job creators. An example is the Clean Power Plan (CPP), proposed by Obama's EPA in June 2014 to reduce the levels of greenhouse gases emitted by coal-, gas- and oil-fired power plants. Reporting on the jobs aspect of this proposal a year later, in June 2015, the Economic Policy Institute criticized the EPA's estimates:

> Estimates made by the Environmental Protection Agency of the likely employment effects . . . . undercount both positive and negative influences on employment. This paper . . . finds that the CPP is likely to lead to a net increase of roughly 360,000 jobs in 2020, but that the net job creation falls relatively rapidly thereafter, with net employment gains of roughly 15,000 jobs in 2030.[120]

In other words, while the EPA's estimates may have been flawed, the CPP would nonetheless create a significant number of jobs. It would, however, cost the fossil-fuels industry money to implement (to put it another way, the industry would be expected to contribute something toward the reduction of its pollution, rather than as usual dumping the cost of cleaning up its messes on the public), and this was clearly a far more important factor than job-creation or public well-being in the minds of Trump and his ever-obliging facilitator, Scott Pruitt. In early October 2017, Pruitt's EPA proposed repealing the plan entirely.

Of course, who cares about greenhouse-gas emissions if you're so scientifically illiterate as to deny the science concerning climate change, and so arrogant as to believe this makes your benighted opinion somehow more valid than that of the people who're actually qualified to judge the matter? This variant of the Dunning–Kruger effect is widespread within the Trump administration, as we've seen, from Trump himself on down through individuals such as Secretary of Energy Rick Perry and, of course, EPA Administrator Scott Pruitt.

# # #

Bearing in mind Pruitt's history of suing the EPA for its efforts to limit pollution and save lives, it's pleasing to note that he hasn't had it all his own way, legally speaking, since taking office. Environmental groups sued over the agency's suspension of an Obama-administration rule concerning the safe disposal of dental amalgam (which contains toxic materials, including mercury); the American Dental Association agreed with the environmental groups. Fifteen states and the District of Columbia sued over Pruitt's delaying of a new regulation to control ozone pollution. Environmental groups and eight states sued over the administration's decision to delay indefinitely a regulation requiring the tracking and reporting of vehicular greenhouse-gas emissions on federal highways. In these and other instances, the administration has had to back down in the face of legal action.

It hardly needs to be pointed out that having repeatedly to go to court to prevent destructive acts by one's own government is no way to run a country. One of the unpredicted consequences of Donald Trump's ascension to the White House in 2017, and the "war on science" that he promptly declared through appointments such as that of Scott Pruitt, was a groundswell of scientists and retired scientists volunteering to run as Democratic candidates at every level, from congressional all the way down to school boards. Some of this movement was coordinated by a new organization, 314 Action (the name comes from the first three digits of π), founded by breast-cancer researcher Shaughnessy Naughton, who himself ran unsuccessfully for Congress in Pennsylvania in 2014 and 2016. Cited in *Mother Jones*,[121] Naughton explained the rationale of his group:

> Traditionally, the attitude has been that science is above politics, and therefore scientists shouldn't get involved in politics, and what that ignores is the fact that politicians are unashamed to meddle in science.

The number of STEM (Science, Technology, Engineering, Math) professionals in Congress, or in the US government as a whole, has typically been almost zero in recent decades; the only science represented in any numbers has been medicine. As any medical educator will tell you, it's today perfectly possible to sail through medical school without gaining any comprehension of science at all, which is one reason you come across creationist physicians—such as Ben Carson, and such as Paul "All that stuff I was taught about evolution and embryology and the Big Bang Theory, all that is lies straight from the pit of Hell" Broun (R–GA)—secure in their beliefs despite a supposedly scientific education.

As a consequence of the dearth of scientists in positions of political power we find some idiotic decisions being made by our elected representatives in the field of science. Leaving aside the imbecilities perpetrated by the likes of Lamar S. Smith (R–TX) as head of the House Science Committee and James Inhofe (R–OK) as chairman of the Senate Committee on Environment and Public Works and a senior member of the Committee on Commerce, Science, and Transportation, consider instead what might have been going through the minds of the congresscritters and senators who elected these profoundly incompetent individuals to such important scientific posts. Are our professional politicians so ignorant that they think science and technology *don't matter*, or ignorant to an even deeper level, where they think individuals such as Smith and Inhofe are in fact knowledgeable in the sciences?

Or, of course, have swaths of them been bought, through "campaign contributions," by special interests in the fossil-fuel and other relevant industries?

Whatever the case, 314 Action seeks to redress the balance. Whether the venture will be successful or not remains to be seen, but at the very least it will surely lay the groundwork for a more rational realignment of representative politics in the future.

Another tool that's being used against corrupt agency and department heads is whistleblowing. Of course, whistleblowers face retaliation. When Joel Clement, an expert on climate policy at the Department of the Interior, spoke out about the dangers climate change presents to the native populations in Alaska, he was in mid-July 2017 reassigned—almost certainly illegally—to what's essentially an accounting office within the department. A couple of months later he resigned, addressing his farewell to the department's boss, Ryan Zinke:

> Retaliating against civil servants for raising health and safety concerns is unlawful, but there are many more items to add to your resume of failure: You and President Trump have waged an all-out assault on the civil service by muzzling scientists and policy experts like myself; you conducted an arbitrary and sloppy review of our treasured National Monuments to score political points; your team has compromised tribal sovereignty by limiting programs meant to serve Indians and Alaska Natives; you are undercutting important work to protect the western sage grouse and its habitat; you eliminated a rule that prevented oil and gas interests from cheating taxpayers on royalty payments; you cancelled the moratorium on a failed coal leasing program that was also shortchanging taxpayers; and you even cancelled a study into the health risks of people living near mountaintop removal coal mines after rescinding a rule that would have protected their health.
>
> You have disrespected the career staff of the Department by questioning their loyalty and you have played fast and loose with government regulations to score points with your political base at the expense of American health and safety. Secretary Zinke, your agenda profoundly undermines the DOI mission and betrays the American people.

That's quite a catalogue. Clement finished his letter thus:

> My thoughts and wishes are with the career women and men who remain at DOI. I encourage them to persist when possible, resist when necessary, and speak truth to power so the institution may recover and thrive once this assault on its mission is over.

THE INCONVENIENT TRUTH

The earth moves closer to the sun every year.* We have more people. You know, humans have warm bodies, so is heat coming off? We're just going through a lot of change, but I think we are, as a society, doing the best we can.

—Pennsylvania State Senator Scott Wagner explains climate change to the masses, June 2017

---

* Someone seems to have told Wagner that our planet's orbit is elliptical. Every year, during the northern hemisphere fall/winter, it moves closer to the sun. But then, during the northern hemisphere winter/spring, it moves further away again. (The seasons are caused by the earth's axial tilt, not the varying distance from the sun.)

Of all the areas in which both Bush II and Trump administrations have abused science, the most prominent—and the most disastrous—is global warming. For decades it has been established science that the global climate is approaching a catastrophic change because of the release of greenhouse gases through the burning of fossil fuels. There may be other contributors towards this approaching cataclysm, but the human contribution is the only one we can do anything about. There is no real question but that this is the case. There are no dissenters among qualified climate scientists except a few mavericks—"few" being a misleadingly strong word for such a small number. The only real debate is over whether the outlook is grim, very grim, horrific, or worse than that.

This fact is inconvenient for those industries whose livelihood currently depends on burning fossil fuels, such as the oil and automobile corporations. It is therefore inconvenient also for politicians whose political or personal welfare depends on the financial contributions of those corporations. Ironically, many corporations, both outside and increasingly within the US, have realized there are healthy profits to be made from exploitation of renewable sources. Yet dinosaur corporations worldwide have entered a state of denial about the harshness of the situation, and, like spoiled children told to stop playing with a live grenade, have mounted vociferous protests against reality—as if reality, like politicians, could be swayed by bullying or bribery.

The Bush II administration's lie that "the jury is still out on climate change" was a constant catchphrase. As a consequence, much of the US public, unlike their counterparts throughout most of the rest of the world, is *still* unaware of just how bleak the future is likely to be unless effective—and drastic—action is taken almost immediately . . . if indeed it's not already too late to escape some of the worst effects. The legislators of the House and Senate have thus seen little political reason to curb those of their actions that, crazily, actually promote increases in the consumption of fossil fuels. Although President Obama, bypassing House and Senate, issued executive orders mandating enforced fuel economies in new automobiles and, most necessarily of all, encouraging the quest to find new, renewable, non-polluting energy sources, he did this late in his presidency, having, as we've seen, squandered years in the doomed attempt to obtain cooperation from a hostile House and Senate; and already, under the Trump administration, suicidal efforts are under way to reverse those initiatives.

It would be bad enough had the members of the Bush and now the Trump administrations confined their corruption of the public understanding of climate science to campaign speeches and the like. It's worse that through their legislative actions they have made the situation more parlous, not less. But they go further, using Stalinist techniques to distort or suppress the climate science they find distasteful in the reports produced by their own scientific advisory bodies.

Let us again look first at the activities of the Bush II administration.

One of the worst such incidents occurred in 2003 when the White House intervened to make a number of fundamental changes to a *Draft Report on the Environment* that had been produced by the EPA. Some of these were:

• Removal of all references to the 2001 report *Climate Change Science: An Analysis of Some Key Questions*, produced for the administration by the NAS's Commission on Geosciences, Environment and Resources. This report, which in strong terms supported the findings of the IPCC that climate change caused by human activities is imminent and its effects on human civilization will be catastrophic unless measures are taken immediately to ameliorate them, was effectively buried on its release through administration members grossly misrepresenting its conclusions. Now they insisted that mention of it be expunged from the EPA report.

• The removal of a record covering global temperatures over the past 1,000 years. Instead, according to a despairing internal EPA memo dated April 29, 2003, on the changes demanded by the administration: "Emphasis is given to a recent, limited analysis [that] supports the administration's favored message."

• Still quoting from that memo, "The summary sentence has been [deleted]: 'Climate change has global consequences for human health and the environment.' . . . The sections addressing impacts on human health and ecological effects are deleted. So are two references to effects on human health. . . . Sentences have been deleted that called for further research on effects to support future indicators."

• The administration inserted reference to the 2003 paper "Proxy Climatic and Environmental Change of the Past 1,000 Years" which, although it appeared in a peer-reviewed journal, *Climate Research*, was in large part funded by the American Petroleum Institute. This paper's conclusions had been comprehensively discredited in the literature and thus, quite correctly, ignored by the EPA.* (For more on this see page 74.)

---

* This was back in the day when at least a few Republicans in the House and Senate had the guts to speak out against blatant falsehood.

• Again from that internal EPA memo: "Uncertainty is inserted (with 'potentially' or 'may') where there is essentially none. For example, the introductory paragraph on climate change . . . says that changes in the radiative balance of the atmosphere 'may' affect weather and climate. EPA had provided numerous scientific citations . . . to show that this relationship is not disputed."

• "Repeated references now may leave an impression that cooling is as much an issue as warming."

The EPA discussed internally the best way of coping with these distortions of scientific fact, and eventually concluded that all it could do was delete entirely its report's section on climate change; anything else would grossly misrepresent the truth and make the EPA and its scientists subject to ridicule. The omission had what one assumes must have been the desired effect: it drew attention to the fact that the White House had, in essence, attempted to make the EPA a mouthpiece for its own ideology. Not just scientists but politicians—Republicans and Democrats alike,** although the administration was swift to smear them as "partisan"—criticized the administration's actions; among the most prominent Republican critics was Russell Train, who had been EPA Administrator under two Republican presidents, Nixon and Ford. The Republican EPA administrator whom the Bush White House had itself appointed, Christine Todd Whitman, soon departed in apparent disgust at the debasement of the truth she'd been expected to institute.

The emasculation of the EPA report is, unfortunately, just one example among many. In March 2005, Rick Piltz resigned from the Climate Change Science Program, alleging that political appointees within that body were acting so as to "impede forthright communication of the state of climate science [and attempting to] undermine the credibility and integrity of the program." On June 8, 2005, the *New York Times* published documents from Piltz's office which had been edited by Philip A. Cooney, a lawyer without scientific training who had represented the oil industry in its fight to prevent restrictions being placed on the emission of greenhouse gases and who was by now, incredibly, Chief-of-Staff of the White House Council on Environmental Quality. The effect of Cooney's edits was, as the *New York Times* summarized, "to produce an air of doubt about findings that most climate experts say are robust." In one instance Cooney removed from a report on the projected impact of global warming on water supplies and flooding a paragraph outlining likely reductions of mountain glaciers;

his marginal note stated that the paragraph was "straying from research strategy into speculative findings/musings."

That same day, June 8, 2005, Bush spokesman Scott McLellan denied charges that the administration was trying to make human-derived climate change appear a matter of scientific uncertainty rather than known reality, and claimed it was standard procedure for political appointees to edit government-sponsored scientific reports before publication—surprising news to many. Two days later Cooney resigned from the CEQ, but he didn't stay unemployed for long: within days he was hired by ExxonMobil.

A new outcry arose in early 2006 when distinguished NASA scientist James Hansen went public with complaints that his discussions of climate science were being impeded and censored by political appointees within the NASA administrative staff. This was not the first time Hansen had been at the center of a storm over politically motivated censorship of science. Back in 1988, under the George H.W. Bush administration, he had testified before Congress that he was 99 percent certain long-term global warming, probably due to overuse of fossil fuels, had already begun, and that the time to start doing something about it was now. In 1989, he discovered this testimony had been "edited," presumably at the behest of the oil and automobile industries, by the Office of Management and Budget to stress "scientific uncertainties."

Now he was up against an even worse case of censorship, motivated not just by political but also, incredibly, by Christian fundamentalist considerations. He had been told by NASA officials that the Public Affairs staff had been instructed to review his forthcoming lectures, papers, newspaper requests for interviews, etc. This was because of Hansen's insistence on speaking about the urgent subject of climate change—a not unreasonable insistence, since his NASA job was to direct the computer simulation of global climate at the Goddard Institute.

Initially the NASA administrative staff came out fighting. Dean Acosta, Deputy Assistant Administrator for Public Affairs, said there were no special restrictions on Hansen that did not apply to all NASA scientists: they were free to discuss scientific findings, but policy statements should be left to policy makers and to appointed political spokesmen—a position that might seem understandable until you start wondering who decides where the line is drawn between scientific and policy statements. Clearly Hansen's justified warnings about climate change were being classified as the latter.

By early February 2006, attention was focused on George C. Deutsch, a 24-year-old public affairs officer at NASA, a Bush appointee whose sole qualification appeared to be that he'd been a part of the 2004 Bush–Cheney re-election campaign. In one specific instance, when National Public Radio wanted to interview Hansen, Deutsch refused permission; his reasoning was, he told a Goddard Institute staffer, that NPR was the country's "most liberal" media voice, while his, Deutsch's, job was to "make the President look good." A few days later it was revealed Deutsch had instructed the designer of the NASA website to add the word "theory" after every mention of the Big Bang:

It is not NASA's place [he memoed], nor should it be, to make a declaration such as this about the existence of the universe that discounts intelligent design by a creator. . . . This is more than a science issue, it is a religious issue. And I would hate to think that young people would only be getting one-half of this debate from NASA. That would mean that we had failed to properly educate the very people who rely on us for factual information the most.

This is NASA, remember?

It emerged that scientists at the climate laboratories of the National Oceanic and Atmospheric Administration (NOAA), accustomed in earlier years to taking journalists' phone calls whenever they wanted, were now permitted to do so only if an interview had been given the green light by Washington and only if a public affairs officer monitored the conversation. Meanwhile, those rare government science employees who disagreed about the need to curb fossil-fuel emissions were permitted to lecture and publish at will.

To his great credit, the Chairman of the House Science Committee, Sherwood Boehlert, despite being a Republican and therefore expected to toe the party line, backed Hansen: "Political figures ought to be reviewing their public statements to make sure they are consistent with the best available science. Scientists should not be reviewing their statements to make sure they are consistent with the current political orthodoxy." (Sadly, Boehlert retired from Congress in 2006.) Unsurprisingly, not all of his GOP colleagues concurred; a spokesman for James Inhofe stated: "It seems that Dr. Hansen, once again, is using his government position to promote his own views and political agenda, which is a clear violation of governmental procedure in any administration." Read that again: Hansen's "political agenda"?

This was to become a theme in the administration's attempts to smear Hansen rather than correct what was obviously an iniquitous situation. Not long afterward, it was discovered by Nick Anthis of the *Scientific Ac-*

tivist blog[122] that Deutsch, the youthful political appointee whose actions had been Hansen's final straw, had "exaggerated" his resume a little when applying for the NASA post. Within hours Deutsch resigned his position. Speaking a day later, on February 9, to Radio WTAW-AM, he said:

> Dr. James Hansen has for a long time been a proponent of a particular global warming agenda . . . What he is willing to do is to smear people and to misrepresent things to the media and to the public to get that message across. What's sad here is that there are partisan ties of his all the way up to the top of the Democratic Party, and he's using those ties, and using his media connections, to push an agenda . . . and anybody who is even perceived to disagree with him is labeled a censor and is demonized and vilified in the media, and the media of course is a willing accomplice here . . . What you do have is hearsay coming from a handful of people who have clear partisan ties, and they're really coming after me as a Bush appointee and the rest of the Bush appointees because this is a partisan issue. It's a culture-war issue, they do not like Republicans, they do not like people who support the President, they do not like Christians, and if you're perceived to be disagreeing with them or being one of those people they will stop at nothing to discredit you.

As ever, the administration's technique, when confronted by criticism, was to accuse the critics of "just playing partisan politics," but it was a little startling to find it in this context—and as startling to discover that Deutsch regarded scientific conclusions as being somehow obedient to agendas. As the *New York Times* remarked in a February 9 editorial, "The shocker was not NASA's failure to vet Mr. Deutsch's credentials, but that this young politico with no qualifications was able to impose his ideology on other agency employees."

It would be wrong to think, though, that the problem was merely one of a young zealot. All this time, further horror stories were emerging from NASA about political manipulation of science, especially in the run-up to the 2004 presidential election. Bush had declared his great visionary goal was for the US to place a man on Mars, and the instructions were passed down through the NASA hierarchies that all NASA news releases should stress the contribution NASA scientists were making toward this "vision statement."

Some of the results were ludicrous. In a December 2004 press release about research on wind patterns and warming in the Indian Ocean, JPL scientist Tong Lee was quoted as saying that some of the methods used in the study would be useful also in space exploration and studying the climate systems on other planets. Queried by his colleagues as to this statement, a startled Lee, learning of it for the first time, demanded that it be removed; NASA's press office duly removed it, but not NASA's public-affairs office in Washington, which retained it on the NASA website.

Finally, in late February 2006, NASA Administrator Michael D. Griffin began to conduct a review of NASA's communications policies that was hailed by scientists there as a genuine attempt to root out at least some of the political censorship that had become endemic in the agency. Hansen himself expressed cautious optimism.

Yet in 2006, Pieter Tans and James Elkins, senior scientists at the NOAA's Boulder, Colorado, laboratory, recounted to federal investigators how between 2000 and February 2005, when Russia ratified the Kyoto Protocol, divisional director David Hofmann had a standing instruction that they were to omit the word "Kyoto" from all their conference presentations. Further, when in late 2005 Tans was organizing the Seventh International Carbon Dioxide Conference, he was instructed by Hofmann that the term "climate change" must not appear in the titles of any of the presentations. (Tans ignored the prohibition.) In response to the latter account, Hofmann said he must have been misunderstood: he had merely been saying that a conference on global CO_2 measurements should stick to that, and not veer off into discussions of greenhouse gases and climate change!

Belatedly driven by the force of public opinion to admit anthropogenic climate change was a reality and something should be done about it, Bush pledged in late 2006 that increased research would be the centerpiece of his new climate strategy. Within weeks it emerged from a National Academy of Sciences study that over the previous two years NASA's earth-sciences budget had been reduced by 30 percent, with no plans for any reversal of the trend—indeed, the reckoning was that by 2010 the number of operative earth-observing instruments in US satellites would fall by 40 percent. This was in the context of a little-changed NASA budget; the funds that should have been spent on this critical work were instead being put toward preparation for a manned lunar base, a media-friendly scheme imposed on NASA by political fiat (and now largely forgotten). As a *New York Times* editorial summarized on January 22, 2007: "Studies that affect the livability of the planet seem vastly more consequential than completing a space station or returning to the Moon by an arbitrary date."

One crucial aspect in assessing global warming is assessing the earth's albedo (reflectivity); through knowledge of the albedo one can infer our planet's net energy balance. To this end, NASA built the Deep Space Climate Observatory, designed to be placed at the Lagrange–1 stationary orbit between the earth and the sun. Thanks to Bush administration intransigence, the project was frozen for six years: the satellite simply sat in a warehouse at Goddard. The Obama administration revived the project but even so, allowing for reactivation time, it was February 2015 before the satellite was finally launched. Well over a decade had been wasted.

# # #

The Bush administration's continued denial of global warming was assisted in no small measure by the chairman (2003–2007 and since 2015) of the Senate's Environment and Public Works Committee, James Inhofe, who has used this position to promote the notion that global warming is simply a grandiose hoax put about by tree-huggers:* "The greatest climate threat we face may be coming from alarmist computer models." It is hard to work out if Inhofe lies about science for reasons of political expediency or simply through stupidity. Or both.

His claim continues to be that the earth is experiencing a natural warming trend that began around 1850, at the end of the four-century cool period known as the Little Ice Age, and that human activities have nothing to do with it; thus it's pointless for us to institute measures like capping greenhouse gas emissions. To back up his case, Inhofe points to the period between the 1940s and the 1970s, when global temperatures fell slightly even as fossil-fuel consumption rose. This was certainly the case in the Northern Hemisphere, but represents only a minor fluctuation in a curve that has gone steadily upward since about the time of the Industrial Revolution; the earth's climate is an incredibly complex mechanism to which numerous factors contribute, so short-term fluctuations are to be expected in any of its trends. This is why there are disagreements in detail between the various computer models Inhofe decries as unreliable and alarmist. What he omits to mention is that they all agree to within a very narrow margin on the general picture. Focusing on disagreements over the timing of the imminent crisis is like arguing over whether one's speeding car is ten or fifteen feet from the edge of the cliff toward which it's hurtling.

On December 6, 2006, Inhofe held a hearing concerning media coverage

---

* This conspiracy theory was also advanced by novelist Michael Crichton in *State of Fear* (2004). Alarmingly, in the wake of the novel's publication Crichton was called to the White House to consult as a supposed expert on the "debate" concerning the science of climate change.

of climate change. Laughably, he claimed that the popular US news media were overhyping the subject, when a steady complaint from scientists and the informed public since before the dawn of the twenty-first century has been that the mainstream news media largely downplay, misrepresent or ignore it. This situation began to improve in 2006, largely due to a rise in public awareness brought about by the publicity surrounding the movie *An Inconvenient Truth* (2006), the climax of a campaign the movie's writer and mainstay, Al Gore, had been waging—despite much derision from rightist pundits—for some thirty years: mainstream news outlets could no longer get away with almost completely ignoring something that most of their audience was talking about.* Alas, the effect was largely temporary: today most of the broadcast media underplay the existential hazard presented by anthropogenic global warming, or even actively work to undermine perception of the reality.

It's to be hoped that the 2017 release of Gore's follow-up, *An Inconvenient Sequel: Truth to Power*, directed by Bonni Cohen and Jon Shenk, will once more influence that situation for the better. Reviewing the movie in *Nature*, Michael E. Mann wrote:[123]

> This sequel is deliciously inconvenient, and for several reasons. It is inconvenient to the vested interests who had hoped that Gore would just give up. Their campaign of vilification was intended both to deter his ongoing outreach efforts and to strike fear into the hearts of others who might consider stepping up to the plate . . . But, as his subtitle promises, Gore is still speaking truth to power . . .
>
> Finally, the film casts an inconvenient light on humanity. It is astonishing that we're still mired in a political debate about whether climate change even exists when, with each passing year of insufficient action, the challenge of averting a catastrophe becomes ever greater.

To return to Inhofe's hearing, it had its hilarious moments. One came when the geologist David Deming, a stalwart of the anti-environmentalist National Center for Policy Analysis, began his closing statement thus:

---

* Following the world premiere of the blockbuster movie *The Day After Tomorrow* (2004) in New York, Gore's slide show about global warming was seen by Laurie David, wife of TV comedian Larry David; Gore had been giving the show here and there all over the world since his disputed loss of the 2000 presidential election. Laurie David was sufficiently inspired by Gore's presentation and passion that she gathered a team to make a feature movie of it. In February 2007, *An Inconvenient Truth*, directed by Davis Guggenheim, received an Oscar.

As far as I know, there isn't a single person anywhere on earth that's ever been killed by global warming. There is not a single species that's gone extinct. In fact, I'm not aware, really, of any deleterious effects whatsoever. It's all speculation.

The extinction over the past few years of numerous species owing to global warming is well documented—as a single example, over 70 species of tree frogs have gone. At the time of writing, there are major fears that polar bears may be driven to extinction within a matter of decades, possibly years, as their Arctic habitat melts. As for human deaths due to global warming, Deming presumably forgot about the tens of thousands of heat-related deaths in Europe, 10,000 in France alone, during the summer of 2003.*

# # #

The political corruption of science under the Bush administration was not confined to the politicians themselves—far from it: propagandists in the broadcast media played a major part in creating and maintaining the whole mess. Some were more egregious than others; listing the scientific distortions perpetrated by, say, radio shoutmeister Rush Limbaugh would fill a book.** To cite just a single media incident as representative of a myriad others, in the January 21, 2006, edition of the TV series *The Journal Editorial Report*, broadcast by Fox News and linked to the *Wall Street Journal*, *WSJ* editorial page editor Paul A. Gigot and his deputy, Daniel Henninger, discussed a report recently released by the Max Planck Institute. Henninger used it as an excuse to attack the Kyoto Accord:

> . . . the eminent Max Planck Institute in Heidelberg, Germany, has just reported in *Nature* magazine that plants, trees, forests emit 10 percent to 30 percent of the methane gas into the atmosphere. This is a greenhouse gas, the sort of stuff the Kyoto Treaty is meant to suppress. So, this is causing big problems for the tree-huggers, if plants, in fact, do cause greenhouse gases,

---

* This conspiracy theory was also advanced by novelist Michael Crichton in *State of Fear* (2004). Alarmingly, in the wake of the novel's publication Crichton was called to the White House to consult as a supposed expert on the "debate" concerning the science of climate change.

** Chapter 16 of my book *Denying Science* (2011) comprises a very brief rundown of the climate-change-intensified weather catastrophes of the single year 2010. The death toll, and the cost in other human suffering, was almost unimaginably high.

and I have just one message for them: the next time you are out for a walk in the woods, breathe the methane.

Fair comment, one might suppose, were it not for the fact that the report's very first sentence specifies that, because of human activities, the atmospheric concentration of methane has approximately tripled since the pre-industrial era, in which context the 10–30 percent contribution of plants is irrelevant; before the advent of industrialization, it was obviously far higher in percentage terms (when human contributions were far smaller). The absolute level of plant methane production has been gradually decreasing during the same period, again thanks to human activities (such as deforestation).

Just to reinforce their point, the report's authors issued a press release on January 18, three days before the Gigot–Henninger distortion, in which they spelled out the truth: "The most frequent misinterpretation we find in the media is that emissions of methane from plants are responsible for global warming." They add that, while reforestation might trivially increase methane production, it at the same time has the important beneficial effect of increasing the amount of plant absorption of CO_2. While the excuse could be made that Gigot and Henninger were guilty of no more than bad journalism—failure to check facts—or plain ignorance, this seems unlikely given their positions at the *Wall Street Journal*, where there are fact checkers aplenty. The most probable explanation is that their presentation was what would later be dubbed "fake news."

That same week, on January 25, CNN weatherman Chad Myers put forward his own hypothesis. Atmospheric temperatures only *seem* to be rising as much as they are because buildings are tending to spring up around the thermometers used by climate scientists, and of course these suddenly mushrooming metropolitan areas are warmer than the surrounding countryside. Climate scientists are obviously too stupid to take account of this effect. Myers did not expand on quite how his hypothesis explains the increased melting of the polar icecaps.

---

* Such as *The Way Things Aren't: Rush Limbaugh's Reign of Error* (1995) by Steve Rendall, Jim Naureckas and Jeff Cohen. Howlers include that it is volcanoes, rather than manmade chlorofluorocarbons, that damage the ozone layer; that it has not been proven that nicotine is addictive; that low levels of dioxin exposure aren't harmful; that the failure rate of condoms in preventing HIV infection is as high as 20 percent; and that there were fewer acres of forest land in the US in the time of the Founders than there are now.

In despair at such stuff, in January 2007 Heidi Cullen, a TV meteorologist and host of *The Climate Code*, called upon the American Meteorological Society to withdraw its customary endorsement from TV and radio weather presenters who tried to plant doubts in viewers' minds about the reality of climate change. She pointed out that if, for example, a meteorologist claimed on air that tsunamis were caused by the weather, the meteorologist would be immediately recognized as incompetent and probably fired. Meteorologists who denied global warming should be treated likewise.

#

The hurricane season of 2017 made it increasingly hard for anyone in the US to continue denying that climate change is increasing the severity of major weather events.* In mid-August, Hurricane Harvey smashed into Texas; the first major hurricane to impact the mainland US since Hurricane Wilma in 2005 (the same year Hurricane Katrina devastated New Orleans), it caused the loss of at least 76 US lives. Less than two weeks later, the even more powerful Hurricane Irma hit Florida; it cost 95 US lives (three of which were in Puerto Rico, four in the US Virgin Islands). Hurricane Jose was of similar strength to Irma, but luckily stayed largely clear of land; never before on record have there been two hurricanes the strength of Irma and Jose operating simultaneously. Yet more powerful still was Hurricane Maria, which ravaged Puerto Rico, essentially wiped out the electrical grid, destroyed the island's already rocky economy, and resulted in at least several hundred deaths, with the toll still climbing—some counts put it at over a thousand. The slowness and inefficiency of the Trump administration's response, its deliberate downplaying of the ongoing crisis, and its self-adulating complacency. have made matters worse. By the end of 2017, months after the hurricane hit Puerto Rico, only about half the island had electrical power and there were continuing large-scale problems with both potable water supplies and cell phone service.

Another casualty of the 2017 hurricane season was Gonzo, the Gulf-

---

* One could make the same observation about the increasing length and severity of the wildfire season. So far as storms are concerned, and using the bills for damage as a metric, 2017 was the worst year in US history since records began. Damages ran to an estimated $306 billion; the next-worst year, 2005, saw a bill (adjusted for inflation) of $215 billion. Of course, these bills take no account of lives lost and other human suffering. We pay quite heavily for our corporations' ongoing profits and our conspiracy-theorizing politicians' whims.

stream IV aircraft that the NOAA uses to gather data on these huge storms by flying over them at high altitude. In the eight days before being grounded for repairs in September, Gonzo had suffered no fewer than three malfunctions. Succeeding administrations have declined to invest in a backup Gonzo. With stronger hurricanes predicted as the pattern for the future, perhaps at last it's time to make that investment, but no one's holding their breath.

Rightwing pundit Ann Coulter tweeted in response to Hurricane Harvey:

> I don't believe Hurricane Harvey is God's punishment for Houston electing a lesbian mayor. But that is more credible than "climate change."*

Prior to the arrival of Hurricane Irma, broadcaster Rush Limbaugh told his fans that he wasn't a meteorologist but "Just as I am the go-to tech guy in my family and here on the staff, when it comes to a hurricane bearing down on south Florida, I'm the go-to guy," and explained his reasoning:

> Now, in the official meteorological circles, you have an abundance of people who believe that manmade climate change is real. And they believe that Al Gore is correct when he has written—and he couldn't be more wrong—that climate change is creating more hurricanes and stronger hurricanes. . . . So there is a desire to advance this climate change agenda, and hurricanes are one of the fastest and best ways to do it. You can accomplish a lot just by creating fear and panic. You don't need a hurricane to hit anywhere. All you need is to create the fear and panic accompanied by talk that climate change is causing hurricanes to become more frequent and bigger and more dangerous, and you create the panic, and it's mission accomplished, agenda advanced. . . . The hurricane doesn't even need to hit in order to for the agenda to be advanced.

His advice to his listeners in southern Florida was to ignore the warnings and sit tight, because Irma was a liberal hoax. Needless to say, a few days later, Limbaugh himself evacuated from Florida in search of safety from the impending hoax.

It's easy enough to laugh at blowhards like Coulter and Limbaugh, but what's not so funny is that (a) far too many members of the US public believe them (obviously we have no way of knowing how many of Lim-

---

* Actor Ron Perlman tweeted back: "Ann, ANN, you're supposed to take the anti-psychotic WITH FOOD! WITH FOOD Ann!"

baugh's audience took his advice and lost their lives in consequence), and (b) the President of the United States is one of them. He has quite solemnly declared that climate change is a hoax invented by the Chinese in order to damage the US economy.

This "rationale" lies behind his selection of nominees to head important scientific departments and agencies such as the Department of Energy, NASA and the EPA. And it lies behind the behavior of those agencies' heads and their equally science-denialist staffers, which is reminiscent of nothing more strongly than those morons—and they do exist—who deliberately rig their pickups to emit additional greenhouse gases in a noxious plume of exhaust fumes. It's as if nothing will satisfy them short of poisoning their own offspring and destroying their own food and living space.

Which would be all well and fine except that they're intent on taking the rest of us with them.

# # #

The Global Change Research Act of 1990 specified, among other things, that thirteen relevant agencies of the US government should collaborate to produce periodic National Climate Assessments (NCAs). NCAs have appeared in the years 2000, 2009, 2014 and 2017; the last of these, appearing in November 2017, discommoded the Trump administration in that it stated plainly what everyone already knew outside the selectively ignorant few, that the science overwhelmingly indicates we're speeding toward climate catastrophe unless we rapidly and radically cut down on our greenhouse-gas emissions and other bad habits, such as deforestation and desertification.

The next National Climate Assessment is scheduled to appear in 2018; bets are open as to whether it will or not, and as to whether it'll contain science or science denial. The auspices are not especially good. In August 2017 the NOAA's acting administrator, political appointee Ben Friedman, announced the disbandment of the Advisory Committee for the Sustained National Climate Assessment, a fifteen-strong panel intended to recommend to policy makers and the private sector the actions that should be taken to ameliorate the worst effects of global warming. For an administration whose approach to climate change is rooted in the pretense that it doesn't exist, the continued existence of this panel was obviously problematic.

In January 2018, however, the panel's former chair, Richard Moss, an adjunct professor in the Department of Geographical Sciences at University of Maryland, announced that the committee was reforming, this time outside the aegis of the federal government. The panel's new incarnation will be based at Columbia University's Earth Institute and financed by a group of entities including New York State. According to New York Governor Andrew Cuomo's announcement of the state's support, "The Advisory Committee will continue its critical work without political interference and provide the guidance needed to adapt to a changing climate."

We'll have to wait to find out how much of an embarrassment this will prove to be for the administration's anti-scientific worldview.

Another reverse that worldview suffered came in the same month, January 2018. A few months earlier Energy Secretary Rick Perry had put forward a scheme for a new rule that, he claimed, would guarantee greater stability for the national grid: preferential treatment would be given to electricity producers who could show themselves capable of stockpiling ninety days' worth of fuel onsite. It was immediately obvious this was merely a cunning plan (as *Blackadder*'s Baldrick would put it) to promote the interests of fossil-fuels producers and nuclear plants, since it's rather hard to stockpile sunshine and wind. The Federal Energy Regulatory Commission unanimously agreed with this assessment of Perry's intention, issuing a binding ruling on January 8 that rejected his scheme.

Back to the drawing board for Perry.

# # #

Back in September 2015 candidate Trump spelled out his thinking on climate change in a radio interview with Hugh Hewitt:*

Well, first of all, I'm not a believer in global warming. And I'm not a believer in man-made global warming. It could be warming, and it's going to start to cool at some point. And you know . . . in the 1920s, people talked about global cooling. I don't know if you know that or not. They thought the Earth was cooling. Now, it's global warming. And actually we've had times where the weather wasn't working out, so they changed it to extreme weather, and they have all different names, you know, so that it fits the bill. But the problem we have, and if you look at our energy costs, and all of the things that we're doing to solve a problem that I don't think in any major fashion exists. I mean, Obama thinks it's the number one problem of the world today. And I think it's very low on the list. So I am not a believer, and I will, unless somebody can prove something to me, I believe there's weather. I believe there's change, and I believe it goes up and it goes down, and it goes up again. And it changes depending on years and centuries, but I am not a believer, and we have much bigger problems. You know, I talk about global warming. You know, to me, the worst global warming, and I mentioned this to you once before, is nuclear warming. That's our global warming. That's what I see, because we have incompetent people, and we have these rogue nations, and not even rogue nations anymore. . . .

Now do you understand?

# # #

By far the most visible of all the Trump administration's actions to harm the climate so far has been the president's decision to withdraw the US from the 2015 Paris Agreement.

This accord, signed by all except three countries in the world,* is a remarkably fluid affair. Nations set their own targets for reductions in greenhouse gas emissions and promise to report their progress, but there's no penalty for failure to achieve those targets. The idea is that the accord works not through mandates but through national honor. And most nations seem to be responding well to the challenge: almost every week, it seems, we learn of a new instance of a nation announcing plans to improve upon its previously stated targets. While a lot of the countries making these announcements are small, low-polluting ones, sometimes it's one of the heavy hitters, such as India or China.

In June 2017, Donald Trump announced that he was withdrawing the US from the accord* on the grounds that to stay in it would damage the US economy. Immediately economists all over the world demonstrated that this reasoning was bunk, but perhaps the best explanation came the following day from Katharine Hayhoe, director of Texas Tech University's Climate Science Center and, lest there be accusations of liberal hoaxing, a conservative Christian. As cited in *Nature*, she said:

---

* The absentees were the Vatican, Nicaragua (because it believed the agreement did not go far enough), and Syria (because of the civil war). In September 2017 Nicaragua changed its mind and signed the Accord; in November 2017 Syria followed suit. Although the Vatican traditionally maintains only an observer role in such international agreements, there have been scattered reports that it too is seeking a way to sign up. Certainly Pope Francis has repeatedly stated his view that it is a Christian duty to counter climate change and thereby save millions of human lives.

The biggest loser from the decision could be the United States itself. Why? Because although the Paris agreement is a climate treaty, a triumph for evidence-based decision-making, it's also much more: a trade agreement, an investment blueprint and a strong incentive for innovation in the energy and the economy of the future.

Earlier this week, India broke its own record for the lowest bids for electricity from solar power. Last month, Ernst & Young listed its most attractive markets for renewables: the United States came third, behind China and India. And earlier this year, China announced a US$360-billion investment in clean energy to create 13 million new jobs. The US announcement shows that it will be doing its best to turn back the clock, while the rest of the world accelerates into the future.

Hayhoe could have added that the decision did great damage to US foreign policy and international diplomacy. Who in the future will trust an administration that, on a matter of such global importance, clearly regards major treaties signed by its predecessor as barely worth the paper they're printed on? Making oneself an international pariah is a further source of economic damage.

Immediately after Trump's June 1 declaration that the US would withdraw from the accord, the governors of three states—Washington, New York and California—announced the formation of a new grouping, the United States Climate Alliance, "a coalition that will convene US states committed to upholding the Paris Climate Agreement and taking aggressive action on climate change."

Another group was Climate Mayors, a coalition of US mayors determined to fight climate change no matter what the Trump administration might do. As of 11pm on June 1, Climate Mayors had 92 members, including many representing major cities. By the last time I checked, in early January 2018, this number had risen to, by my count, 393. The credo of the organization reads, in part:

> We will continue to lead. We are increasing investments in renewable energy and energy efficiency. We will buy and create more demand for electric cars and trucks. We will increase our efforts to cut greenhouse gas emissions, create a clean energy economy, and stand for environmental justice. And if the President wants to break the promises made to our allies enshrined in the historic Paris Agreement, we'll build and strengthen relationships around the world to protect the planet from devastating climate risks.[124]

---

* The earliest the withdrawal can actually take effect is November 2020.

Internationally, Trump's announcement of the US withdrawal was greeted with a mixture of, from the US's allies, sorrow and derision, and, from the country's economic and/or political rivals, ill concealed glee. Here, for example is an excerpt from the editorial in the June 3, 2017 *South China Morning Post*:

China has Leadership Role in Fight against Climate Change
China's stature as a nation with global interests at heart has been amplified with US President Donald Trump's decision to pull his country out of the Paris climate agreement. Along with the EU, Beijing now has a leadership role in helping ensure the world brings down temperatures to avoid rising seas and severe weather conditions.

The German tabloid newspaper *Berliner Kurier* was a tad more succinct in its June 2 headline:

Earth to Trump: Fuck You

In his speech announcing the withdrawal, Trump made the mistake of saying: "I was elected to represent the citizens of Pittsburgh, not Paris"—a phrase that must have sounded so folksily appealing when it rolled off the speechwriter's pen that no one paused to check it for meaning. Bill Peduto, Pittsburgh's mayor, immediately retorted that Pittsburgh wholeheartedly supported the Paris Agreement. In the 2016 presidential election, over 75 percent of Pittsburgh's voters chose Hillary Clinton. Furthermore, while the city's economy used to be based heavily on coal and steel, it has turned itself around and now has a thriving industry based on technology and renewables. Not just the city's mayor but its primary newspaper, the *Post-Gazette*, and it seems the bulk of its population are strongly in favor of green energy and action on climate change.

At *Climate News Network*, Alex Kirby commented:

I was elected to represent the citizens of Pittsburgh, not Paris"—a wonderfully clear statement of [Trump's] inability to recognize that the Earth shares one atmosphere.

Our Responsibility

Other governments, in other times, have corrupted science for short-term political gain or for longer-term ideological reasons. In the last three chapters we've looked at just three of the worst offenders. There are examples of governments imposing their scientific dogma on the populace from earlier centuries, but they're relatively rare outside theocracies: the political corruption of science is very much a modern phenomenon. There are, too, smaller-scale examples than these of modern governments attempting to pervert one or another aspect of science; one thinks immediately of the South African government's mercifully brief flirtation with "traditional remedies" as a counter to the spread of AIDS. But such episodes are dwarfed by the systematic onslaught mounted by the three regimes treated here.

It might be expected that, with the internet offering ever-greater freedom of communication, such onslaughts would become progressively harder to mount. The world's experience during the twenty-first century has been the opposite. The propaganda technique known as "fake news," deliberate falsehood put out under the guise of genuine news in order to bamboozle and confuse the voting public, has corrupted the communal understanding of science as much as it has any other field of public knowledge . . . as too has the appropriation of the term "fake news" to be brayed moronically at any piece of genuine news that happens to be inconvenient to one's ideology.

Unless we, the public of every nation, maintain constant vigilance, we can expect authoritarian regimes everywhere to recognize the benefits— however illusory those benefits might in fact be—of corrupting science at its roots. If we let them get away with it, we can indeed expect the arrival of, in every sense of the term, a new dark age.

ENDNOTES

1. In *Lying Truths* (1979), edited by Ronald Duncan and Miranda Weston-Smith.

2. An extensive discussion of the case appears in Horace Freeland Judson's *The Great Betrayal* (2004).

3. "The Effect of Cyclooxygenase-2 Inhibition on Analgesia and Spinal Fusion," *Journal of Bone and Joint Surgery,* March 2005.

4. "Evaluating the Analgesic Efficacy of Administering Celecoxib as a Component of Multimodal Analgesia for Outpatient Anterior Cruciate Ligament Reconstruction Surgery" by S.S. Reuben, E.F. Ekman and D. Charron, *Anesthesia & Analgesia,* July 2007.

5. "Perpetuation of Retracted Publications Using the Example of the Scott S. Reuben Case: Incidences, Reasons and Possible Improvements" by Helmar Bornemann-Cimenti, Istvan S. Szilagyi and Andreas Sandner-Kiesling. Abstract at https://link.springer.com/article/10.1007/s11948-015-9680-y.

6. http://retractionwatch.com/the-retraction-watch-leaderboard/.

7. Published online at http://stanford.edu/~dbroock/broockman_kalla_aronow_lg_irregularities.pdf.

8. "The Economics of Reproducibility in Preclinical Research" by Leonard P. Freedman, Iain M. Cockburn and Timothy S. Simcoe, *PLOS Biology,* June 9, 2015. https://doi.org/10.1371/journal.pbio.1002165.

9. "The Truth Wears Off: Is There Something Wrong with the Scientific Method?" by Jonah Lehrer, *New Yorker,* December 13, 2010. http://www.newyorker.com/magazine/2010/12/13/the-truth-wears-off.

10. "1,500 Scientists Lift the Lid on Reproducibility" by Monya Baker, *Nature,* May 25, 2016. http://www.nature.com/news/1-500-scientists-lift-the-lid-on-reproducibility-1.19970.

11. *PLOS Medicine,* August 30, 2005. https://www.ncbi.nlm.nih.gov/pmc/articles/PMC1182327/.

12. https://www.omicsonline.org/open-access/the-refutation-of-the-climate-greenhouse-theory-and-a-proposal-for-ahopeful-alternative.php.

13. "Editor of New 'Sham Journal' Is Climate Science Denier with Ties to Heartland Institute" by Graham Readfearn, June 11, 2017. https://www.desmogblog.com/2017/06/11/new-sham-journal-OMICS-climate-science-denier-ties-heartland-institute.

14. "Defending Science" by Rudy Baum, *Chemical & Engineering News,* June 9, 2008. http://cen.acs.org/articles/86/i23/Defending-Science.html.

15. "Microbes and the Days of Creation" by Dr. Alan Gillen, *Answers Research Journal,* January 16, 2008. https://answersingenesis.org/days-of-creation/microbes-and-the-days-of-creation/.

16. See https://www.splcenter.org/fighting-hate/extremist-files/group/alliance-defending-freedom.

17. The DOAJ website is at https://doaj.org/.

18. Both are cited in "No More 'Beall's List'" by Carl Straumsheim, *Inside Higher Ed*, January 18, 2017. https://www.insidehighered.com/news/2017/01/18/librarians-list-predatory-journals-reportedly-removed-due-threats-and-politics.

19. "Journal Accepts Bogus Paper Requesting Removal from Mailing List" by Michael Safi, November 25, 2014. https://www.theguardian.com/australia-news/2014/nov/25/journal-accepts-paper-requesting-removal-from-mailing-list.

20. "iOS Just Got a Paper on Nuclear Physics Accepted at a Scientific Conference" by Christoph Bartneck, October 20, 2016. http://www.bartneck.de/2016/10/20/ios-just-got-a-paper-on-nuclear-physics-accepted-at-a-scientific-conference.

21. "FTC Charges Academic Journal Publisher OMICS Group Deceived Researchers," August 26, 2016. https://www.ftc.gov/news-events/press-releases/2016/08/ftc-charges-academic-journal-publisher-omics-group-deceived.

22. "Who's Afraid of Peer Review?" by John Bohannon, *Science*, October 4, 2013.

23. Learned Publishing, September 19, 2016. The paper is, alas, behind a paywall.

24. "Predatory Journals Hit by 'Star Wars' Sting." http://blogs.discovermagazine.com/neuroskeptic/2017/07/22/predatory-journals-star-wars-sting/.

25. "Predatory Journals Recruit Fake Editor" by Piotr Sorokowski, Emanuel Kulczycki, Agnieszka Sorokowska and Katarzyna Pisanski, *Nature*, March 22, 2017. http://www.nature.com/news/predatory-journals-recruit-fake-editor-1.21662.

26. Reprinted in *The Sociology–Philosophy Connection* (1999).

27. "What the Social Text Affair Does and Does Not Prove" by Alan Sokal, *A House Built on Sand* (1998) edited by Noretta Koertge—see Bibliography.

28. http://www.skeptic.com/downloads/conceptual-penis/23311886.2017.1330439.pdf.

29. "The Conceptual Penis as a Social Construct: A Sokal-Style Hoax on Gender Studies," *eSkeptic*, May 19, 2017. http://www.skeptic.com/reading_room/conceptual-penis-social-contruct-sokal-style-hoax-on-gender-studies/

30. "The Conceptual Penis as a Social Construct: A Sokal-Style Hoax on Gender Studies," op cit.

31. "Cogent Criticisms: A Point-by-Point Reply to Criticisms of the 'Conceptual Penis' Hoax" by James Lindsay & Peter Boghossian, *eSkeptic*, June 14, 2017. http://www.skeptic.com/reading_room/reply-to-10-popular-criticisms-of-conceptual-penis-hoax-in-cogent-social-sciences.

32. "Some Thoughts on 'The Conceptual Penis as a Social Construct' Hoax" by Alan Sokal, *eSkeptic*, June 7, 2017. https://www.skeptic.com/reading_room/thoughts-on-the-conceptual-penis-as-social-construct-hoax/.

33. Cited in Torres, op cit.

34. "Detection of B-Mode Polarization at Degree Angular Scales by BICEP2" by P.A.R. Ade et al., *Physical Review Letters*, June 19, 2014. https://journals.aps.org/prl/abstract/10.1103/PhysRevLett.112.241101.

35. Cited in "Cosmic Inflation: Confidence Lowered for Big Bang Signal" by Jonathan Amos, BBC, June 20, 2014. http://www.bbc.com/news/science-environment-27935479.

36. "Unconventional Wisdom: The Lessons of Oakland" by John J. Lentini, David M. Smith and Richard W. Henderson, *Fire & Arson Investigator*, June 1993.

37. *Fire & Arson Investigator*, April 1999.

38. http://www.creationmoments.com/content/baconian-method-science. Amusingly, there have been some minor ass-covering changes made since the first edition of this book was released.

39. http://www.creationmoments.com/content/age-earth.

40. "Darwinian Assumptions Leave 'No Doubt' About Extraterrestrial Life" by Casey Luskin, *Evolution News & Science Today*, October 22, 2010.

41. "SETI Astronomer Says Life's 'Not All That Special' Even as His Own Program Suggests Otherwise" by David Klinghoffer, *Evolution News & Science Today*, November 25, 2011.

42. https://evolutionnews.org/2013/10/the_anxious_sea/.

43. "Curiosity Rover Finally Scratches the Surface," *Answers in Genesis*, March 16, 2013. https://answersingenesis.org/astronomy/life-on-mars/curiosity-rover-finally-scratches-surface/.

44. "NASA Discovers Planet Within a Habitable Zone," *Answers in Genesis*, December 17, 2011. https://answersingenesis.org/astronomy/extrasolar-planets/nasa-discovers-planet-within-habitable-zone/.

45. http://www.icr.org/planet-earth.

46. https://www.documentcloud.org/documents/3417915-IBT-Howard-Johnson-Transcript.html. For a much more complete account of the affair, see Rebekah Wilce's "Transcript from Secret Meeting Illustrates EPA Collusion with the Chemical Industry," *Truth Out*, July 28, 2017. http://www.truth-out.org/news/item/41429-poison-papers-snapshot-hojo-transcript-illustrates-epa-collusion-with-the-chemical-industry.

47. You can find more detail in "Breaking the Silence" by the Sierra Club's Joanne Lauck, March 1999. http://www.sciencegroup.org.uk/ifgene/lapperev.htm.

48. Cited in "Science Writer Wins Libel Appeal," BBC, April 1, 2010. http://news.bbc.co.uk/go/pr/fr/-/2/hi/uk_news/8598472.stm.

49. "The Russian 'Firehose of Falsehood' Propaganda Model: Why It Might Work and Options to Counter It" by Christopher Paul and Miriam Matthews, RAND International Security and Defense Policy Center, 2016.

50. "Smoking Conspiracy: Secondhand Smoke." http://www.smokingaloud.com/ets.html.

51. "VIDEO FLASHBACK: Obama EPA Chief Ignorant of Basic Climate Facts." https://junkscience.com/2017/08/video-flashback-obama-epa-chief-ignorant-of-basic-climate-facts/.

52. "Unnatural Science," *New York Times Magazine*, July 30, 2010. http://www.nytimes.com/2010/08/01/magazine/01FOB-medium-t.html.

53. "The Religious and Political Origins of Evangelical Protestants' Opposition to Environmental Spending" by Philip Schwadel and Erik Johnson, *Journal for the Scientific Study of Religion*, March 2017. http://onlinelibrary.wiley.com/doi/10.1111/jssr.12322/full.

54. Cited in "On its 100th Birthday in 1959, Edward Teller Warned the Oil Industry about Global Warming" by Benjamin Franta, *The Guardian*, January 1, 2018. https://www.the-guardian.com/environment/climate-consensus-97-per-cent/2018/jan/01/on-its-hundredth-birthday-in-1959-edward-teller-warned-the-oil-industry-about-global-warming.

55. http://www.miamiherald.com/latest-news/article163066413.ece/binary/Miami_Dade_Real_%20Estate_Study_2017.pdf

56. "In Florida, Officials Ban Term 'Climate Change'" by Tristram Korten, March 8, 2015.

57. "Crop-Damaging Temperatures Increase Suicide Rates in India." http://www.pnas.org/content/early/2017/07/25/1701354114.

58. This list is derived from Michael Mann's discussion in Chapter 12 of his book *The Hockey Stick and the Climate Wars* (2012).

59. See "Climate Change in the American Mind: May 2017" by Anthony Leiserowitz, Edward Maibach, Connie Roser-Renouf, Seth Rosenthal and Matthew Cutler. The summary is at http://climatecommunication.yale.edu/publications/climate-change-american-mind-may-2017/, where there's also a link to download a PDF of the full report.

60. "US Scientist Sees New Ice Age Coming, by Victor Cohn, *Washington Post*, July 9, 1971.

61. Government Response to the House of Commons Science and Technology Committee 8th Report of Session 2009–10: The disclosure of climate data from the Climatic Research Unit at the University of East Anglia, Presented to Parliament by the Secretary of State for Energy and Climate Change by Command of Her Majesty, Stationery Office, September 2010.

62. http://comments.americanthinker.com/read/42323/731481.html.

63. "Large Losses of Total Ozone in Antarctica Reveal Seasonal ClO_x/NO_x Interaction," May 1985. Abstract at https://www.researchgate.net/publication/246650409_Large_Losses_of_Total_Ozone_in_Antarctica_Reveal_Seasonal_ClO_xNO_xInteraction.

64. Both these DuPont quotes come from the 1997 Greenpeace document "DuPont: A Case Study in the 3D Corporate Strategy." https://web.archive.org/web/20120406093303/http://archive.greenpeace.org/ozone/greenfreeze/moral97/6dupont.html.

65. Cited in "DuPont: A Case Study in the 3D Corporate Strategy," Greenpeace, *op cit.*

66. "Fixing Flint's Contaminated Water System Could Cost $216m, Report Says" by Brian Felton, *Guardian,* June 6, 2016. https://www.theguardian.com/us-news/2016/jun/06/flint-water-crisis-lead-pipes-infrastructure-cost.

67. "Flint Water Crisis May Cost City $400 Million in Long-Term Social Costs" by Josh Sanburn, *Time*, August 8, 2016. http://time.com/4441471/flint-water-lead-poisoning-costs.

68. https://www.freep.com/story/news/local/michigan/flint-water-crisis/2016/01/21/flint-red-flag-2015-report-urged-corrosion-control/79119240. See also http://scitechconnect.elsevier.com/flint-water-crisis-corrosion-pipes-erosion-trust.

69. https://www.freep.com/story/news/local/michigan/flint-water-crisis/2016/01/21/flint-red-flag-2015-report-urged-corrosion-control/79119240/ and http://scitechconnect.elsevier.com/flint-water-crisis-corrosion-pipes-erosion-trust/.

70. "Crackdown on East Chicago Air Polluter Stalls under Trump EPA" by Michael Hawthorne, July 24, 2017. http://www.chicagotribune.com/news/local/breaking/ct-epa-east-chicago-coke-oven-met-20170716-story.html.

71. "Lead Foot in Mouth," July 24. 2017.

72. "Lead Found in Water at More Than Half of School Drinking Water Taps in Bergen County," *Environment New Jersey*, July 17, 2017. http://www.environmentnewjersey.org/news/nje/lead-found-water-more-half-school-drinking-water-taps-bergen-county.

73. Issue dated October 1, 1962. I found a copy, albeit riddled with typos that I assume weren't present in the original (and have silently corrected here), at http://shipseducation.net/pesticides/library/darby1962.htm.

74. Published in the Winter 2017 issue of the Chemical Heritage Foundation's magazine *Distillations*, but posted online months earlier.

75. "Sucrose in the Diet and Coronary Heart Disease" in *Atherosclerosis*, September–October 1971. Cited by Robert H. Lustig in his introduction to the 2012 edition of Yudkin's Pure, White and Deadly.

76. "Nonnutritive Sweeteners and Cardiometabolic Health: A Systematic Review and Meta-analysis of Randomized Controlled Trials and Prospective Cohort Studies" by Meghan B. Azad et al., *CMAJ*, July 17, 2017. http://www.cmaj.ca/content/189/28/E929.

77. "Coca-Cola's Secret Influence on Medical and Science Journalists," April 5 2017.

78. *Guardian*, April 7, 2016.

79. By Aseem Malhotra, Rita F. Redberg and Pascal Meier. http://dx.doi.org/10.1136/bjsports-2016-097285.

80. "Associations of Fats and Carbohydrate Intake with Cardiovascular Disease and Mortality in 18 Countries from Five Continents (PURE): A Prospective Cohort Study" by Mahshid Dehghan et al., *Lancet*, August 29, 2017. http://www.thelancet.com/journals/lancet/article/PIIS0140-6736%2817%2932252-3/fulltext.

81. http://apps.who.int/iris/bitstream/10665/204871/1/9789241565257_eng.pdf.

82. As summarized by Legal Sleuth at http://legalsleuth.com/r-t-vanderbilt-talc-asbestos-mesothelioma/.

83. "Pharmaceutical Industry Sponsorship and Research Outcome and Quality: Systematic Review" by J. Lexchin et al., *BMJ*. "Scope and Impact of Financial Conflicts of Interest in Biomedical Research: A Systematic Review" by J.E. Bekelman et al., *JAMA*. "Pharmaceutical Company Funding and its Consequences: A Qualitative Systematic Review" by S. Sergio, in CCT. All three cited by Ben Goldacre in *Bad Pharma* (2012).

84. You can find Grassley's minority report at https://www.grassley.senate.gov/sites/default/files/about/upload/Senator-Grassley-Report.pdf.

85. "'It Said the Drug was the Best Thing Since Sliced Bread. I Don't Think It Is.'" February 7, 2002.

86. "The Myth of Drug Expiration Dates" by Marshall Allen. *ProPublica*, July 18, 2017. https://www.propublica.org/article/the-myth-of-drug-expiration-dates.

87. "The Myth of Drug Expiration Dates," op cit.

88. "Stability of Active Ingredients in Long-Expired Prescription Medications" by Cantrell et al., November 26, 2012.

89. "Effect of Rosiglitazone on the Risk of Myocardial Infarction and Death from Cardiovascular Causes" by S.E. Nissen and K. Wolski.

90. https://projects.propublica.org/docdollars/.

91. The 114-page report is available for free PDF download at https://www.cdc.gov/drugresistance/threat-report-2013/index.html.

92. "What Drives Inappropriate Antibiotic Use in Outpatient Care?" June 28, 2017. http://www.pewtrusts.org/en/research-and-analysis/issue-briefs/2017/06/what-drives-inappropriate-antibiotic-use-in-outpatient-care.

93. "Prevalence of Inappropriate Antibiotic Prescriptions Among US Ambulatory Care Visits, 2010–2011," by Katherine Fleming-Dutra et al., *JAMA*, May 3, 2016.

94. "British Supermarket Chickens Show Record Levels of Antibiotic-Resistant Superbug" by Fiona Harvey, *Guardian*, January 18, 2018. https://www.theguardian.com/environment/2018/jan/15/british-supermarket-chickens-show-record-levels-of-antibiotic-resistant-superbugs.

95. "Peripheral Modifications of [Ψ [CH2NH]Tpg4]Vancomycin with Added Synergistic Mechanisms of Action Provide Durable and Potent Antibiotics" by Akinori Okanoa, Nicholas A. Isleya and Dale L. Boger, *PNAS*, May 23, 2017. Abstract at http://www.pnas.org/content/114/26/E5052.abstract.

96. See, for example, the Pew Charitable Trusts' *Antibiotic Resistance and Food Animal Production: a Bibliography of Scientific Studies (1969–2012)* (2012). https://web.archive.org/web/20120811001709/ http://www.pewhealth.org/uploadedFiles/PHG/Content_Level_Pages/Issue_Briefs/HHIFBibliographyFinal%20with%20TOC%20_071712.pdf.

97. For further discussion, see "Meat Industry Still Denying Antibiotic Resistance" by Tom Philpott, *Mother Jones*, September 2011.

98. GAO-17-192: "More Information Needed to Oversee Use of Medically Important Drugs in Food Animals."

99. "Sanderson Farms: Spreading Deception & Antibiotic Resistance," June 1, 2017. https://www.nrdc.org/experts/sanderson-farms-spreading-deception-antibiotic-resistance.

100. "The CEO of a Huge Chicken Company Denied that Antibiotic Use is a Problem," May 25, 2017. https://www.buzzfeed.com/venessawong/sanderson-ceo-denies-role-antibiotic-resistance?utm_term=.oig18XnoQ#.fj6qEVpMO.

101. You can read it in full at http://files.shareholder.com/downloads/ABEA-6BBVPE/4520 728307x0xS1193125-17-16859/812128/filing.pdf.

102. "Volkswagen Dieselgate Reexamined." https://skeptoid.com/episodes/4586.

103. "How Trump Signed a Global Death Warrant for Women" by Sarah Boseley, *Guardian*, July 21, 2017. https://www.theguardian.com/global-development/2017/jul/21/trump-global-death-warrant-women-family-planning-population-reproductive-rights-mexico-city-policy.

104. "A Guide to the Bush Administration's Environmental Doublespeak," Sierra Club, October 2004. Reposted at http://www.southshore.com/baedd.htm.

105. http://spj.org/pdf/news/obama-letter-final-08102015.pdf.

106. "In Washington Speak, Censorship Is Called 'Transparency'" by Charles Seife, January 10, 2018. https://www.scientificamerican.com/article/in-washington-speak-censorship-is-called-ldquo-transparency-rdquo.

107. https://en.wikipedia.org/wiki/List_of_executive_actions_by_Barack_Obama.

108. "Proof in Pudding: EPA Toxics Nominee Dourson Has Consistently Recommended 'Safe' Levels for Chemicals that Would Weaken Health Protections" by Richard Denison, EDF, September 22, 2017. http://blogs.edf.org/health/2017/09/22/proof-in-pudding-epa-toxics-nominee-dourson-has-consistently-recommended-safe-levels-for-chemicals-that-would-weaken-health-protections/.

109. Cited in "EPA Chief Met with Dow Chemical CEO before Deciding not to Ban Toxic Pesticide" by Associated Press, *Los Angeles Times*, June 30, 2017. http://www.latimes.com/business/la-fi-epa-pesticide-dow-20170627-story.html.

110. "Fired U.S. Attorney Preet Bharara Said to Have Been Investigating HHS Secretary Tom Price" by Robert Faturechi, *ProPublica*, March 17, 2017. https://www.propublica.org/article/preet-bharara-fired-investigating-tom-price-hhs-stock-trading.

111. "The Pathological Cruelty of Trump's Medicaid Work Requirement," January 11, 2018. http://nymag.com/daily/intelligencer/2018/01/the-cruelty-of-trumps-medicaid-work-requirement.html.

112. "Private Emails, Private Jets and Mr. Trump's Idea of Public Service," editorial, September 28, 2017. https://www.nytimes.com/2017/09/28/opinion/editorials/tom-price-plane-trump-cabinet.html.

113. See the Americans for Tax Fairness special report "The Pharma Big 10: Price Gougers, Tax Dodgers," December 7, 2017. https://americansfortaxfairness.org/pharma-big-10-price-gougers-tax-dodgers.

114. Press release, "Alex Azar to Face Senate Finance Committee on Tuesday; Public Citizen Experts Available for Comment," January 8, 2018. https://www.citizen.org/media/press-releases/alex-azar-face-senate-finance-committee-tuesday-public-citizen-experts.

115. "Medicaid Block Grant Would Slash Federal Funding, Shift Costs to States, and Leave Millions More Uninsured" by Edwin Park. https://www.cbpp.org/research/health/medicaid-block-grant-would-slash-federal-funding-shift-costs-to-states-and-leave.

116. "Charmaine Yoest's Cheerful War on Abortion" by Emily Bazelonnov, *New York Times*, November 2, 2012. http://www.nytimes.com/2012/11/04/magazine/charmaine-yoests-cheerful-war-on-abortion.html.

117. https://web.archive.org/web/20100823232059/http://www.drbfitz.com:80/index.php?pid=15.

118. Cited in "Trump's New CDC Chief Reportedly Promoted a Sketchy Anti-Aging Therapy" by Kiera Butler, *Mother Jones*, July 10, 2017. http://www.motherjones.com/politics/2017/07/trumps-new-cdc-chief-promoted-a-sketchy-anti-aging-therapy/.

119. Cited by *On the Issues* at http://www.ontheissues.org/House/Jim_Bridenstine_Energy_+_Oil.htm.

120. "A Comprehensive Analysis of the Employment Impacts of the EPA's Proposed Clean Power Plan" by Josh Bivens, June 9, 2015. http://www.epi.org/publication/employment-analysis-epa-clean-power-plan/.

121. "Donald Trump's War on Scientists Has Had One Big Side Effect" by Tim Murphy, July 31, 2017.

122. www.scientificactivist.blogspot.com.

123. "Climate change: Al Gore Gets Inconvenient Again" by Michael E. Mann, *Nature*, July 27, 2017. http://www.nature.com/nature/journal/v547/n7664/full/547400a.html.

124. http://www.climate-mayors.org/.

BIBLIOGRAPHY

Agin, Dan. *Junk Science: An Overdue Indictment of Government, Industry, and Faith Groups that Twist Science for Their Own Gain*. New York: Thomas Dunne Books, 2006

Alterman, Eric, and Green, Mark. *The Book on Bush: How George W. (Mis)leads America*. New York: Viking, 2004

Ayala, Francisco J. Darwin's Gift to Science and Religion. Washington DC: Joseph Henry Press, 2007

Babbage, Charles. *Reflections on the Decline of Science in England, and on Some of Its Causes*. London: Fellowes, 1830

Bartholomew, Robert E., and Radford, Benjamin. *Hoaxes, Myths, and Manias: Why We Need Critical Thinking*. Amherst: Prometheus, 2003

Beinecke, Frances, with Deans, Bob. *Clean Energy Common Sense: An American Call to Action on Global Climate Change*. Lanham MD: Rowman & Littlefield, 2010

Bergman, Jerry. *The Criterion: Religious Discrimination in America*. Colorado Springs CO: Onesimus, 1984

Bernasconi, Robert (ed). *American Theories of Polygenesis*. London: Thoemmes, 2002

Berra, Tim. *Evolution and the Myth of Creationism: A Basic Guide to the Facts in the Evolution Debate*. Redwood City CA: Stanford University Press, 1990

Blaser, Martin J. *Missing Microbes: How the Overuse of Antibiotics is Fueling Our Modern Plagues*. New York: Holt, 2014

Bloch, Sidney, and Reddaway, Peter. *Soviet Psychiatric Abuse: The Shadow Over World Psychiatry*. London: Gollancz, 1984

Block, N.J., and Dworkin, Gerald (eds). *The IQ Controversy: Critical Readings*. New York: Pantheon, 1976

Bloomberg, Michael, and Pope, Carl. *Climate of Hope: How Cities, Businesses, and Citizens Can Save the Planet*. New York: St. Martin's Press, 2017

Broad, William, and Wade, Nicholas. *Betrayers of the Truth: Fraud and Deceit in the Halls of Science*. New York: Simon & Schuster, 1982

Broad, William J. *Teller's War: The Top-Secret Story Behind the Star Wars Deception*. New York: Simon & Schuster, 1992

Brock, David. *The Republican Noise Machine: Right-Wing Media and How It Corrupts Democracy.* New York: Crown, 2004

Brown, James Robert. *Who Rules in Science?: An Opinionated Guide to the Wars.* Cambridge MA: Harvard University Press, 2001

Browne, Janet. *Darwin's Origin of Species: A Biography.* New York: Atlantic Monthly Press, 2006

Bunge, Mario. "In Praise of Intolerance to Charlatanism in Academia." *Annals of the New York Academy of Sciences.* Vol 775, June 24, 1996

Centers for Disease Control and Prevention. *Antibiotic Resistance Threats in the United States, 2013.* Atlanta: CDC, 2013

Cirincione, Joseph. "A Brief History of Ballistic Missile Defense." Carnegie Endowment for International Peace, updated 2000 from 1998 conference paper "The Persistence of the Missile Defense Illusion"

Close, Frank. *Too Hot to Handle: The Race for Cold Fusion.* London: W.H. Allen, 1990

Cohen, Stewart J., and Waddell, Melissa W. *Climate Change in the 21st Century.* Montreal & Kingston: McGill–Queen's University Press, 2009

Colborn, Theo, Dumanoski, Dianne, and Myers, John Peterson. *Our Stolen Future: Are We Threatening Our Fertility, Intelligence, and Survival?— A Scientific Detective Story.* New York: Dutton, 1996

Collins, Paul. *Banyard's Folly: Thirteen Tales of Renowned Obscurity, Famous Anonymity, and Rotten Luck.* New York: Picador, 2001

Cornwell, John. *Hitler's Scientists: Science, War and the Devil's Pact.* New York: Viking Penguin, 2003

Darwin, Charles. *The Origin of Species by Means of Natural Selection; or The Preservation of Favoured Races in the Struggle for Life.* London: John Murray, 6th edition, 1872

Dawkins, Richard. *The Blind Watchmaker: Why the Evidence of Evolution Reveals a Universe Without Design.* New York: Norton, 1986

Dawkins, Richard. *Unweaving the Rainbow: Science, Delusion and the Appetite for Wonder.* Boston: Houghton Mifflin, 1998

Dean, Jodi. *Aliens in America: Conspiracy Cultures from Outerspace to Cyberspace.* Ithaca NY: Cornell University Press, 1998

de Camp, L. Sprague. *The Ragged Edge of Science.* Philadelphia: Owlswick, 1980

Dennett, Daniel. *Darwin's Dangerous Idea: Evolution and the Meanings of Life.* New York: Simon & Schuster, 1995

Denworth, Lydia. *Toxic Truth: A Scientist, a Doctor, and the Battle over Lead.* Boston: Beacon Press, 2008

Dewdney, A.K. *Yes, We Have No Neutrons: An Eye-Opening Tour through the Twists and Turns of Bad Science.* Hoboken: Wiley, 1997

Diamond, Jared. *Collapse: How Societies Choose to Fail or Succeed.* New York: Viking Penguin, 2005

Diamond, John. *Snake Oil, and Other Preoccupations.* New York: Vintage, 2001

Dowie, Mark. *Losing Ground: American Environmentalism at the Close of the Twentieth Century.* Cambridge MA: MIT Press, 1995

Druyan, Ann, and Soter, Steven. "The Clean Room." *Cosmos,* episode 7, aired April 20/21, 2014

Dwyer, William M. *What Everyone Knew About Sex.* London: Macdonald, 1973

Dyer, Gwynne. *Climate Wars: The Fight for Survival as the World Overheats.* Oxford: Oneworld, 2010

Ecklund, Elaine Howard. *Science vs Religion: What Scientists Really Think.* New York: Oxford University Press, 2010

Erickson, George A. *Time Traveling with Science and the Saints.* Amherst: Prometheus, 2003

Evans, Christopher. *Cults of Unreason.* London: Harrap, 1973

Eve, Raymond A., and Harrold, Francis B. *The Creationist Movement in Modern America.* Boston: Twayne, 1991

Eysenck, Hans. *Uses and Abuses of Psychology.* Harmondsworth: Penguin, 1953

Feder, Kenneth L. *Frauds, Myths, and Mysteries: Science and Pseudoscience in Archaeology.* Mountain View CA: Mayfield, 3rd edition, 1999

Finkel, Elizabeth. *Stem Cells: Controversy at the Frontiers of Science.* Sydney: ABC Books, 2005

Fitzgerald, A. Ernest. *The High Priests of Waste.* New York: Norton, 1972

Flank, Lenny. *Deception by Design: The Intelligent Design Movement in America.* St Petersburg FL: Red and Black, 2007

Forrest, Barbara, and Gross, Paul R. *Creationism's Trojan Horse: The Wedge of Intelligent Design.* Oxford: Oxford University Press, 2004

Frankfurt, Harry G. *On Bullshit.* Princeton NJ: Princeton University Press, 2005

Friedlander, Michael W. *At the Fringes of Science.* Boulder CO: Westview, 1995

Fritz, Ben, Keefer, Bryan, and Nyhan, Brendan. *All the President's Spin: George W. Bush, the Media, and the Truth.* New York: Touchstone, 2004

Futuyma, Douglas J. *Science on Trial: The Case for Evolution.* New York: Pantheon, 1982

Gardner, Martin. *Fads and Fallacies in the Name of Science.* New York: Dover, 1952; revised and expanded edition, 1957

Gardner, Martin. *Science: Good, Bad and Bogus.* Amherst: Prometheus, 1989

Gelbspan, Ross. *Boiling Point: How Politicians, Big Oil and Coal, Journalists, and Activists Have Fueled the Climate Crisis—and What We Can Do to Avert Disaster.* New York: Basic Books, 2004

Glick, Thomas F. (ed). *The Comparative Reception of Darwinism*. Austin: University of Texas Press, 1975

Goldacre, Ben. *Bad Science*. London: Fourth Estate, revised edition, 2009

Goldacre, Ben. *Bad Pharma: How Drug Companies Mislead Doctors and Harm Patients*. New York: Faber, 2012

Goldin, Claudia, and Katz, Lawrence F. *The Race Between Education and Technology*. Cambridge MA: Belknap Press, 2008

Gore, Al. *An Inconvenient Truth: The Planetary Emergency of Global Warming and What We Can Do About It*. New York: Rodale, 2006

Gould, Stephen Jay. *The Mismeasure of Man*. New York: Norton, 1981

Grant, John. *Discarded Science: Ideas that Seemed Good at the Time*. London: AAPPL, 2006

Grant, John. *Corrupted Science: Fraud, Ideology and Politics in Science*. Wisley: AAPPL, 1st edition, 2007

Grant, John. *Denying Science: Conspiracy Theories, Media Distortions, and the War Against Reality*. Amherst: Prometheus, 2011

Grant, John. *Debunk It!: How to Stay Sane in a World of Misinformation*. San Francisco: Zest, 2014

Gratzer, Walter. *The Undergrowth of Science: Delusion, Self-Deception and Human Frailty*. Oxford: Oxford University Press, 2000

Haffner, Sebastian (trans. Ewald Osers). *The Meaning of Hitler*. Cambridge MA: Harvard University Press, 1979

Hahn, Otto (trans. Ernst Kaiser, Eithne Wilkins). *My Life: The Autobiography of a Scientist*. New York: Herder & Herder, 1970

Healy, David. *Let Them Eat Prozac: The Unhealthy Relationship Between the Pharmaceutical Industry and Depression*. New York: New York University Press, 2004

Healy, David. *Pharmageddon*. Berkeley: University of California Press, 2012

Heard, Alex. *Apocalypse Pretty Soon: Travels in End-Time America*. New York: Norton, 1999

Hendricks, Stephenie. *Divine Destruction: Wise Use, Dominion Theology, and the Making of American Environmental Policy*. New York: Melville House, 2005

Hoggan, James, with Littlemore, Richard. *Climate Cover-Up: The Crusade to Deny Global Warming*. Vancouver: Greystone, 2009

Holton, Gerald. *Science and Anti-Science*. Cambridge MA: Harvard University Press, 1993

Hounam, Peter, and McQuillan, Steve. *The Mini-Nuke Conspiracy: How Mandela Inherited a Nuclear Nightmare*. London: Viking, 1995

Huber, Peter W. *Galileo's Revenge: Junk Science in the Courtroom*. New York: Basic Books, revised edition, 1993

Hurley, Dan. *Natural Causes: Death, Lies and Politics in America's Vitamin and Herbal Supplement Industry.* New York: Broadway, 2006

Irvine, William. *Apes, Angels, and Victorians.* New York: McGraw–Hill, 1955

Jacobsen, Annie. *Phenomena: The Secret History of the U.S. Government's Investigations into Extrasensory Perception and Psychokinesis.* New York: Little, Brown, 2017

Judson, Horace Freeland. *The Great Betrayal: Fraud in Science.* New York: Harcourt, 2004

Kaminer, Wendy. *Sleeping with Extra-Terrestrials: The Rise of Irrationalism and Perils of Piety.* New York: Pantheon, 1999

Kaplan, Esther. *With God on Their Side: How Christian Fundamentalists Trampled Science, Policy, and Democracy in George W. Bush's White House.* New York: New Press, 2004

Keen, Andrew. *The Cult of the Amateur: How Today's Internet is Killing Our Culture.* New York: Doubleday/Currency, 2007

Kessler, David A. *The End of Overeating: Taking Control of the Insatiable American Appetite.* New York: Rodale, 2009

Keys, Ancel. *Seven Countries: A Multivariate Analysis of Death and Coronary Heart Disease.* Cambridge MA: Harvard University Press, 1980

Kitcher, Philip. *Abusing Science: The Case Against Creationism.* Cambridge MA: MIT Press, 1982

Koertge, Noretta (ed). *A House Built on Sand: Exposing Postmodernist Myths About Science.* Oxford: Oxford University Press, 1998

Koestler, Arthur. *The Case of the Midwife Toad.* London: Hutchinson, 1971

Kohn, Alexander. *False Prophets: Fraud and Error in Science and Medicine.* Oxford: Blackwell, revised edition, 1988

Kolbert, Elizabeth. "The Climate of Man." *New Yorker.* April 25, 2005, May 2 2005, May 9 2005

Krupp, Fred, and Horn, Miriam. *Earth: The Sequel: The Race to Reinvent Energy and Stop Global Warming.* New York: Norton, 2008

Lear, Linda. *Rachel Carson: Witness for Nature.* New York: Holt, 1997

Lewis, Sophie C. *A Changing Climate for Science.* London: Palgrave Macmillan, 2017

Lustig, Robert H. *Sugar: The Bitter Truth.* San Francisco: University of California Television, 2009 (lecture available on YouTube and at www.uctv.tv)

Lustig, Robert H. *Fat Chance: Fructose 2.* San Francisco: University of California Television, 2013 (lecture available on YouTube and at www.uctv.tv)

Lustig, Robert H. *Fat Chance: Beating the Odds Against Sugar, Processed Food, Obesity, and Disease.* New York: Hudson Street, 2013

MacDougall, Curtis D. *Hoaxes*. New York: Dover, 1958

McRare, Ron. *Mind Wars: The True Story of Secret Government Research into the Military Potential of Psychic Weapons*. New York: St. Martin's Press, 1984

Mann, Michael E. *The Hockey Stick and the Climate Wars: Dispatches from the Front Lines*. New York: Columbia University Press, 2012

Mann, Michael E., and Toles, Tom. *The Madhouse Effect: How Climate Change Denial is Threatening Our Planet, Destroying Our Politics, and Driving Us Crazy*. New York: Columbia University Press, 2016

Martin, Brian. *Information Liberation: Challenging the Corruptions of Information Power*. London: Freedom Press, 1998

Michaels, David. *Doubt Is Their Product: How Industry's Assault on Science Threatens Your Health*. New York: Oxford University Press, 2008

Millar, Ronald. *The Piltdown Men*. London: Gollancz, 1972

Miller, Arthur I. *Empire of the Stars: Obsession, Friendship, and Betrayal in the Quest for Black Holes*. Boston: Houghton Mifflin, 2005

Miller, Mark Crispin. *The Bush Dyslexicon: Observations on a National Disorder*. New York: Norton, 2001

Mitroff, Ian I., and Bennis, Warren. *The Unreality Industry: The Deliberate Manufacturing of Falsehood and What It Is Doing to Our Lives*. New York: Oxford University Press, 1989

Monbiot, George. Heat: *How to Stop the Planet Burning*. London: Allen Lane, 2006

Monmonier, Mark. *Drawing the Line: Tales of Maps and Cartocontroversy*. New York: Henry Holt, 1995

Mooney, Chris. *The Republican War on Science*. New York: Basic Books, 2005

Mooney, Chris, and Kirshenbaum, Sheril. *Unscientific America: How Scientific Illiteracy Threatens Our Future*. New York: Basic Books, 2009

Moore, Kate. *The Radium Girls: The Dark Story of America's Shining Women*. Naperville: Sourcebooks, 2017

Moore, Kathleen Dean, and Nelson, Michael P. (eds). *Moral Ground: Ethical Action for a Planet in Peril*. San Antonio TX: Trinity University Press, 2010

Morton, Eric. "Race and Racism in the Works of David Hume." *Journal of African Philosophy*, vol. 1, no. 1, 2002

Numbers, Ronald L. *The Creationists: The Evolution of Scientific Creationism*. New York: Knopf, 1992

O'Donnell, Michael. *Medicine's Strangest Cases*. London: Robson, 2002

Offit, Paul A. *Autism's False Profits: Bad Science, Risky Medicine, and the Search for a Cure*. New York, Columbia University Press, 2008

Offit, Paul A. *Killing Us Softly: The Sense and Nonsense of Alternative Medicine*. London: Fourth Estate, 2013

Oreskes, Naomi, and Conway, Erik M. *Merchants of Doubt: How a Handful of Scientists Obscured the Truth on Issues from Tobacco Smoke to Global Warming.* New York: Bloomsbury, 2010

Otto, Shawn Lawrence. *Fool Me Twice: Fighting the Assault on Science in America.* Emmaus: Rodale, 2011

Otto, Shawn [Lawrence]. *The War on Science: Who's Waging It, Why It Matters, What We Can Do About It.* Minneapolis: Milkweed, much revised edition of the above, 2016

Park, Robert. *Voodoo Science: The Road from Foolishness to Fraud.* Oxford: Oxford University Press, 2000

Park, Robert L. *Superstition: Belief in the Age of Science.* Princeton NJ: Princeton University Press, 2008

Pennock, Robert T. (ed). *Intelligent Design Creationism and Its Critics: Philosophical, Theological, and Scientific Perspectives.* Cambridge MA: MIT Press, 2001

People for the American Way. *A Right Wing and a Prayer: The Religious Right and Your Public Schools.* Washington DC: People for the American Way, 1997

Podolsky, Scott H. *The Antibiotic Era: Reform, Resistance, and the Pursuit of a Rational Therapeutics.* Baltimore: Johns Hopkins University Press, 2015

Porter, Roy. *Madness: A Brief History.* Oxford: Oxford University Press, 2002

Porter, Roy, and Hall, Lesley. *The Facts of Life: The Creation of Sexual Knowledge in Britain 1650–1950.* New Haven: Yale University Press, 1995

Press, Bill. *Spin This! All the Ways We Don't Tell the Truth.* New York: Pocket, 2001

Pringle, Heather. *The Master Plan: Himmler's Scholars and the Holocaust.* New York: Hyperion, 2006

Quaratiello, Arlene R. *Rachel Carson: A Biography.* Westport: Greenwood, 2004

Radner, Daisie, and Radner, Michael. *Science and Unreason.* Belmont CA: Wadsworth, 1982

Rampton, Sheldon, and Stauber, John. *Weapons of Mass Deception: The Uses of Propaganda in Bush's War on Iraq.* New York: Tarcher/Penguin, 2003

Rees, Martin. *Just Six Numbers: The Deep Forces that Shape the Universe.* London: Weidenfeld & Nicolson, 1999

Regal, Brian. *Human Evolution: A Guide to the Debates.* Santa Barbara: ABC–CLIO, 2004

Reilly, Lucas. "The Most Important Scientist You've Never Heard Of." *Mental Floss,* May 17 2017. http://mentalfloss.com/article/94569/clair-patterson-scientist-who-determined-age-earth-and-then-saved-it.

Rendall, Steve, Naureckas, Jim, and Cohen, Jeff. *The Way Things Aren't: Rush Limbaugh's Reign of Error.* New York: Norton, 1995

Rice, Berkeley. *The C-5A Scandal: A $5 Billion Boondoggle by the Military–Industrial Complex*. Boston: Houghton Mifflin, 1971

Richelson, Jeffrey T. *The Wizards of Langley: Inside the CIA's Directorate of Science and Technology*. Boulder: Westview, 2001

Ronson, Jon. *The Men Who Stare at Goats*. New York: Picador, 2004

Rorvik, David. *In His Image: The Cloning of a Man*. Philadelphia & New York: Lippincott, 1978

Ross, Benjamin, and Amter, Steven. *The Polluters: The Making of Our Chemically Altered Environment*. New York: Oxford University Press, 2010

Sagan, Carl. *Billions & Billions: Thoughts on Life and Death at the Brink of the Millennium*. New York: Random, 1997

Sagan, Carl. *The Demon-Haunted World: Science as a Candle in the Dark*. London: Headline, 1996

Sagan, Carl. *Pale Blue Dot: A Vision of the Human Future in Space*. New York: Random House, 1994

Salomon, Jean-Jacques (trans. Noël Lindsay). *Science and Politics*. Cambridge MA: MIT Press, 1973

Schneider, Andrew, and McCumber, David. *An Air that Kills: How the Asbestos Poisoning of Libby, Montana, Uncovered a National Scandal*. New York: Putnam, 2004

Schneider, Andrew, and McCumber, David. *An Air that Still Kills: How a Montana Town's Asbestos Tragedy is Spreading Nationwide*. Seattle: Cold Truth, much revised edition of the above, 2016

Scott, Eugenie C. *Evolution vs. Creationism: An Introduction*. Berkeley: University of California Press, 2004

Segal, Sanford L. *Mathematicians Under the Nazis*. Princeton: Princeton University Press, 2003

Shanks, Niall. *God, the Devil, and Darwin: A Critique of Intelligent Design Theory*. Oxford: Oxford University Press, 2006

Shattuck, Roger. *Forbidden Knowledge: From Prometheus to Pornography*. New York: St. Martin's Press, 1996

Sheehan, Helena. *Marxism and the Philosophy of Science: A Critical History*. Amherst: Humanity Books, 2nd edition, 1993

Shermer, Michael. *Why People Believe Weird Things: Pseudoscience, Superstition, and Other Confusions of Our Time*. New York: W.H. Freeman, 1997

Shermer, Michael. *The Borderlands of Science: Where Science Meets Nonsense*. Oxford: Oxford University Press, 2001

Shulman, Seth. *Undermining Science: Suppression and Distortion in the Bush Administration*. Berkeley: University of California Press, 2006

Singer, Peter. *A Darwinian Left: Politics, Evolution and Cooperation*. New Haven CT: Yale University Press, 1999

Singh, Simon, and Ernst, Edzard. *Trick or Treatment: Alternative Medicine on Trial*. London: Bantam, 2008

Singham, Mano. *God vs. Darwinism: The War Between Evolution and Creationism in the Classroom*. Lanham MD: Rowman & Littlefield, 2009

Sladek, John. *The New Apocrypha: A Guide to Strange Sciences and Occult Beliefs*. St Albans: Hart-Davis, MacGibbon, 1974

Sokal, Alan, and Bricmont, Jean. *Fashionable Nonsense: Postmodern Intellectuals' Abuse of Science*. New York: Picador, 1998. Cut translation of *Impostures Intellectuelles*. Paris: Odile Jacob, 1997

Specter, Michael. *Denialism: How Irrational Thinking Hinders Scientific Progress, Harms the Planet, and Threatens Our Lives*. New York: Penguin, 2009

Specter, Michael. "The Denialists: The Dangerous Attacks on the Consensus about H.I.V. and AIDS." *New Yorker*. March 12, 2007

Steingraber, Sandra. *Living Downstream: An Ecologist Looks at Cancer and the Environment*. Reading, MA: Addison–Wesley, 1997

Stern, Robin. *The Gaslight Effect*. New York: Morgan Road, 2007

Taibbi, Matt. *The Great Derangement: A Terrifying True Story of War, Politics, and Religion at the Twilight of the American Empire*. New York: Spiegel & Grau, 2008

Taubes, Gary. *The Case Against Sugar*. New York: Knopf, 2016

Taverne, Dick. *The March of Unreason: Science, Democracy, and the New Fundamentalism*. Oxford: Oxford University Press, 2005

Teicholz, Nina. *The Big Fat Surprise: Why Butter, Meat, and Cheese Belong in a Healthy Diet*. New York: Simon & Schuster, 2014

Thompson, Damian. *Counterknowledge: How We Surrendered to Conspiracy Theories, Quack Medicine, Bogus Science, and Fake History*. New York: Norton, 2008

Tilton, George R. *Clair Cameron Patterson: 1922–1995*. Washington DC: National Academies Press, 1998

Toumey, Christopher P. *God's Own Scientists: Creationists in a Secular World*. New Brunswick NJ: Rutgers University Press, 1994

Tucker, Todd. *The Great Starvation Experiment: The Heroic Men who Starved so that Millions Could Live*. New York: Free Press, 2006

Union of Concerned Scientists. "Scientific Integrity in Policymaking: An Investigation into the Bush Administration's Misuse of Science." Cambridge MA: March 2004

Union of Concerned Scientists. "Scientific Integrity in Policymaking: Further Investigation of the Bush Administration's Misuse of Science." Cambridge MA: July 2004

US Global Change Research Program (Thomas R. Karl, Jerry M. Melillo, and Thomas C. Peterson, eds). *Global Climate Change Impacts in the United States*. Cambridge: Cambridge University Press, 2009

Van Strum, Carol. *A Bitter Fog: Herbicides and Human Rights*. Alfred, New York: Jericho Hill Interactive, 2nd edition, 2014

Vorzimmer, Peter J. *Charles Darwin: The Years of Controversy—The Origin of Species and its Critics 1859–82*. London: University of London Press, 1972

Wanjek, Christopher. *Bad Medicine: Misconceptions and Misuses Revealed, from Distance Healing to Vitamin O*. Hoboken: Wiley, 2003

Webb, George E. *The Evolution Controversy in America*. Lexington KY, University Press of Kentucky, 1994

Weinberger, Sharon. *Imaginary Weapons: A Journey Through the Pentagon's Scientific Underworld*. New York: Nation Books, 2006

Wertheim, Margaret. *Pythagoras' Trousers: God, Physics, and the Gender Wars*. New York: Times Books, 1995

Westfall, Richard S. *Never at Rest: A Biography of Isaac Newton*. Cambridge: Cambridge University Press, 1980

Wheen, Francis. *How Mumbo-Jumbo Conquered the World: A Short History of Modern Delusions*. London: Fourth Estate, 2004

Wicker, Christine. *Not in Kansas Anymore: A Curious Tale of How Magic is Transforming America*. San Francisco: HarperSanFrancisco, 2005

Wynn, Charles M., and Wiggins, Arthur W. *Quantum Leaps in the Wrong Direction: Where Real Science Ends . . . and Pseudoscience Begins*. Oxford: Oxford University Press, 2nd edition, 2016

Young, James Harvey. *The Medical Messiahs: A Social History of Health Quackery in Twentieth-Century America*. Princeton: Princeton University Press, 2nd edition, 1992

Yudkin, John. *Pure, White and Deadly: How Sugar is Killing Us and What We Can Do to Stop It*. New York: Penguin, 2013; revised edition of a book published in 1972 (US title: *Sweet and Dangerous*) and first revised/expanded in 1986

INDEX